Management Principles of Sustainable Industrial Chemistry

Management Principles of Sustainable Industrial Chemistry

Contributors :
Timothy Gonzalez,
Ebrahim Ghafar-Zadeh, *et al.*

AURIS REFERENCE LTD.
London, UK

Management Principles of Sustainable Industrial Chemistry
Contributors : Timothy Gonzalez *and* Ebrahim Ghafar-Zadeh, *et al.*

Auris Reference Ltd., UK

www.aurisreference.com

United Kingdom

Printed in 2017 for Sale in the Indian Subcontinent

Management Principles of Sustainable Industrial Chemistry

ISBN: 978-1-78154-516-4

British Library Cataloguing in Publication Data
A CIP record for this book is available from the British Library

Exclusively distributed by CBS Publishers & Distributors Pvt. Ltd.

Sales & Distribution Rights only for India, Pakistan, Bangladesh, Sri Lanka, Nepal and Bhutan.This book is not to be sold outside these territories.

PREFACE

Sustainable Industrial Chemistry is the branch of chemistry which applies physical and chemical processes towards the transformation of raw materials into products that are of benefit to humanity. Since the time of Industrial Revolution, human intellect throughout the civilised world has been driving this Chemical Revolution. Chemical Revolution has been a continuing series of interconnected developments, growing like a string of firecrackers, which show no sign of letting up.

Sustainable Chemistry has succeeded almost too well. Some of its products have become so common and necessary that they are inevitable for the human beings, while, some others are viewed with contempt by many. On the other hand, accidents and poor waste management together are sensationalized by media coverage.

The global industry globalization process, the emphasis was on supply chain management, and customer-oriented business process design. It seeks for methods to reduce waste, energy and use of resources.

The aspects of Management Principles of Sustainable Industrial Chemistry are presented significantly in this book. The order of the techniques included is conventional and would be liked by students. The chapters have been arranged in a conventional way, as it may be easy for students to pass from one to another with continuity.

This page left intentionally blank.

CONTENTS

This page left intentionally blank.

List of Contributors

Mohamad Hajj-Hassan

Department of Electrical and Computer Engineering, McGill University, 3480 University Street, Montreal, Quebec, Canada H3A 2A7; E-Mails: mohamad.hajjhassan@mail.mcgill.ca (M. H. H); Ebrahim.ghafarzadeh@mail.mcgill.ca (E. G. Z.)

Timothy Gonzalez

Department of Chemistry, McGill University, 801 Sherbrooke Street West, Montreal, Quebec, Canada H3A 2K6; E-Mails: timothy.gonzalez@mail.mcgill.ca (T. G.); hagop.djeghelian@mail.mcgill.ca (H. D.); mark.andrews@mcgill.ca (M. A.);

Ebrahim Ghafar-Zadeh

Department of Electrical and Computer Engineering, McGill University, 3480 University Street, Montreal, Quebec, Canada H3A 2A7; E-Mails: mohamad.hajjhassan@mail.mcgill.ca (M. H. H); Ebrahim.ghafarzadeh@mail.mcgill.ca (E. G. Z.)

Hagop Djeghelian

Department of Chemistry, McGill University, 801 Sherbrooke Street West, Montreal, Quebec, Canada H3A 2K6; E-Mails: timothy.gonzalez@mail.mcgill.ca (T. G.); hagop.djeghelian@mail.mcgill.ca (H. D.); mark.andrews@mcgill.ca (M. A.);

Vamsy Chodavarapu

Department of Electrical and Computer Engineering, McGill University, 3480 University Street, Montreal, Quebec, Canada H3A 2A7; E-Mails: mohamad.hajjhassan@mail.mcgill.ca (M. H. H); Ebrahim.ghafarzadeh@mail.mcgill.ca (E. G. Z.)

Mark Andrews

Department of Chemistry, McGill University, 801 Sherbrooke Street West, Montreal, Quebec, Canada H3A 2K6; E-Mails: timothy.gonzalez@mail.mcgill.ca (T. G.); hagop.djeghelian@mail.mcgill.ca (H. D.); mark.andrews@mcgill.ca (M. A.);

Daniel Therriault

Department of Mechanical Engineering, Ecole Polytechnique de Montreal, Case postale 6079, Succursale Centre-ville, Quebec, Canada H3C 3A7; E-Mail: daniel.therriault@polymtl.ca (D. T.)

Chapter 1

INTRODUCTION

CONTRIBUTION OF THE CHEMICAL INDUSTRY TO SUSTAINABLE DEVELOPMENT

The chemical industry is a key industry. Its products and services are instrumental in meeting the needs of mankind. It is present in all areas of life, from food and clothing, housing, communications, transport - right through to leisure activities. In addition, it helps to solve the problems of other sectors of industry, such as the energy sector, information technologies, environmental industries and the waste disposal sector, as examples.

Due to its size, the chemical industry is an important supplier to a broad range of downstream industries and is, as well, a customer of a broad range of products and services from other industries. It follows, therefore, that the chemical industry plays a major role in providing/supporting performance improvements, research and development progress and, last but not least, employment in other industries.

In itself, it is a large-scale provider of jobs and makes a significant contribution to wealth creation and, hence, to the financing of both public works and the exercise of public responsibilities. Since living standards are determined to a large degree by material considerations, it is clear that the chemical industry with its unique capabilities is in a position to make a decisive contribution to Sustainable Development.

Commitment by the world chemical industry to the concept of Sustainable Development requires words to be transposed into company-specific action programmes in order to provide a framework for all those working in the sector. Its "Responsible Care" initiative, self-monitoring systems and other voluntary programmes such as Sustainable Technology (SUSTECH), Education-Industry Partnerships, Energy Efficiency Programmes are also part of this framework. Thereby, companies are also confronted with new challenges and must act responsibly. They must take account of the consequences of their actions upon society and future generations.

The global chemical industry believes that the key to improving the performance of the industry is both its commitment to achieving environmentally sound Sustainable Development and improved performance and transparency. Under the concept of "Responsible Care", chemical companies are committed, in all aspects of safety, health and protection of the environment, to seek continuous improvement in performance, to educate all staff and work with customers and communities regarding product use and overall operation. Through these efforts the industry is improving its efficiency, reducing risks to health and the environment and making better products which, in turn, help individual and industry customers.

The Chemical Industry's Leadership in Innovation

The very notion of Sustainable Development will require new approaches in a number of areas. Innovation at all levels and in all fields of activity is the most effective instrument for ensuring that the economic, and environmental goals, as well as those of society, are being advanced.

The chemical industry's contribution is to continue innovation of new products that meet customer needs and manufacturing processes that reduce risks to health and the environment. This contribution is based upon the knowledge and experience the industry has acquired from applying innovation not only to making, handling and use of chemical compounds, but also to reprocessing, recycling and solving environmental problems. The challenge facing the chemical industry is to maximize innovation, which can contribute to society meeting its goals for Sustainable Development.

The chemical industry is firmly convinced that leadership in innovation represents the best way of attaining Sustainable Development. For the individual company, this means:

- a consistent orientation towards products, technologies and solutions which offer the greatest promise for the future
- development of new integrated environmental technologies
- a close cooperation with the customers of the chemical industry
- adaptation to the conditions of global competition
- bringing the most promising products quickly on the market
- strengthening the R&D effort which requires resources which can only be financed from profitable earnings
- actively contributing ideas and suggestions to the policy debates taking place in society
- improving process yield (efficiency).

Approach to the Economic Goal of Sustainable Development

The internationalization of the economy at large, in conjunction with a growing trend towards global competition, is becoming more and more apparent. This is being manifested by:

- an increase of imports and exports of goods as well as services
- growing outward and inward flows of direct investment
- an ever increasing exchange of technology transfers
- globalization of monetary and financial schemes.

The inter-relation of economic systems is complex, with a variety of relationships among countries. Multi-national chemical companies apply common standards in spreading investment capital and stimulating markets around the globe, thus setting the scene for the world market. What they need, in order to play a constructive role in Sustainable Development, is, first and foremost, freedom and fairness in international trade. Trade as an engine of economic growth is essential for Sustainable Development. A climate needs to be fostered within which such growth may take place on the basis of a clear set of rules with predictable consequences, by which investors may be guided in their long-term decision-making process. This includes bringing to a halt the growing intervention by governments in industry and their ever increasing demands to raise income by taxation, thus imposing a disproportionate load on the business community.

Wealth creation and profits are fundamental to Sustainable Development. They sustain economies (not just the chemical industry), and contribute, *via* reinvestment and R&D, to new technologies and environmental improvements. Profits are needed to create flexible company structures oriented towards economic, environmental and society-related requirements.

The chemical industry is a major industrial sector and an essential contributor to welfare and employment on a global scale. In order to maintain this position under the imperative of Sustainable Development, the long-term future of the industry must be rooted in a dynamic policy, whereby continual innovation and re-engineering of companies result in an increase of productivity and, thus, keeping up international competitiveness as a pre-requisite of sustainable job creation.

Because it operates on a global scale, the chemical industry makes an important contribution to the attainment of economic goals. The advancement of the world economy along Sustainable Development lines will be influenced positively by:

- the transfer to other countries or parties (particularly in the developing world) of efficient technologies, together with high environmental, health and safety standards
- comparable standards worldwide, applying to the industry's operations in other countries
- monitoring to ensure that equivalent and acceptable standards are being achieved worldwide
- improving education and training and furthering the knowledge of staff and customers by means of training courses, again organised at the industry's plants and offices in other countries.

Whilst the chemical industry thus feels it has an important role to play in contributing to the establishment of an appropriate economic framework, the power to decide this lies with the regulators.

Approach to the Environmental Goal of Sustainable Development

In order to take environmental protection further, environmental standards need to be set to serve as parameters - established within the framework of a market economy. It is then up to each industry to determine how these standards are to be attained within its sector. For the chemical industry, the following parameters are relevant:

- Scientific-environmental monitoring systems are of particular importance. Such scientific- environmental monitoring systems must be systematically and scientifically backed up.
- The chosen integrated approach to environmental protection and waste minimisation must be developed in a consistent manner. The aim must be to integrate environmental protection considerations into products and processes as early as possible in the development phase. Integrated environmental protection also enhances plant and product safety and allows waste disposal to be improved and made more efficient. This parameter clearly meets the Product Stewardship Code of the chemical industry's "Responsible Care" initiative.
- Energy costs are crucial to the chemical industry. Since a policy of economically and environmentally optimised materials and energy use is in the best interest of the individual company, the company itself should assume the responsibility for implementation. As well as ensuring an environmentally and economically sound attitude towards scarce resources - and helping to husband those resources - the policy also makes provision, as said, for the re-use of recycling of used substances and products. The chemical industry is already working on these aspects in close cooperation with its customers and consumer organisations.
- Competitive access to energy is necessary to the chemical industry. Since the reservoir of natural resources is finite, their proper management is a crucial pillar of Sustainable Development. The chemical industry is contributing to this goal by continuously reducing the environmental impact of its activities. Perhaps, more importantly, the products of the chemical industry's innovation enable other industries to use resources more efficiently with less environmental impact.

In order to achieve these aims, industry requires a worldwide politico-economic framework applied by Governments, which **facilitates** decision-taking within companies and not a dirigiste approach. Any attempt to impose centralised control over entire substance chains from cradle to grave would simply not work.

Approach to the Society-Related Goal of Sustainable Development

The chemical industry considers a functioning, pragmatic social partnership to be a key factor in Sustainable Development. This form of cooperation is the best way of allowing a company to compete effectively in the global market place and prosper, thus making it possible to ensure the future, safeguard jobs in the long term, give proper recognition to performance, be consistent in labour protection and show concern for the problems of developing countries.

Current needs of society must be defined by society itself acting through the various audiences involved, who will be affected by the definition, in a genuine process of **consensus-building**. Public discussion can frequently tend to lack focus. What is needed is a constructive form of debate within society. The chemical industry has to demonstrate its willingness to be publicly responsible for its contribution to Sustainable Development. The results of efforts to achieve **consensus** will be influenced to a significant extent by the differences in the way the partners involved in the debate perceive the risks. Whether these perceptions are "rational" or not, they need to be addressed. For this reason, the chemical industry attaches considerable importance to environmental reporting and risk communication and recognises that there is an urgent need to debate these issues openly within society.

Efforts to move towards Sustainable Development will be to no avail unless consensus can be reached within society. The task of politicians and the political parties must therefore be to encourage cohesion within society and to marshal the forces of society behind this goal.

The Policy Challenges

The chemical industry is convinced that advances along the path of Sustainable Development are possible only in the context of a market economy and by way of a collective learning process involving all members of society. The chemical industry therefore calls for policies which guarantee the maintenance of the market economy framework. Within this framework, the best means of attaining our goals is *via* a combination of legislation, voluntary measures and market forces. Those who will be directly affected must participate in the process, thereby guaranteeing that realistic targets are set and that sufficient time is provided for their implementation.

Establishing an **environmental policy** compatible with a market economy is not a matter of drafting detailed regulations but of setting transparent, attainable environmental policy goals. In practice, the fixing of environmental standards must be the outcome of a process of opinion-forming and consensus-seeking within society. The long-term interests of society, however, depend upon this process being as well informed as possible about current science. This gives them the status of environmental conventions which, as a set of strategies, cannot be static but must respond to changes and trends within both science and society.

Economic policies must create an environment conducive to structural change. There can be no artificial propping up of economically outdated structures. The effects on the distribution of incomes and on employment, however, must be given proper consideration at the national level. Society must be given time to adjust to structural change.

Given the nature and importance of the problems, a "go-it-alone" national approach is not enough. There should be no unilateral measures applied to achieve an environmental objective outside one country's borders. We would therefore wish to see national policies to foster Sustainable Development implemented in most countries worldwide. Countries must strive for harmonization or, at least, convergence of environmental goals. This can only be achieved through international cooperation.

The most important tasks of economic and social policies include the need to:

- Secure the framework for the market economy by discouraging state interventions
- Promote freedom and fairness in international trade
- Improve international competitiveness by
  - o facilitating restructuring of industries
  - o reducing burdens causing uncompetitive costs (labour, energy)
  - o fostering the scientific and technological base
- Promote rational regulatory and environmental policies based on sound science whilst allowing voluntary approaches
- Develop a clear set of rules with predictable consequences in the long run.

Moreover, the structural change involved in the move towards Sustainable Development calls for close cooperation within and between all major policy areas - *i.e.* measures must be co-ordinated, non-contradictory and in line with the overall goals. Those responsible for individual policy areas will need to set their own priorities accordingly.

The task of **research and development policy** should be to co-ordinate aims and focal points for research. In future, the success of research and technology policies will to a large degree depend on the extent to which a consensus can be arrived at between innovative potential - whether it be that of companies, research institutes or universities - and the concerns of the other groups and forces within society, and hence give impetus to creative competition. In order to be successful, activities in these areas must be integrated into a worldwide network of scientific and information exchange.

Sustainable Development presupposes the existence of an efficient education and training system. **Education and training policy** must help to pass on the relevant knowledge and an awareness of the problems on the basis of proven scientific knowledge, fostering an ability and willingness to make the necessary changes in economy and society. The concept of Sustainable Development has a strong appeal for young people and is well suited also to the international context.

CRITERIA FOR APPRAISING GOALS AND INSTRUMENTS

Because the three stated aspects of Sustainable Development (economic, environmental, society-related) are inseparable, attainment of the goals will generally call for a package of instruments. Since action limited to one country cannot do justice to the global goals of Sustainable Development, international harmonisation of the political and policy framework will be decisive for success.

A fundamental prerequisite is agreement not only on the specific goals and the manner in which they are to be attained through national policies, but also - and above all - on international programmes and decision-making processes. However, it serves no purpose to determine the means unless the end is also made clear - *i.e.* there is no point in encouraging or supporting the use of specific instruments without defining beforehand the goal one is seeking to reach.

Moreover, conflicts between different goals must be resolved before individual measures can be put into effect. Given the complexity of the issues, it will be necessary to undertake case-by-case studies of appropriate combinations of goals and measures.

The goals, methods and instruments must satisfy the following criteria:

- The goals must be set realistically and should neither overestimate nor overstrain the capabilities or absorption capacity of the system (practicability).
- In an overall appraisal on the basis of economic, technical, environmental criteria and social acceptability, the effort involved must be in proportion to the desired result (proportionality).
- The goals and measures must be designed in such a way that they can be tailored to meet changes in requirements (flexibility).
- There must be some provision for retracting certain goals or reversing individual instruments, if new facts come to light or if they produce unintended adverse effects (reversibility).
- The goals and instruments must not run counter to the principle of equal treatment nor, even more importantly, jeopardise equal opportunities in international competition (competitive neutrality) and should be taken at the most appropriate level.

The means and methods selected must comply with the laws of the market and not interfere with fundamental rights. Therefore, there can be no alternative to responsible action not only on the part of companies but also by the individual citizen within the framework of the goals set by Governments.

CHEMICALS AND WASTE

Chemicals

Environmentally sound management of "toxic chemicals". A substantial use of chemicals is essential to meet the social and economic goals of the world community, and these can be used with a high degree of safety when best practices are followed. However, much remains to be done. Two of the major problems identified, particularly in developing countries, are

1. Lack of sufficient scientific information for the risk assessment, and
2. Lack of resources of assessment of chemicals for which data are at hand.

Gross chemical contamination, with grave damage to human health, genetic structures and reproductive outcomes and the environment, has been continuing within some of the world's most important industrial areas, and restoration will require major investment as well as the development of new techniques.

Six programme areas, as follows:

- expanding and accelerating international assessment of chemical risks;
- harmonization of classification and labelling of chemicals;
- information exchange on toxic chemicals and chemical risks;
- establishment of risk reduction programmes;
- strengthening of national capabilities and capacities for management of chemicals; and
- prevention of illegal international traffic in toxic and dangerous products.

In paragraph 23 of the Johannesburg Plan of Implementation, Member States renewed the commitment, to sound management of chemicals throughout their life cycle and of hazardous wastes for sustainable development as well as for the protection of human health and the environment, inter alia, aiming to achieve, by 2020, that chemicals are used and produced in ways that lead to the minimization of significant adverse effects on human health and the environment, using transparent science-based risk assessment procedures and science-based risk management procedures, taking into account the precautionary approach, as set out in principle 15 of the Rio Declaration on Environment and Development, and support developing countries in strengthening their capacity for the sound management of chemicals and hazardous wastes by providing technical and financial assistance.

The focus provided on the sound management of chemicals led to a number of developments which culminated in 2006 with the adoption of the Dubai Declaration on International Chemicals Management and the formal establishment of the Strategic Approach to International Chemicals Management (SAICM).

At its second, fifth, eighteenth and nineteenth sessions, held in 1994, 1997, 2010 and 2011 respectively, the Commission on Sustainable Development (CSD) had substantive discussions on sound management of chemicals throughout their life cycle.

In "The Future We Want", the outcome document of Rio+20 (2012), Member States reaffirmed their commitment to achieve, by 2020, the sound management of chemicals throughout their life cycle and of hazardous waste in ways that lead to minimization of significant adverse effects to human health and the environment, as set out in the JPOI. They also called for the effective implementation and strengthening of the SAICM.

Waste (Hazardous)

Effective control of the generation, storage, treatment, recycling and reuse, transport, recovery and disposal of hazardous wastes is "of paramount importance for proper health, environmental protection and natural resource management, and sustainable development." Prevention of the generation of hazardous wastes and the rehabilitation of contaminated sites are the key elements, and both require knowledge, experienced people, facilities, financial resources and technical and scientific capacities.

Among the overall targets are the following:

1. Preventing or minimizing the generation of hazardous wastes as part of an overall integrated cleaner production approach;
2. Eliminating or reducing to a minimum transboundary movements of hazardous waste;
3. Ratifying the Basel Convention on the Control of Transboundary Movements of Hazardous Wastes and their Disposal;
4. Ratifying and full implementation of the Bamako Convention on the Ban of the Import into Africa and the Control of Transboundary Movement and Management of Hazardous Wastes within Africa; and
5. Eliminating the export of hazardous wastes to countries that prohibit such imports.

Managing hazardous wastes was discussed by the Commission on Sustainable Development (CSD) at its second, fifth, seventh, eighteenth and nineteenth sessions and by the General Assembly at its nineteenth Special Session.

Member States renewed the commitment to sound management of chemicals throughout their life cycle and of hazardous wastes for sustainable development as well as for the protection of human health and the environment.

In 2004, the Ministerial Statement on Partnerships for Meeting the Global Waste Challenge was adopted at the seventh meeting of the Conference of the Parties to the Basel Convention on the Transboundary Movement of Hazardous Wastes and their Disposal. The Statement recognizes the environmentally sound management of hazardous wastes as part of the wider issues of water protection, improved sanitation, solid waste management and economic and social development. It calls for the reduction of the impacts of hazardous wastes on human health and the environment and promotes a fundamental shift in emphasis from remedial measures to preventive measures such as reduction at source, reuse, recycling and recovery. It recognizes the importance of mobilizing new and additional financial resources to build partnerships to meet the global waste challenge worldwide.

In "The Future We Want", the outcome document of Rio+20 (2012), Member States reaffirmed their commitment to achieve, by 2020, the sound management of chemicals throughout their life cycle and of hazardous waste in ways that lead to minimization of significant adverse effects to human health and the environment.

Waste (Solid)

The solid wastes include all domestic refuse and non-hazardous wastes such as commercial and institutional wastes, street sweepings and construction debris and, in some countries, human wastes. Hazardous waste is frequently intermixed with other waste, posing particular management challenges.

The Commission on Sustainable Development (CSD) at its second, fifth, sixth, seventh, eighteenth and nineteenth sessions. At the World Summit on Sustainable Development in 2002, Governments reaffirmed the importance of solid waste management. They called for priority attention to be given to waste prevention and minimization, reuse and recycling. They also called for the development of environmentally sound disposal facilities, including technology to convert waste into energy.

Waste (Radioactive)

The safe and environmentally sound management of radioactive wastes. The radiological and safety risk from radioactive waste varies, from very low for short-lived, low-level waste, to very large for high-level waste. Annually, about 200,000 m^3 of low-level and intermediate-level waste and 10,000 m^3 of high-level waste (as well as spent nuclear fuel destined for final disposal) are generated worldwide from nuclear power production, and these volumes are increasing.

The ensure that radioactive waste is safely managed, transported, stored and disposed of, with a view to protecting human health and the environment, within the wider framework of an interactive and integrated approach to radioactive waste management and safety.

The Commission on Sustainable Development (CSD) considered the safety of radioactive wastes during its fifth session in 1997, seventh session in 1999, in relation to transboundary movement of this waste, and again during its ninth session in 2001, in relation to nuclear energy technologies.

As a result of its deliberations on this issue, the World Summit on Sustainable Development in 2002 stressed the importance of effective liability measures for international maritime transportation and other transboundary movement of radioactive material, radioactive waste and spent nuclear fuel, and encouraged Governments to examine and improve measures and internationally agreed regulations regarding the safe handling, transport and disposal of this waste.

GREEN CHEMISTRY ON THE RISE

The adoption of green chemistry as one of the primary methods of pollution prevention is a fairly recent phenomenon. It is certainly reasonable to question why this fairly straightforward approach is only now taking hold. The answer is found in a combination of economic, regulatory, scientific, and even social factors, which coalesced in the 1990s to give rise to green chemistry. Since then, green chemistry has found implementation and commercialization on a wide scale.

Since the early 1960s, environmental statutes and regulations have proliferated at an exponential rate. With these regulations came costs, restrictions on the use of chemicals, and increased testing of chemical substances to determine their hazards. This provided powerful incentives for industry to find replacements, substitutes, or alternatives. The toxicity testing required by many of these statutes generated new knowledge and a new awareness about the types and degrees of hazard associated with many chemicals.

As this collective knowledge grew in scientific and industrial circles, there was corresponding growth in the public's demand for more information about chemicals that are present in their communities. For example, in the USA, this culminated in the 1980s with the passage of the Emergency Planning and Community Right-to-Know Act (EPCRA), which made public relevant data on chemicals being released to the air, water, and land by industry. As a consequence, industry has been confronted by tremendous pressure, not only to reduce the release of toxic chemicals to the environment, but also to reduce the use of hazardous chemicals overall. Thus, it is not surprising that since 1990, sustainable chemistry has been an official focus of the U.S. Environmental Protection Agency, involving a great deal of activity in research, symposia, and education.

In the same year in Italy, the Consorzio Interuni-versitario "La Chimica per l'Ambiente" (Interuniversity Consortium Chemistry for the Environment), or INCA, was established with the aim of uniting academic groups concerned with chemistry and the environment. One of its main focus areas became pollution prevention through the development of cleaner reactions, products, and processes. INCA organized its first meeting in Venice, Processi Chimici Innovativi e Tutela dell'Ambiente (Innovative Chemical Processes and Environmental Protection), in February 1993.

However, it was only between 1996 and 1997 that the term green chemistry was first used. The definition "green chemistry" or "sustainable chemistry" has been the subject of a long debate. Both expressions have been used for the same or very similar meanings, but each has its supporters and detractors. The word "green" is brightly evocative, but may assume unintended connotations, whereas "sustainable" can be paraphrased as "chemistry for a sustainable environment," and may be perceived as a less focused and less incisive description of the discipline. Other terms have been proposed, such as "chemistry for the environment," but this combination of words does not capture the economic and social implications of the concept of sustainability.

This definition is not only straightforward, but it also marks a significant departure from the manner in which environmental issues have been considered or ignored in the up-front design of the molecules and molecular transformations that are at the heart of the chemical enterprise. Paul Anastas, then of the EPA, and

John C. Warner developed 12 principles of green chemistry, which illustrate the definition in a practical sense. The principles cover such concepts as:

Invention and design imply that the impacts of chemical products and chemical processes are included in the design criteria. This definition inextricably links hazard considerations to performance criteria.

Use and generation signify that green chemistry must focus not only on those undesirable substances that might be inadvertently produced in a process, but also on all substances that are part of the process. In this sense, green chemistry recognizes that there are significant consequences associated with the use of hazardous substances, including regulatory, handling and transport, and liability issues, to name a few.

Hazardous means that green chemistry is a way of dealing with risk reduction and pollution prevention by addressing the intrinsic hazards of the substances rather than those circumstances and conditions of their use that might increase their risk.

In August 1996, IUPAC began its involvement in the green chemistry field with the creation of the Working Party on Synthetic Pathways and Processes in Green Chemistry. A year later, in September 1997, the First International Conference on Challenging Perspectives on Green Chemistry was held in Venice. The meeting was sponsored by IUPAC and co-sponsored by ACS, EPA, and UNESCO.

Another important step in the development of green chemistry was the foundation in 1997 of the Green Chemistry Institute (GCI) in the USA. After more than a year of planning by individuals from industry, government, and academia, GCI was incorporated in 1997 as a not-for-profit corporation devoted to promoting and advancing green chemistry. In January 2001, GCI joined the American Chemical Society in an effort to address global issues at the intersection of chemistry and the environment.

From its earliest days, the Institute has sought to be the premier change agent that has the knowledge, expertise, and capabilities to catalyze the movement of the chemical enterprise toward sustainability through the application of green chemistry principles. To fulfill its mission, GCI supports research, works to integrate green chemistry into all levels of chemical education, aides companies with industrial implementation, hosts conferences, and coordinates efforts with an international network of green chemistry advocates. The GCI also provides national recognition for outstanding student contributions to furthering the goals of green chemistry.

GCI offers two well-known awards established in memory of Kenneth G. Hancock and Joseph Breen. Hancock, former director of the Division of Chemistry at the National Science Foundation, was one of the earliest proponents of green chemistry and "environmentally benign chemical synthesis and processing." The Hancock Memorial Award in Green Chemistry allows undergraduate and graduate students an opportunity to compete for a prestigious award in recognition of their studies and/or research in green chemistry.

GCI and ACS also established the Joseph Breen Memorial Fund through the ACS International Endowment Fund in 2000. Breen was a founder of the worldwide green chemistry movement and the first director of the Green Chemistry Institute. This fund commemorates his commitment and accomplishments, and each year awards one or more fellowships.

In the late 1990s and at the beginning of the new millennium, interest in green chemistry spread all over the world. In 1998, following an EPA proposal, the OECD (Organization for Economic Co-operation and Development) issued a directive to develop a program on sustainable chemistry. The USA and Japan were nominated co-leaders in the field of research and development, while Italy was appointed leader of education. In particular, INCA was chosen as coordinator of the Educational Acts on Green Chemistry for OECD.

A year later, in 1999 the Royal Society of Chemistry (UK) introduced a new journal entirely dedicated to sustainable chemistry: Green Chemistry. Around the same time, similar projects were initiated in many other countries. Following are just a few examples:

- Foundation of the Green Chemistry Institute of Spain in February 2000
- Creation of the Green and Sustainable Chemistry Network in Japan in 1988 (officially operating since 2000), which in 2007 co-organized the first Asian-Oceania Conference on Green and Sustainable Chemistry
- The foundation of the Centre of Green Chemistry of Monash University in Australia, operating since January 2000

In addition, the German government has begun to dedicate special attention to renewable chemistry. In 2002, in connection with the Johannesburg's "World Summit for Sustainable Development," the government launched an international conference to promote the use of renewable energy in industrial and global development processes. The conference was held in Bonn and gathered 3600 participants.

In order to address the increasing need for green chemistry education, the Carnegie Group (the biannual meeting of the G8 Ministers for Research), which was held in Victoria, Canada and in New York (2–3 December 2005), founded a research and training network on green-sustainable chemistry called the International Green Network (IGN). The network was unanimously approved, following a proposal by the Italian minister for Education, Research, and University. INCA, located in Venice, Italy, was selected as the hub of IGN. INCA and the other research centers dedicate a space within their institutes to the IGN. The IGN aims to provide know-how, coordination and sponsorship for scientific collaborations, proper training for the new generation of chemists, and support for sustainable use of chemistry in developing nations. INCA organized the kick-off meeting of the IGN on 1 December 2005 (Marghera, Venice). Delegates of all the G8 countries participated, as well as observers from other nations and prominent invited speakers. The purpose of the event was to officially inaugurate IGN, illustrate its goals (both scientific and political), and gather feedback from delegates. A document presented at the meeting, and approved shortly afterward by the Carnegie Group in New York, outlines the goals and mission of IGN.

Another green chemistry network was founded in December 2005: the Mediterranean Countries Network on Green Chemistry (MEGREC). The network was created to further facilitate increasing collaboration between the European regions of the Mediterranean and North Africa on green chemistry issues. The official ceremony was held during the first board meeting at the University of Belgrade.

The MEGREC founding institutions include the Suez Canal University, Ismailia (Egypt); University Institute of Science and Technology, (Barcelona, Spain); Fez University (Morocco); Belgrade University (Serbia); Athens University (Greece); and INCA, Venice (Italy).

Among the objectives of the MEGREC program are the creation of a university master course (second level) on green chemistry and the recruitment of young researchers who will conduct research in the laboratories of partner institutions.

In Latin American countries, the scientific community is focusing intensively on green chemistry by developing a scientific collaboration called Chemistry for Clear Reactions and Processes: Green Chemistry. This joint project has already provided fellowship opportunities at an Italian university for researchers from Chile, Argentina, Brazil, Uruguay, Costa Rica, and Venezuela.

In 2006, INCA and the German Chemical Society organized the first international IUPAC conference dedicated to green-sustainable chemistry. The meeting, held 10–15 September 2006, was organized under the auspices of IUPAC, the Italian Ministry of Research, Federal Ministry of Environment, Nature Conservation and Nuclear Safety, and the German Federal Environmental Ministry. The wide selection of topics offered at the conference attracted industrial researchers and representatives, university researchers, politicians, and students. The enormous effort of the organization committee fully paid off with over 450 participants attending from 42 countries. Upon conclusion of the conference, the organization committee extended an invitation to all delegates to participate in the Second International IUPAC Conference on Green Chemistry to be held aboard a luxury cruiseliner that will travel from Moscow to St. Petersburg. The conference is scheduled to take place in the last week of September 2008.

Another important step in the history of green chemistry has been realized with the introduction of REACH (Registration, Evaluation and Authorisation of Chemicals Regulation), which was formally adopted on 18 December 2006 by the European Council of Environment Ministers. This new regulation aims to improve the protection of human health and the environment through improved assessment of chemical substances. Thus, the REACH Regulation gives greater responsibility to industry as manufacturers and importers will be required to compile information on the properties of their substances and to register the information in a central database. This regulation ultimately calls for a progressive substitution of the most dangerous chemicals when suitable and greener alternatives have been identified, which is the main goal of green chemistry itself. REACH entered into force on 1 June 2007.

SUSTAINABLE CHEMISTRY SUCCESS

With personal care and home products increasingly in the spotlight, expectations have grown formore sustainable chemicals and formulations. Target, Walmart and states such as Minnesota,California and Washington are demanding that companies take a new approach to managing chemicals in these products.

To help guide companies managing this added pressure, my consulting firm, Pure Strategies, evaluated how leaders in the formulated products marketplace have constructed their goals, strategies, tools and partnerships. What is at the core of these programs — and what makes themsuccessful?

We found the following four essential parts to enhanced sustainable chemistry programs:

Strategy and Alignment

A strategy with leadership support is the core of any company's efforts to move toward sustainable chemistry. That's why the first questions to answer concern basic strategy: What issues are we trying to solve? What approach do we want to take? Issues range from meeting market demands to bringing new innovations in product design to customers. Strategies include engaging with the U.S. Environmental Protection Agency's Design for the Environment program. Other approaches involve determining what chemicals of concern are in products and prioritizing these chemicals for replacement.

Our review of industry front-runners identified strategies that address specific ingredient objectives as part of their sustainability programs, with a protective approach to health and environmental considerations across the product life cycle. This moves from restricting chemicals of concern to improving overall ingredient profiles for intentionally added ingredients, known by-products and known contaminants by:

- Understanding and assessing chemicals and their hazard profiles.
- Prioritizing chemicals and avoiding chemicals of concern.
- Improving the chemical hazard profile in products.
- Disclosing chemical composition.

Beyond ensuring safety, a key piece of the strategy is to provide appropriate organizational alignment. The leading companies interact and share information across functions to empower more groups across the organization to engage on key objectives. Often, R&D, procurement and regulatory groups work together to advance on the targets — especially when working to improve the profile of product ingredients.

The company's goal is to increase the proportion of ingredients in its products that have a lower impact on the environment and human health. It is also committed to disclosing product ingredients. SC Johnson scientistscreating a new product or a reformulation are armed with chemical profile information when making ingredient selections to ensure that environmental and health ratings are equal to or higher than the original formula. By including R&D in raw material evaluation, in addition to having a safety team, the company provides greater alignment across the organization. In this way, it goes beyond regulatory requirements to continually improve products.

Competencies

To meet strategic objectives in sustainable chemical management, companies need several critical competencies. These include expertise in evaluating ingredient hazards; proficiency in developing improved formulations; and the ability to assess the performance, cost, compatibility and safety of alternatives. To support these internal competencies, leaders have tools in place that:

- Provide clear guidance to R&D teams to assist in the selection of preferable ingredients.
- Include a substance database with environmental and health hazard information.
- Integrate these two elements in a development tool that shows the environmental and health impact of possible formulations.

Effective use of external resources is important here, as well. Strong supplier relationships are needed to assess, communicate and assure the composition of raw materials. Suppliers must increasingly understand and be willing to share details on trace ingredients, impurities and contaminants. For example, are preservatives or stabilizers included with ingredients? Are extractions, distillation or other activities conducted in PVC piping that may lead to phthalate contamination? These partnerships also help to solve issues and innovate solutions.

As an example, Seventh Generation has developed product and ingredient guidelines to ensure that products are not acutely toxic and are free of chronic toxicants. The company also has a scorecard system to evaluate performance. Additionally, it has strong supplier relationships to provide the highest level of transparency and assurance of ingredient composition. The company co-develops solutions and new ingredients to enable its top-of-the-class program. For example, the company worked for two years with its suppliers to eliminate 1,4-dioxane from its products and introduced liquid dish and liquid laundry detergents with no detectable levels of 1,4-dioxane in 2010.

Measurement and Tracking

Performance measurement and tracking provide structure and accountability for organizations leading in chemicals management. This includes a system to evaluate conformance to the company's stated policy and progress against its objectives. Performance tracking encompasses internal product evaluation, testing and scorecards. Leaders also engage with independent reviewers and analysis to assure their standards are being met.

For instance, all of the ingredients in Method's products undergo a comprehensive evaluation by an independent environmental research institute, EPEA. This material research process assures the company that its products are healthy and environmentally safe. Method also leverages the third-party certification program Cradle to Cradle to provide an additional, rigorous assessment of all ingredients, packaging materials and design, and processes.

External Engagement

Communication and collaboration with external stakeholders are the final elements of an effective chemical management program. Consumer education about ingredient sustainability and safety that empowers the public and encourages responsible product use has become a standard practice. Some form of ingredient disclosure, typically online, is becoming commonplace. The differentiator in communication is that leaders disclose ingredients on the packaging and provide consumers with more details, such as the composition of fragrance mixtures.

External engagement also includes other stakeholders. Companies in the forefront contribute to industry initiatives and support external centers of excellence. We found that involvement in thought leadership organizations such as the Green Chemistry in Commerce Council (GC3) or work with advocacy groups was limited to leading companies such as Seventh Generation and Method.

In the case of Seventh Generation, it discloses all ingredients both on the packaging and on its website, so consumers can access the information right in the store aisle. While fragrance information is often hidden in the personal care and household cleaning industries, the company discloses all fragrances.

Seventh Generation additionally holds a strong leadership role within the industry and chairs the American Cleaning Institute's (ACI) Sustainability Committee. The company helped develop ACI's successful Ingredient Communication Initiative, leading the way for many ACI members (including most major players in the cleaning products market) to disclose the ingredients in their products voluntarily beginning in 2010.

Leaders also engage with their customers, working to set standards for the industry, as well as government policy makers to help them establish better policies and regulation to protect consumers and the environment.

An Apple a Day

Firms with successful sustainable chemistry programs not only protect workers, consumers and the environment — they also protect their reputations. The core of their approach is a solid strategy with alignment across the company. The meat of a leading program combines critical expertise, clear guidance and tools with a solid measurement and tracking strategy.

Finally, ingredient disclosure and consumer empowerment, as well as engagement with important external stakeholders, puts the final polish on leading programs. Just as an apple a day keeps the doctor away, adopting best practices for sustainable chemical management keeps programs healthy and strong.

CHEMICALS IN THE CONTEXT OF SUSTAINABLE DEVELOPMENT GOALS

As an outcome of the UN Conference on Sustainable Development (UNCSD, or Rio+20), the United Nations has launched a process to develop a set of Sustainable Development Goals (SDGs). This process should ultimately become part of a single global development agenda beyond 2015, building on the Millennium Development Goals (MDGs) and with sustainable development at its core.

Chemicals and waste will be discussed in the SDG Open Working Group meeting in January 2014, under the topic of sustainable consumption and production. Although chemicals and waste are validly considered a cross-cutting issue, the challenge with this approach is that the substantive nature of the problem can become diluted. Chemicals are inextricably linked to our lives, present in almost every industry, from health, transportation, agriculture, construction, textile, and consumer products, just to name a few. In 2010, global chemical output was valued at US$4.12 trillion, making the chemical industry a major player in the global economy. However, the essential contribution made by chemicals to modern societies and economies may present a potential threat to sustainable development if chemicals are not managed soundly. More importantly, the costs of not taking action towards the sound management of chemicals is often higher than the costs of implementing measures to manage chemicals in ways that minimize adverse effects to human health and the environment.

The Global Chemicals Outlook (GCO) explores a number of challenges related to chemicals, including: the increasing variety and complexity of chemicals; replacement of traditional plant, animal and/or ceramic-based products; and the increased length and intricacy of chemical supply chains and waste streams that expose serious gaps, lapses and inconsistencies in government and international policies and corporate practices. The GCO also presents evidence suggesting that the costs of inaction to a country due to the negative consequences of unsound chemicals management on human health and the environment often are significantly higher than the costs of action. Furthermore, the WHO has estimated that chemical-induced disease contributes up to 25% of the Global Burden of Disease.

The link between sustainable development and sound chemicals management is clearly demonstrated by the experience with phasing out lead in gasoline. 'The Global Benefits of Phasing out Leaded Fuel' (December 2011) study, for example, confirmed that the phase-out of leaded petrol contributed US$2.4 trillion (4% global GDP) to the global economy. This monetary saving is calculated by measuring social benefits such as heightened IQ levels and reduced criminality, as well as health savings from afflictions such as cardiovascular disease. Ending the use of leaded petrol also translated to 1.2 million fewer deaths per year. In addition, it is expected that these health and associated financial savings may increase as connections between lead and additional health problems such as kidney disease become clearer.

Agreed to at the highest international political level in 2006, the Strategic Approach to International Chemicals Management (SAICM) supports achievement of a central goal agreed at the 2002 Johannesburg World Summit on Sustainable Development to ensure that, by 2020, chemicals are produced and used in ways that minimize significant adverse impacts on the environment and human health. The Dubai Declaration (2006) that gave birth to SAICM recognizes that the sound management of chemicals is critical to achieving sustainable development. SAICM, as a unique global initiative involving all the relevant sectors and stakeholders, has been touted as the overarching governing mechanism dealing with the complexity of chemical safety issues at the national, regional and global levels in the context of sustainable development. While the multilateral chemical-related conventions are contributors to sound chemicals management, they only cover specific chemicals and selected aspects of chemical safety.

SAICM recognizes the complexity of sound chemicals management and the adverse effects that various chemicals interactions can have on human health and the environment, and provides a multi-stakeholder forum for discussing these complex issues, identifying emerging issues, and finding feasible and attainable solutions to global problems. As the international community focuses on agreeing the SDGs that will drive the global sustainable development agenda, the platform provided under SAICM, including its 20 indicators of progress, becomes an important forum to discuss how to integrate the sound management of chemicals into the SDGs.

Looking ahead to 2020, chemicals production and consumption are increasing around the world. The relative shift of production from highly industrialized to developing countries, the rapid expansion of chemicals production and use in developing countries, and the economic implications of this increase under unsound chemicals management are global realities.

The UN's post-2015 agenda will be critical to ensuring technical and financial resources are directed towards addressing the most basic chemical challenges that stakeholders currently face and to taking strategic action on these challenges. With only seven years remaining to reach the Strategic Approach 2020 goal of sound chemicals management, it is all the more imperative for concrete collaborative actions to turn aspirations into practical implementation and delivery on the ground. The fourth session of the International Conference on Chemicals Management, scheduled to convene in 2015, will take important decisions defining the chemicals and waste agenda for the five year period in the lead up to 2020, in particular through the 'Overall Orientation and Guidance towards the SAICM 2020 Goal'. These decisions could jump-start the implementation process on global goals for sound chemical management for the post-2015 period.

SUSTAINABILITY: THE EMERGING CONTEXT OF BUSINESS OPERATIONS

You can think of sustainability as a feelgood issue that you can support when it's convenient to do so. Or you can think of it as the emerging context in which businesses must operate as customers engage more with the concept of being green, as competitors adopt sustainability practices for business advantage, and as governments regulate behavior to reduce pollution and ensure long-term stocks of raw materials, or to seek more positive relationships between society and industry.

In PwC's 14th Annual Global CEO Survey (2011), CEOs saw big opportunities in making their enterprises responsive to social expectations. Whether they primarily sell to businesses, government, or consumers, half of the CEOs anticipated changes to their business strategies because customers are factoring environmental and social responsibility practices into purchasing decisions.

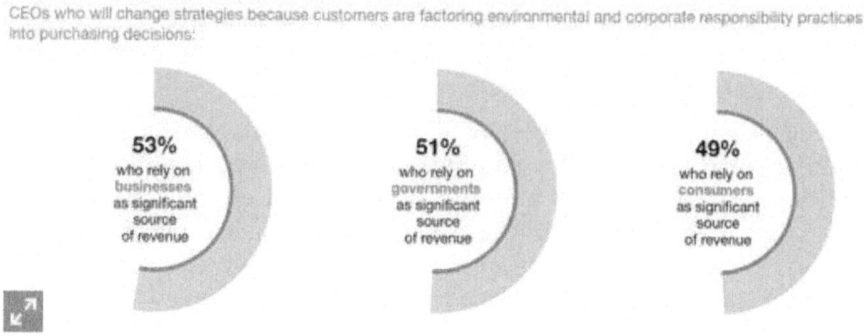

Fig. : CEOs see big opportunities to make their companies more responsive to society's expectations.

Failure to broadly adopt sustainable business practices will tarnish corporate reputations, which happened inthe 1990s when the extensive use of child labor by apparel and other manufacturers hurt their reputations.

The lack of genuine sustainability practices will cause scarcity of essential materials (through pollution, depletion, and waste), from foodstuffs to fuels, from wood to water. This scarcity will raise the costs of inputs and limit the product and service options available. Failure to act now presents risks to the future growth and long-term viability of businesses.

Conversely, the adoption of real sustainability practices will burnish corporate reputations and, most importantly, allow continued economic and business success for the long term.

But how do businesses make sustainability part and parcel of how they operate, how they source, how they manufacture, how they distribute, and how they service? Addressing sustainability will require a behavior shift.

"A lot of companies still think of sustainability as: 'I have my business plan. How do I layer sustainability thinking on top of that?'" Kepler says. "Our view is that sustainable development is a combination of how you achieve economic value, along with how you integrate the environmental and social impact into strategy and operations." This prompts the need to internalize the concerns around environmental and social impact, which if kept external result in an approach where sustainability becomes a separate activity, layered on top of existing operations.

Dow created a new metric called the Sustainable Chemistry Index (SCI), which aggregates the economic performance of sales with the environmental and social impact, and is regularly measured and reported. This measure integrates the life cycle analysis and impact of products with business thinking and decisions, so ongoing actions result in a more sustainable manufacturer and solution provider. For example, business unit leaders include a review of their SCI performance as part of the business strategy review.

Similarly, SAP established 400 sustainability key performance indicators (KPIs) and embedded them in existing processes to create visibility of sustainability performance. "Every employee can go to the portal and see the computing stats, review the energy consumption trends in buildings, and discover printing data for every printer," Graf says.

The next step is to provide feedback on individual actions. "We are also looking at a project to see if we can break down our overall footprint on an individual basis," Graf says. "Such visibility changes behavior."

Among its methods, Intel links individual compensation to environmental performance, promotes employee sustainability groups, funds innovative environmental projects, and rewards employees who deliver significant sustainable impact. "A key role and challenge for the Intel IT sustainability program office is to start embedding sustainability oriented decision making, activities, and best practices throughout our daily business activities, so we can transition sustainability from a program to a mindset," says Chris Peters, IT director of the Industry Engagement Group at Intel.

Managing the Commons: Extending to Industry Value Chains

Sustainability is embedded when every part of the business makes decisions that intelligently weigh the economic, environmental, and social impacts on the long-term ability to sustain the business.

The sustainability efforts of any one entity can have a limited impact and can put a business at a disadvantage if others in the value chain, upstream or downstream, do not act. Often the actual impact of any individual sustainability behavior is unclear, beyond those activities that have a direct cost, such as the purchase of paper and electricity.

The good news is that awareness of sustainability issues, coupled with customer pressure and government regulation, is causing entire business networks—value chains and cross industry groups—to act together to adapt their processes to support sustainability. This joint action is becoming increasingly possible as standards emerge to measure and value the sustainability attributes involved, such as energy efficiency, carbon emissions, water usage, resource renewability, labor practices, trade practices, and social impacts.

These standards—and the underlying metrics—set the stage for understanding the impact of sustainability attributes, which means that individual businesses can make decisions that have a known impact across the ecosystem. For example, in 2007 Nike created the Considered Apparel Index to score the environmental attributes of its apparel. In 2010, it released a web interface to this index, called the Environmental Apparel Design Tool, which designers use early in the product creation process to inform design and development decisions and to reduce the environmental impacts related to materials, manufacturing waste, and garment treatments. The tool allows suppliers to understand Nike's approach to scoring apparel products for sustainability while providing feedback to Nike.

"A key part of the strategy is transparency," says Tom Sedory, IT director of strategy and sustainability at Nike. "Our efforts have to be visible to the industry to let others see, evaluate, use, and develop further."

The Journey to Embed Sustainability in Business

Few people would argue against protecting the environment or responsible citizenship. But few businesses make decisions with environmental and social impact as a routine concern. "Sustainability should be embedded as a natural extension to the existing processes," suggests Jon Chorley, chief sustainability officer of Oracle Corporation. "It needs to transform from a project-based orientation to an operations orientation."

To achieve an operations orientation, PwC believes that companies will move through a maturity path along a continuum that spans compliance, obligation, efficiency, and leadership. In the process, they will embed sustainability in operations.

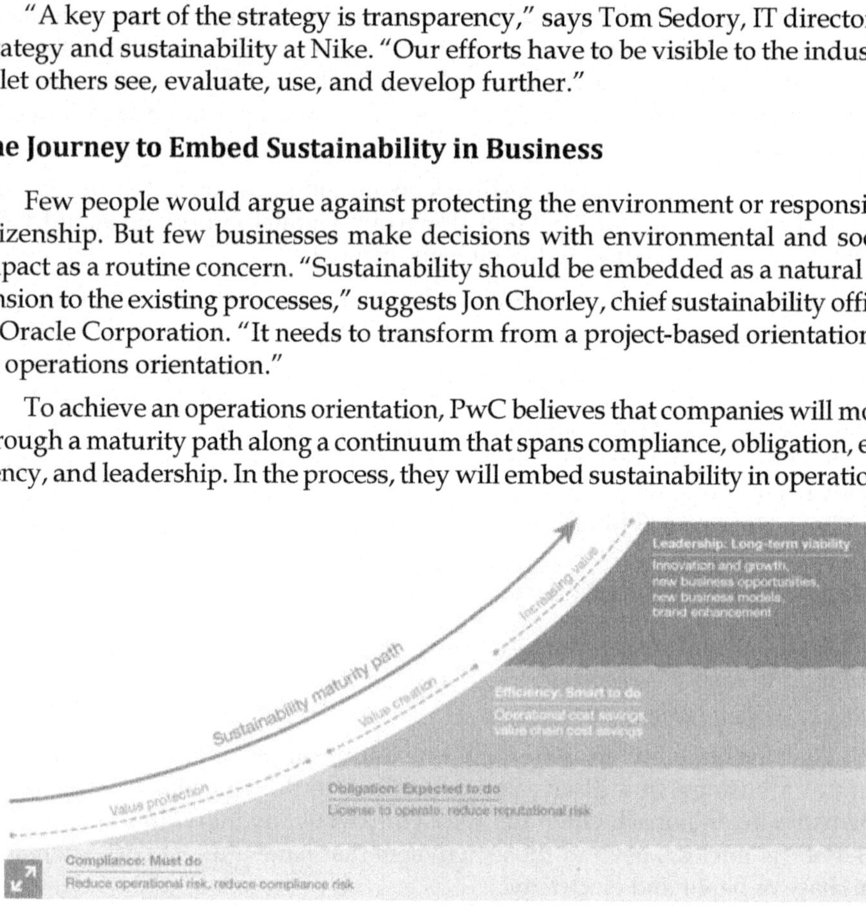

Fig. : The sustainability maturity path: as organizations embed sustainable practices in operations, they move along the continuum from compliance to obligation to efficiency to leadership. Information technology is an enabler of this journey.

The Must Do : Complying with Regulations

Compliance with government regulations is a key external driver and the legally required must-do task. Regulations large and small will constrain decisions and options. Some are indirect, such as recent US federal regulations mandating greater light bulb efficiency that alter the products available for purchase. Others are quite direct, such as the carbon cap-and-trade schemes deployed in Europe, California, and Australia, and China's recent tightening of permissible emissions by commercial vehicles.

It can be difficult to assess the direct value for complying with regulations beyond avoiding civil penalties. But the economic and other impacts used to justify the regulations can be helpful in setting a common valuation for various sustainability activities, which then can create direct enterprise value. While compliance alone has been an adequate corporate response in the past, social attitudes toward sustainable practices will have more impact in the future. Just being compliant will not be enough, because regulations will be slow to catch up to market expectations and competitive reality.

The Obligated to Do: License to Operate

Dealing with stakeholders' perceptions of corporate obligation to act is the second stage of maturity. It is typically associated with the large-scale impacts of industry and what to do about them. Larger enterprises attract the highest expectations, or obligations, to act.

"When multinationals have broad footprints across the world and operatein as many countries as they do, expectations are associated with the companies that are beyond just making a profit for their shareholders back in some other part of the world," says Gary Niekerk, director of global citizenship at Intel.

"Energy [and other resource] management will likely become the basis for organizational and national competitiveness, with financial and market advantages accruing to those that systematically and strategically optimize their use of energy and resources."

—Pat House, C3

Even if their practices are not much different from others or not under their direct control, industry leaders are expected to "do the right thing" or risk reputational damage and lost business if they don't. Many original equipment manufacturers (OEMs) and brands that have a large impact on the market have been a subject of criticism for the labor or manufacturing practices of their suppliers. Enterprises that have taken appropriate actions have benefited from bolstering their reputations. Meeting the obligation by promoting sustainable methods becomes a source of differentiation from competitors. In meeting the obligations, companies can turn environmental and social challenges into new business opportunities. By seizing these opportunities, they shape the world and advance toward leadership.

The Smart to do: Efficient Use of Resources

Efficiency has been a key internal driver for sustainable practices and is the lowhanging fruit when moving beyond compliance and obligation. Waste of resources is costly. Reuse and reduction of waste reduces cost and creates new revenue streams. Efficiency in all forms reduces costs, and so it is no surprise that many organizations have promoted sustainability efforts that result in reduced direct costs.

"A key role and challenge for Intel IT... is to start embedding sustainability-oriented decision making, activities, and best practices throughout our daily business activities."

—Chris Peters, Intel

Common examples include reduced energy usage in the data center, reduced water usage in manufacturing and office facilities, reduced paper usage in the office, and reduced material usage in packaging. These efficiencies reduce costs while also aiding sustainability through the reduced use of resources. Dow, for example, has saved $9.8 billion since 1994 from energy-efficiency efforts that required an investment of less than $2 billion.

Such efficiencies make sense with or without a sustainability impact, but pairing their contributions to the bottom line with their environmental and social impact creates an even greater incentive to support sustainability.

Leadership: Embedding Sustainability

The fourth and most mature stage of becoming a sustainable business is market leadership. Sustainability leaders embed real, measurable, ongoing commitments to sustainability practices as a strategic differentiator, going beyond the immediate benefits of compliance, obligations, and efficiency. Sustainability is embedded when every part of the business makes decisions that intelligently weigh the economic, environmental, and social impacts on the long-term ability to sustain the business.

When embedded, sustainability actually becomes a key driver of innovation and growth. "We've evolved our use of the term corporate responsibility. We now talk about it as sustainable business and innovation, and it is a key to how we think about driving long-term growth," Nike's Sedory says.

In the 2011 PwC survey, 64 percent of the CEOs said that environmentally friendly products and services are an important part of their innovation strategy. When linked to business strategy, sustainability becomes a core enterprise value. By becoming embedded in operations, sustainability objectives can become a part of long-standing measures of enterprise performance, such as profitability, innovation, growth, and industry leadership.

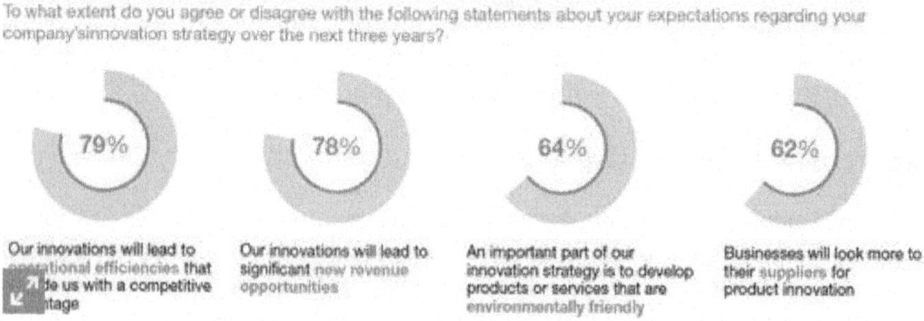

Fig. : Of the CEOs surveyed, 64 percent see environmentally friendly products and services as an important part of their overall innovation strategy.

The leadership opportunity is not limited to enterprises, but extends to countries. "Energy [and other resource] management will likely become the basis for organizational and national competitiveness, with financial and market advantages accruing to those that systematically and strategically optimize their use of energy and resources," predicts Pat House, senior vice president of strategy at C3.

"Technology is accelerating the use of sustainability as a driver of growth, particularly information technology, as it allows greater monitoring, independent verification, transparency, and accuracy of resource usage and its impact." —Alan McGill, PwC

Compliance is an "outside-in" approach in regard to where the goals are established and how individual stakeholders relate to the topic. Engaging with sustainability to meet obligations, then for efficiency, and ultimately for market leadership transcribes a path that converts outside concerns into inside engagement. It becomes a meaningful component of every business decision.

The Crucial Role of Information in Embedding Sustainability

IT is playing an important role in the transition from compliance to market leadership. "Technology is accelerating the use of sustainability as a driver of growth, particularly information technology, as it allows greater monitoring, independent verification, transparency, and accuracy of resource usage and its impact," says Alan McGill, a partner in PwC UK's Sustainability & Climate Change practice.

Sustainability has emerged as a key social issue due to one broadly accepted reality: environmental resources once believed to be limitless are increasingly seen as limited. Any limited resource needs to be managed for efficient use. "We used to live in a world where energy was cheap and information was expensive. Now, information is a resource that is becoming limitless, and energy and other environmental resources are becoming constrained," says SAP's Graf.

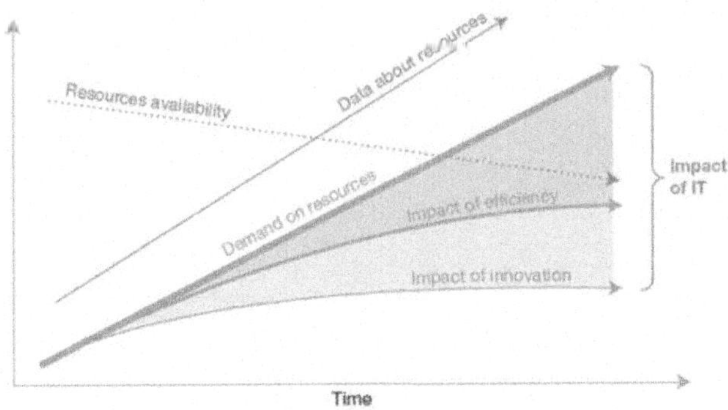

Fig.: As resource availability is reducing, the data about resources is increasing. Sustainability provides the context for using data and information technology to advance on efficiency and innovation.

In an increasingly digitized world, the sustainability journey is about using the limitless nature of information to become better at managing constrained resources. "Information technology transforms energy usage into something that is manageable," says Amit Chatterjee, founder and board member of Hara Software. IT is accustomed to managing the flow of materials in enterprise resource planning, supply chain management, and other systems. What has been missing is the flow of sustainability-specific information about those materials across the full value chain.

Over the long term, the role of IT is to provide information to help strike the balance between the availability of resources and the demand for resources. Achieving such balance has the potential to eliminate volatility in the supply or prices of necessary resources, avoid the unintended consequences of business operations, and address many of the grand challenges facing the world today.

Closing the Loop with Information

Many enterprises have started to look at the life cycle of their products and services and track the environmental and social impact by taking a cradle-to-cradle or cradle-to-grave view of them— the closed-loop resource system. Coupled with this system is the life cycle of information—from its collection to analysis for setting targets, to changes in business practices, and to reporting on performance—making up the closed-loop information system. When these two closed-loop systems— resource and information—conceptually interact, they become the basis for embedding sustainability in operations. The integration of operational information with sustainability impact enables employees to make decisions that weigh the environmental, social, and economic concerns with the long term ability to sustain the business. How adopting the closed loop framework makes advancement on a sustainability maturity path possible.

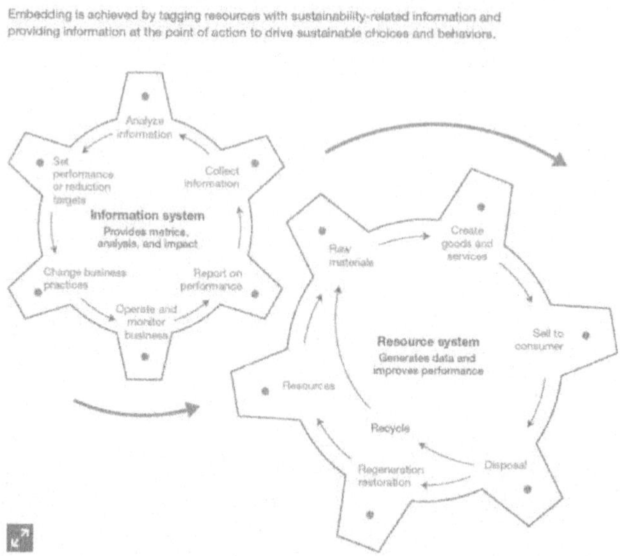

Fig. : The coupling of two closed-loop systems—one on resources and the other on information—provides a framework for embedding sustainability.

On this path, the role of information spans the following key objectives to demonstrate readiness for embedding sustainability.

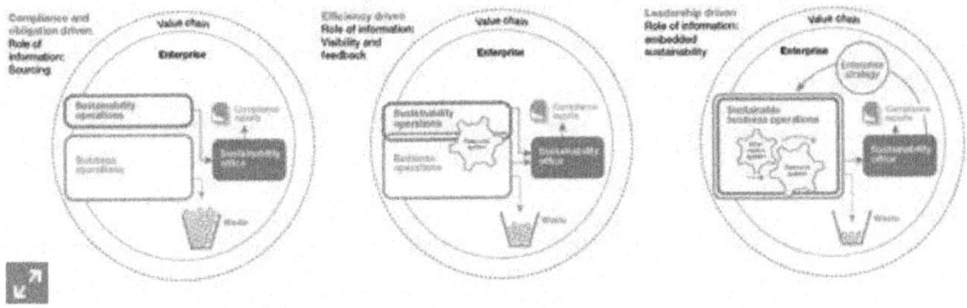

Fig. : The closed-loop systems enable the sustainability maturity journey.

Synthesize: Define and Collect Information

"Sustainability should be embedded as a natural extension to the existing processes. It needs to transform from a project based orientation to an operations orientation."

—Jon Chorley, Oracle

Much sustainability activity today is driven by the need to reduce the environmental footprint. While there is an economic basis to do so, the underlying metrics are unlikely to be adequate for a future in which sustainability is strategic and drives the top line and the bottom line. Industry specific metrics are needed to integrate economic, environmental, and social dimensions across the life cycle of products and services. For example, Nike defined the Considered Apparel Index, the Outdoor Industry Association defined the Eco Index, and Dow defined the Sustainable Chemistry Index.

"Sustainable chemistry is a high-level way to look at life cycle management, the attributes of your business model, and the products you make, and then to evaluate its level of sustainability," Kepler says. "SCI touches all parts of our business. It looks at all [environmental, social, and profit impact] of them collectively and shows where you have an impact and where you can make changes," he adds. At Dow, SCI has become a basis for assessing the sustainability performance of products, processes, business units, business models, and individuals.

Table : Dow's Sustainable Chemistry Index blends the following eight factors to determine products that have sustainable chemistry advantages.

1. Renewable/recycled content	5. Social need
2. Resource management	6. Manufacturing/transportation
3. Life cycle benefit	7. Product application
4. Manufacturing efficiency	8. Public policy/end of life

Source: http://www.dow.com/sustainability/goals/chemistry.htm

Engage: Information at the Point of Action

Sustainability performance depends on individual actions taken by all employees. Raising awareness to drive employee engagement is essential to embed sustainability values. IT plays an important role here. "When you're developing and reporting and really working on being transparent, a whole new internal conversation occurs around all the indicators. It raises awareness and forces companies to look at how they're performing," says Ryan Whisnant, director of sustainability at SunGard.

Many people use e-mail signatures that promote awareness for reducing printing and paper usage. It would be more powerful if every time somebody goes to print, a pop-up says, "This print is going to cost $x. Do you really want to do that?" Most printers are not set up to provide that prompt, but policy based tools can allow this process level sustainability reinforcement. By bringing the information about the environmental impact to the point of action, individual action will more frequently be the sustainable choice.

"Acting sustainably – at a corporate and individual level – becomes the new normal if we can understand and control the impacts of our choices upstream and downstream on stakeholders we may never meet, as well as the 'boomerang' benefit on us at the point of action."

—Nic Delaye, PwC

SAP uses information this way. "When employees book a flight, we will tell them what the carbon footprint of that flight will be. Employees are empowered to take this additional information into consideration when making a decision to travel," Graf says. SAP has embarked on a project to break down its total environmental footprint to the individual level. Such visibility changes behavior, because individuals can compare against location or enterprise averages. "All these approaches are positively influencing behavior," Graf notes.

This type of embedding can go much further when the metrics and desired sustainability impact results are known across the value chain. "Acting sustainably—at a corporate and individual level—becomes the new normal if we can understand and control the impacts of our choices upstream and downstream on stakeholders we may never meet, as well as the 'boomerang' benefit on us at the point of action," says Nic Delaye, director in PwC's Sustainable Business Solutions practice. For example, Nike's Environmental Apparel Design Tool is used by designers to surface information about the environmental impacts of material, dye, and related choices in the value chain.

Integrate: Modeling the Unintended Consequences

Environmental and social considerations were historically seen as external factors, and most enterprises did not understand or model them. As a result, questions about footprint, end-of-life of products, and other considerations were never raised. "Such questions usually are not asked until after the product is developed and it's successful, and the result at times is unintended consequences on society or the environment," Dow's Kepler says. "Frankly, they're unintended many times because you didn't sit down and model market adoption impacts up front."

By having a model of the relationships among the various dimensions during the creation of products and business initiatives, many unintended consequences can be avoided or planned for. IT has a large role to play in such modeling.

"We are also looking at a project to see if we can break down our overall footprint on an individual basis. Such visibility changes behavior." —Peter Graf, SAP

In some cases, that means working with existing enterprise applications and data sources and applying them to financial, manufacturing, and other business processes. For example, the evaluation process for building a new campus could include energy-efficiency ratings, green tax incentives, off-gassing ratings, improved employee productivity and retention, reduced employee health costs, and sustainable-source warranties in materials used. The building industry has standard metrics and warranties in these cases. As standards evolve for, say, measuring toxicity impact, water usage, and carbon emissions, those factors could be added to existing processes and information systems as part of the evaluation criteria used within a business and across its value chain.

Such modeling and visibility should not be limited to the enterprise. They must extend to the value chain. When resources are tagged with sustainability related information, it's possible to know much more about the relative performance of value chain partners. "Buyers of goods and services can place demands on their suppliers and can actually know whether those demands are being met in ways they could never do before," Delaye says. Such visibility is essential, because when enterprises create sustainability targets, they often make assumptions about the performance they can get from value chain partners.

The Emerging Practice of Embedding with Financial Metrics

Much of the information for sustainability is nonfinancial, such as tons of greenhouse gas emissions, gallons of water used, tons of waste disposed, and so on. However, it's possible to translate this information into financial metrics. This appliedvaluation concept is familiar to all businesses. After all, money is a set of metrics and shared valuations for the basket of economic activities—goods, services, labor, intellectual property, goodwill, and so on. And financial metrics are the most fundamental measurement used across business decisions and processes.

Existing financials usually do not include the cost of inefficiency, the cost of environmental damage, and the present and future value of the environment and its resources that businesses and people use to sustain themselves. However, metrics are beginning to emerge that include these effects, opening the door to more programmatic sustainability decisions. The most complete accounting for these assets also incorporates people, culture, social infrastructure (such as education), and the macroeconomic impacts of enterprise behavior.

"Taking the nonfinancial drivers of sustainability that are critically important to a business and putting a valuation on them in economic terms means that people will understand the impact of sustainability concerns on their business much better," says PwC's McGill. "The organization will be better positioned to know where the risks and opportunities for innovation are and where they should focus their scarce resources and avoid potential disruptions to business."

Athletic apparel manufacturer PUMA has taken this approach to embedding sustainability, using emerging modeling and valuation methods to develop an environmental profit and loss statement. PUMA released the first environmental profit and loss statement in May 2011.

In many ways, emerging sustainability developments are redefining the notion of externalities and the definition of cost. "Once this is done, it's possible to use cost as an honest measure that also includes considerations for sustainability," forecasts Chorley of Oracle.

ENVIRONMENT, HEALTH AND QUALITY OF LIFE

Environment, Health, Life Expectancy and Social Inequalities are Linked

The environment plays a crucial role in people's physical, mental and social well-being. Despite significant improvements, major differences in environmental quality and human health remain between and within European countries. The complex relationships between environmental factors and human health, taking into account multiple pathways and interactions, should be seen in a broader spatial, socio-economic and cultural context.

In 2006, life expectancy at birth in the EU-27 was among the highest in the world — almost 76 years for men and 82 years for women. Most of the gain in life expectancy in recent decades has been due to improved survival of people above the age of 65, while before 1950 it was mostly due to a reduction in premature deaths (*i.e.* death below the age of 65). On average, men are expected to live almost 81% of their lives free of disability, and women 75%. There are, however, differences between genders, and between Member States.

The degradation of the environment, through air pollution, noise, chemicals, poor quality water and loss of natural areas, combined with lifestyle changes, may be contributing to substantial increases in rates of obesity, diabetes, diseases of the cardiovascular and nervous systems and cancer — all of which are major public health problems for Europe's population. Reproductive and mental health problems are also on the rise. Asthma, allergies, and some types of cancer related to environmental pressures are of particular concern for children.

The World Health Organization (WHO) estimates the environmental burden of disease in the pan-European region at between 15 and 20% of total deaths, and 18 to 20% of disability-adjusted life years (DALYs) (A), with a relatively higher burden in the eastern part of the region. The preliminary results of a study conducted in Belgium, Finland, France, Germany, Italy and the Netherlands, indicate that 6 to 12% of the total burden of disease could be attributed to nine selected environmental factors, out of which particulate matter, noise, radon, and environmental tobacco smoke were leading. Due to uncertainties, the results need to be interpreted with caution as an indicative ranking of environmental health impacts only.

The significant differences in the quality of the environment across Europe depend on the varying pressures related, for example, to urbanisation, pollution and natural resource use. Exposures and associated health risks, as well as the benefits of pollution reduction and of a natural environment, are not uniformly distributed within populations. Studies show that poor environmental conditions affect vulnerable groups especially. The evidence is scarce, but shows that deprived communities are more likely to be affected; for example, in Scotland, mortality rates in people aged under 75 in the 10% most deprived areas were three times higher than those in the 10% least deprived.

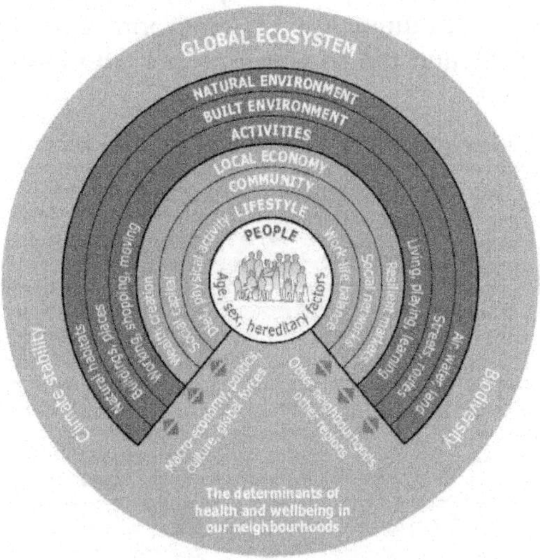

Fig. : The health map.

Better understanding of differences in the social distribution of environmental quality can be helpful for policy, since specific population groups, such as those on low incomes, children, and the elderly, may be more vulnerable — mostly due to their health, economic and educational status, access to health care, and lifestyle factors that affect their adaptation and coping capacities.

Environmental Burden of Disease — Estimating the Impacts of Environmental Factors

The environmental burden of disease (EBD) represents the proportion of ill health attributed to exposure to environmental factors. Use of the EBD approach allows: comparison of health losses due to different risk factors; setting priorities; and evaluating the benefits of specific measures. However, the results are likely to underestimate the overall environmental burden as they focus on single risk factors and health outcomes, rather than taking full account of complex causal pathways. Estimates of similar issues may vary, depending on the underlying assumptions, methods and data used; and, for many risk factors EBD estimates are not yet available.

Attribution of the role of the environment in the development of diseases, and the development of novel assessment approaches aimed at taking the inherent complexity and uncertainty of environment and health interactions into account, remain a subject of intense debate.

An Environment not Giving Rise to Harmful Effects on Health

The main European policies aim to provide an environment in which the level of pollution does not give rise to harmful effects on human health and the environment, and vulnerable population groups are protected. They are the 6th Environment Action Programme, the EU Environment and Health Strategy and Action Plan 2004–2010, and the pan-European WHO Environment and Health process.

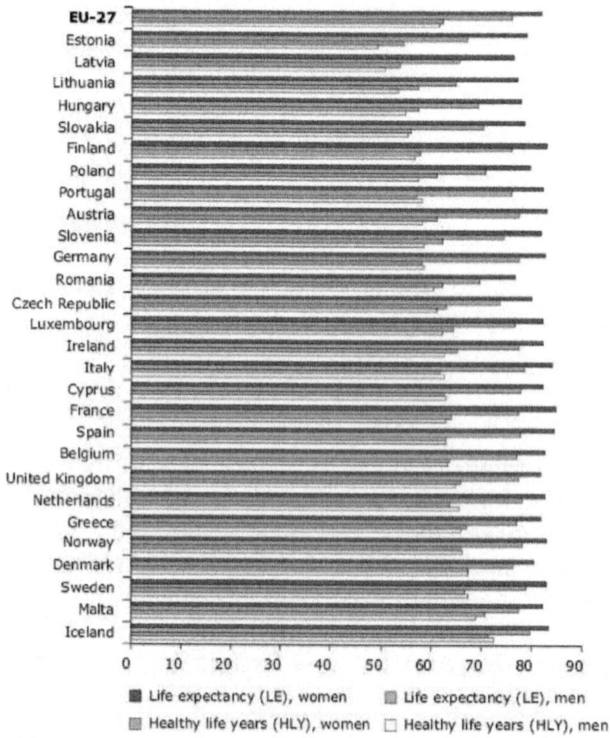

Fig. : Life expectancy and healthy life years at birth in EU-27, Iceland and Norway in 2007, by gender.

Note: Healthy life years (HLY) at birth – the number of years a person at birth is expected to live in a healthy condition. Life expectancy (LE) at birth – the number of years a newborn child is expected to live, assuming that the age-specific mortality levels remain constant. Data coverage: no HLY data for Bulgaria, Switzerland, Croatia, Liechtenstein, and the former Yugoslav Republic of Macedonia. Time coverage: 2006 data used for LE for Italy and EU-27.

Several areas for action have been identified, related to air and noise pollution; water protection; chemicals, including harmful substances such as pesticides; and improving the quality of life, especially in urban areas. The Environment and Health process aims at achieving a better understanding of the environmental threats to human health; reducing the disease burden caused by environmental factors; strengthening EU capacity for policymaking in this area; and identifying and preventing new environmental health threats.

While EU policy emphasis is on reducing pollution and the disturbance of crucial services provided by the environment, there is also a growing recognition of the benefits of the natural, biologically diverse environment to human health and well-being.

Furthermore, it is worth noting that most health-related pollution policies are targeted to the outdoor environment. A somewhat neglected area in this regard is the indoor environment – considering that European citizens spend up to 90% of their time indoors.

Indoor Environment and Health

The quality of indoor environment is affected by ambient air quality; building materials and ventilation; consumer products, including furnishings and electrical appliances, cleaning and household products; occupants' behaviour, including smoking; and building maintenance (for example, energy saving measures). Exposure to particulate matter and chemicals, combustion products, and to dampness, moulds and other biological agents has been linked to asthma and allergic symptoms, lung cancer, and other respiratory and cardiovascular diseases.

Recent assessments of the sources of, exposure to and policies related to indoor air pollution have analysed the benefits of different measures. The highest health benefits are linked to smoking restrictions. Building and ventilation policies that control indoor exposure to particulate matter, allergens, ozone, radon and noise from outdoors offer high long-term benefits. Better building management, prevention of moisture accumulation and mould growth, and prevention of exposure to exhausts from indoor combustion can bring substantial medium to long-term benefits. Substantial short to medium term benefits result from harmonised testing and labelling of indoor materials and consumer products.

For Some Pollutants Ambient Air Quality has Improved, but Major Health Threats Remain

In Europe, there have been successful reductions in the levels of sulphur dioxide (SO_2) and carbon monoxide (CO) in ambient air, as well as marked reductions in NOX. Also, lead concentrations have declined considerably with the introduction of unleaded petrol. However, exposure to particulate matter (PM) and ozone (O_3) remain of major environment-related health concern, linked to a loss of life expectancy, acute and chronic respiratory and cardiovascular effects, impaired lung development in children, and reduced birth weight.

Over the past decade, ozone concentrations have frequently and widely exceeded health- and ecosystem-related target values. The Clean Air for Europe (CAFE) programme estimated that at current levels of ground-level ozone, exposure to concentrations exceeding the health-related target value (B) is associated with more than 20000premature deaths in EU-25 (C) annually.

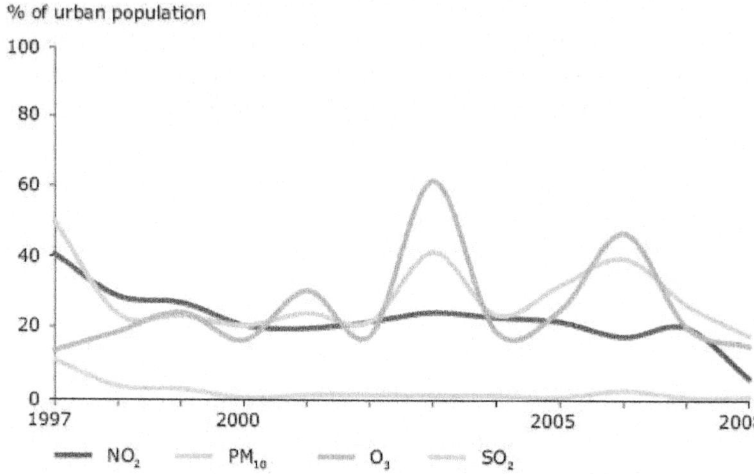

Fig. : Percentage of urban population in areas where pollutant concentrations are higher than selected limit/target values, EEA member countries, 1997–2008.

Note: Only urban and sub-urban background monitoring stations are included. Since O_3 and the majority of PM10 are formed in the atmosphere, meteorological conditions have a decisive influence on the airborne concentrations. This explains at least partly inter-annual variations and for example the high O_3 levels in 2003, a year with extended heat waves during summer.

In the period 1997 to 2008, 13 to 62% of Europe's urban population was potentially exposed to ambient air concentrations of fine and coarse particulate matter (PM10) in excess of the EU limit value set for the protection of human health. However, particulate matter has no threshold concentration, thus adverse health effects can also occur below the limit values.

The fine-particulate fraction (PM2.5) represents a particular health concern because these can penetrate the respiratory system deeply and be absorbed into the bloodstream. An assessment of the health impacts of exposure to PM2.5 in EEA-32 countries in 2005 indicated that almost 5 million lost life years could be attributed to this pollutant. Reducing such exposure has recently been shown to bring measurable health gains in the United States of America, where life expectancy increased most in the regions with the largest reductions in PM2.5 over the past 20 years.

PM10 and PM2.5 concentrations are indicators of complex mixtures of pollutants and are used as proxies for the particulate characteristics responsible for the effects. Other indicators, such as black smoke, elemental carbon, and the number of particles, might provide a better link to the sources of pollution which need mitigation in response to specific health effects. This could be beneficial for targeted abatement strategies and setting air quality standards.

Evidence is increasing that the chemical properties and composition of particles, along with their mass, are important for health impacts. For example, benzo(a)pyrene (BaP), which is a marker of carcinogenic polycyclic aromatic hydrocarbons, is emitted mainly from the burning of organic material and mobile sources. High levels of BaP occur in some regions, such as the Czech Republic and Poland. The increasing wood burning in homes in some parts of Europe may become an even more prominent source of such hazardous pollutants. Climate change mitigation strategies may also play a role, by stimulating use of wood and biomass as domestic energy sources.

The 6th EAP sets the long-term objective of achieving levels of air quality that do not give rise to unacceptable impacts on, and risks to, human health and the environment. Its subsequent Thematic Strategy on air pollution set interim objectives through the improvement of air quality by 2020. The Air Quality Directive has set legally binding limits for PM2.5 and for organic compounds such as benzene. It has also introduced additional PM2.5 objectives, based on the average exposure indicator (AEI) to determine a required percentage reduction to be attained in 2020.

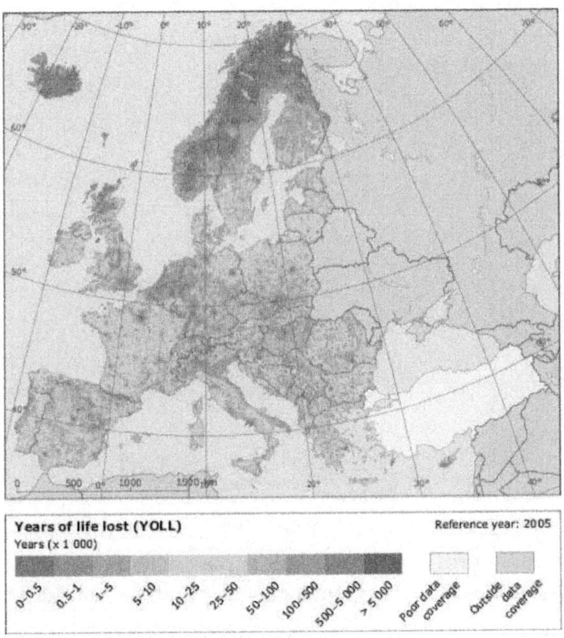

Map : Estimated years of life lost (YOLL) in reference year 2005 attributable to long-term PM2.5 exposure.

Furthermore, several international bodies are discussing the setting of targets for 2050 in relation to the long-term environmental objectives of European policies and international protocols.

Road Traffic is a Common Source of Several Health Impacts, Especially in Urban Areas

Air quality is worse in urban areas than in rural areas. Yearly average PM10 concentrations in the European urban environment have not changed significantly over the past decade. The main sources are road traffic, industrial activities, and the use of fossil fuels for heating and energy production. Motorised traffic is the major source of the PMfractions responsible for adverse health effects, which also come from non-exhaust PM emissions, for example, brake and tyre wear or re-suspended particles from pavement materials.

Meanwhile, road traffic injuries, with an estimated more than 4million incidents in the EU every year, remain an important public health issue. There were 39000 fatalities in the EU in 2008; 23% of fatal accidents in built-up areas affected people under the age of 25. Transport sources also account for a substantial proportion of human exposure to noise, which has negative impacts on human health and well-being. Data delivered in accordance with the Directive on Environmental Noise are available through the Noise Observation and Information Service for Europe.

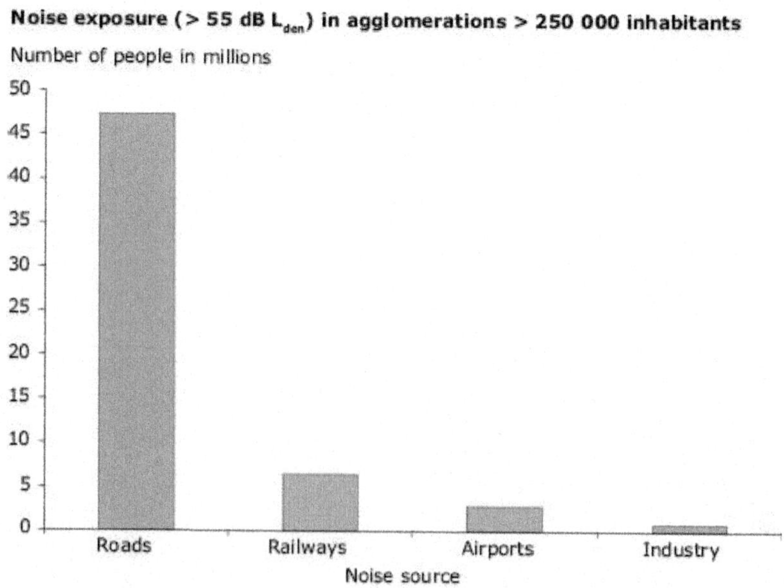

Fig. : The reported long-term (yearly average) exposure to day-evening-night noise (Lden) of more than 55dB in EU-27 agglomerations with more than 250000 inhabitants.

Approximately 40% of the population living in the largest cities in the EU-27 may be exposed to long-term average road traffic noise levels(I) exceeding 55 decibels (dB), and at night, almost 34million people may be exposed to long-term average road noise levels(I) exceeding 50 dB. The WHO night noise guidelines for Europe recommend that people should not be exposed to night noise greater than 40 dB. Night-time noise levels of 55 dB, described as 'increasingly dangerous to public health', should be considered as an interim target in situations where the achievement of the guidelines is not feasible.

According to a German Environmental Survey for Children, children from families of low socio-economic status are more heavily exposed to traffic, and annoyed by road traffic noise, during the day, as compared with children with higher socio-economic status. Urban air quality and noise often share a common source and may cluster spatially. There are examples, such as Berlin, of successful integrated approaches to reducing both local air pollution and noise levels.

Better Wastewater Treatment Has Led to Improved Water Quality, but Complementary Approaches May be Needed for the Future

Wastewater treatment, and the quality of both drinking and bathing water have improved significantly in Europe over the past 20 years, but continued efforts are needed to further improve the quality of water resources.

Human health can be affected through a lack of access to safe drinking water, inadequate sanitation, the consumption of contaminated freshwater and seafood, as well as exposure to contaminated bathing water. The bio-accumulation of mercury and some persistent organic pollutants, for example, can be high enough to raise health concerns in vulnerable population groups such as pregnant women.

Understanding of the relative contribution of different exposure routes is, however, incomplete. The burden of water-borne diseases in Europe is difficult to estimate and most likely underestimated.

The Drinking Water Directive (DWD) sets quality standards for water 'at the tap'. The majority of the European population receives treated drinking water from municipal supply systems. Thus, health threats are infrequent and occur primarily when contamination of the water source coincides with a failure in the treatment process.

While the DWD addresses water supplies serving more than 50people, a European data exchange and reporting system applies only to supplies for more than 5000 people.

In a 2009 survey, the compliance rate with drinking water standards in smaller supplies was 65%, while for larger ones exceeded 95%. In 2008, 10 out of 12 outbreaks of waterborne diseases reported in the EU-27 were linked to the contamination of private wells.

Implementation of the Urban Wastewater Treatment Directive (UWWTD) remains incomplete in many countries. However, EU-12 Member States have staggered transition periods for full implementation ranging up to 2018. The UWWTD addresses agglomerations with a population of 2000 or more; thus potential public health risks linked to sanitation exist in some rural areas of Europe. For these areas, complementary, 'low-technology' solutions are available.

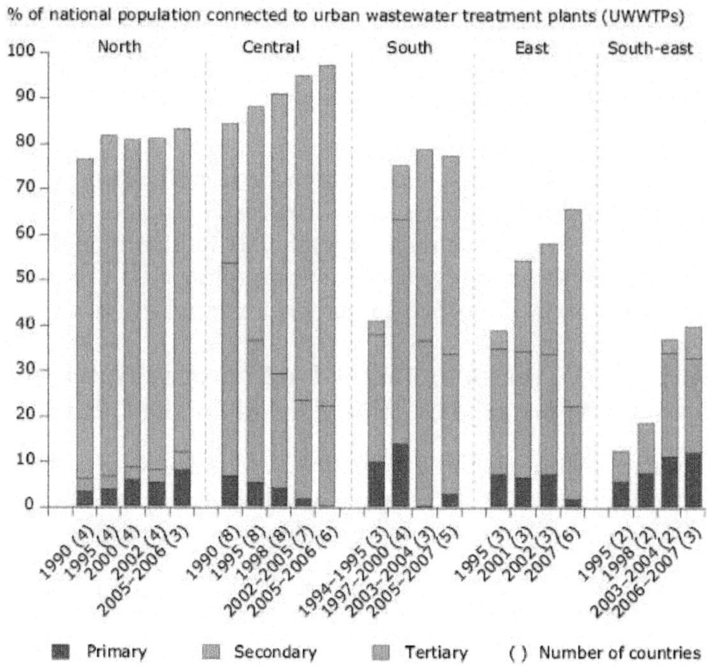

Fig. : Regional variation in wastewater treatment between 1990 and 2007.

Note: Only countries with data for virtually all of each period were included, the numbers of countries are given in parentheses. Regional percentages have been weighted by country population.

North: Norway, Sweden, Finland and Iceland.

Central: Austria, Denmark, England and Wales, Scotland, the Netherlands, Germany, Switzerland, Luxembourg and Ireland. For Denmark no data have been reported to the joint questionnaire since 1998. However, according to the European Commission, Denmark has achieved 100% compliance with secondary treatment and 88% compliance with more stringent treatment requirements (with respect to load generated) under the UWWTD. This is not accounted for in the figure.

East: Czech Republic, Estonia, Hungary, Latvia, Lithuania, Poland, Slovenia, Slovakia.

South-east: Bulgaria, Romania and Turkey.

The implementation of the UWWTD has led to an increasing proportion of Europe's population being connected to a municipal treatment works. The assoc iated improvements in wastewater treatment have resulted in a decline in the discharges of nutrients, microbes and some hazardous chemicals to receiving waters, and substantial improvement in the microbial quality of Europe's inland and coastal bathing waters.

Whilst wastewater treatment has improved, both point and diffuse pollutant sources are still significant in parts of Europe and health risks remain. For example, algal blooms linked to excessive nutrient levels, particularly during extended periods of hot weather, are associated with toxin-producing cyanobacteria – which, in turn, can cause allergic reactions, skin and eye irritation and gastroenteritis in exposed people. Large populations of cyanobacteria can occur in European water bodies used for drinking water, aquaculture, recreation and tourism.

Looking ahead, major investment will be needed to maintain existing wastewater treatment infrastructures. In addition, the discharge of some pollutants in treated effluent can raise environmental concerns, for example, endocrine-disrupting chemicals or pharmaceuticals. While wastewater treatment at municipal plants will continue to play a critical role, complementary approaches, such as tackling pollutants at source need to be explored more extensively.

New legislation related to chemicals (such as the Registration, Evaluation, Authorisation and Restriction of Chemical regulation (REACH) and the Environmental Quality Standards (EQS) Directive are likely to help drive such a source control approach. In combination with the full implementation of the Water Framework Directive, this should lead to a reduced emission of pollutants to water, leading to healthier aquatic ecosystems and reducing risks to human health.

Pesticides in the Environment Have Potential for Unintended Impacts to Wildlife and Humans

Pesticides disrupt essential biological processes, for example through affecting nerve transmission or mimicking hormones. Thus, human health concerns related to exposure *via* water, food, or close proximity to spraying have been raised. Due to their intrinsic properties, pesticides can also be harmful to organisms in the wider environment, including freshwater organisms.

Mixtures of pesticides are common both in the human food supply and in the aquatic environment. Though assessment of mixture toxicity has been a challenge, a single-chemical approach is likely to underestimate ecological risk, including impacts of mixture of pesticides on fish and amphibians.

The EU Thematic Strategy on the sustainable use of pesticides sets objectives to minimise the hazards and risks to health and the environment stemming from the use of pesticides, and to improve controls on the use and distribution of pesticides. Full implementation of the associated Pesticides Directive will be required to support the achievement of good chemical status under the Water Framework Directive.

Information on pesticides in surface and ground waters in Europe is limited; however, the reported levels, including pesticides classified as priority substances, can exceed environmental quality standards. Some pesticide impacts are not captured by routine monitoring programs – for example fatal exposure of aquatic species to short-term contamination during rainfall events immediately after pesticide application to cropland. These limitations combined with growing concerns about potential adverse effects strengthen the case for a more precautionary approach to their use in agriculture, horticulture and to control unwanted plant growth in public spaces close to where people live.

New Chemical Regulation May Help, but the Combined Effects of Chemicals Remain an Issue

Water, air, food, consumer products, and indoor dust can play a role in human exposure to chemicals through ingestion, inhalation or contact through skin. Of particular concern are persistent and bio-accumulative compounds, endocrine-disrupting chemicals and heavy metals used in plastics, textiles, cosmetics, dyestuffs, pesticides, electronic goods and food packaging. Exposure to these chemicals has been associated with declining sperm counts, genital malformation, impaired neural development and sexual function, obesity and cancer.

Chemicals in consumer goods may also be of concern when products become waste, as many chemicals migrate easily to the environment and can be found in wildlife, ambient air, indoor dust, wastewater and sludge. A relatively new concern in this context is waste electrical and electronic equipment, which contains heavy metals, flame retardants or other hazardous chemicals. Brominated flame retardants, phthalates, bisphenol A, and perfluorinated chemicals are most often discussed because of their suspected health effects and ubiquitous presence in the environment and in humans.

Possible combined effects of exposure to a mixture of chemicals found at low levels in the environment or in consumer goods, especially in vulnerable young children, are receiving particular attention. Furthermore, some adult diseases are linked to early-life or even prenatal exposures. The scientific understanding of mixture toxicology has recently been advanced significantly, not least as a result of EU-funded research.

While concerns about chemicals are growing, data for chemical occurrence and their fate in the environment, as well as for exposures and associated risks, remain scarce. There remains a need to establish an information system on concentrations of chemicals in various environmental compartments and in humans. New approaches and use of information technology offer the scope to do this effectively.

Furthermore, there is increasing recognition that cumulative risk assessment is necessary to avoid underestimation of risks that might occur under the current paradigm of considering substances on a chemical-by-chemical basis. The European Commission has been asked to take account of 'chemical cocktails' and to apply the precautionary principle in considering effects of chemical combinations when drafting new legislation.

Good management plays a crucial role in preventing and reducing exposures. A combination of legal, market-based and information-based instruments to support consumer choices is critical, given public concerns about the possible health effects of exposure to chemicals in consumer products. For example, Denmark has published guidelines on how to reduce children's exposure to chemical cocktails, focusing on phthalates, parabens, and polychlorinated biphenyls (PCBs). In the EU rapid alert system for non-food dangerous products, operating since 2004, chemical risks represented 26% of almost 2000notifications in 2009.

The Registration, Evaluation, Authorisation and Restriction of Chemical regulation (REACH) aims to improve the protection of human health and the environment from the risks of chemicals. Manufacturers and importers are required to gather information on the properties of chemical substances and propose risk management measures for safe production, use and disposal – and to register the information in a central database. REACH also calls for the progressive substitution of the most dangerous chemicals once suitable alternatives have been identified. However, the regulation does not address simultaneous exposure to multiple chemicals.

The efforts to better protect human health and the environment through safer chemical substitutes need to be complemented by a systemic approach to chemicals assessment. Such assessments should include not only toxicity and eco-toxicity, but also address the starting material, water and energy use, transport, release of CO_2 and other emissions, as well as waste generation through the life cycle of different chemicals. Such a 'sustainable chemistry' approach requires new, resource-efficient production processes and the development of chemicals that use fewer raw materials and are of high quality, with limited impurities to reduce or avoid waste – however, there is no comprehensive legislation on sustainable chemistry in place as yet.

Climate Change and Health is an Emerging Challenge for Europe

Nearly all the environmental and social impacts of climate change may ultimately affect human health through altering weather patterns, and through changes in water, air and food quality and quantity, ecosystems, agriculture, livelihoods and infrastructure. Climate change can multiply risks and existing health problems: potential health effects depend largely on populations' vulnerability and their ability to adapt.

The heat wave in Europe in summer 2003, with a death toll exceeding 70000, highlighted the need for adaptation to a changing climate. The elderly and people with particular diseases are at higher risk, and deprived population groups are more vulnerable. In congested urban areas with high soil sealing and heat absorbing surfaces, the effects of heat waves can be exacerbated due to insufficient nocturnal cooling and poor air exchange. For populations in the EU, mortality has been estimated to increase by 1 to 4% for each degree increase of temperature above a (locally-specific) cut-off point. In the 2020s, the estimated increase in heat-related mortality resulting from projected climate change could exceed 25000per year, mainly in central and southern European regions.

An anticipated impact of climate change on the spread of water-, food-and vector-borne (K) diseases in Europe emphasises the need for tools to address such threats to public health. Transmission patterns of communicable diseases are also influenced by ecological, social and economic factors, such as changing land-use patterns, declining biological diversity, alterations in human mobility and outdoor activity, as well as access to health care and population immunity. This can be exemplified by the shift in the distribution of ticks, vectors of the lyme disease and tick-borne encephalitis. Other examples include the extended range in Europe of the Asian tiger mosquito, avector of several viruses, with a potential for further transmission and dispersion under the changing climate conditions.

Climate change may also exacerbate existing environmental problems, such as particulate emissions and high ozone concentrations, and pose additional challenges to providing sustainable water and sanitation services. Climate-related changes in air quality and pollen distribution are expected to affect several respiratory diseases. Systematic assessments of the resilience of water supply and sanitation systems to climate change and inclusion of its impacts in water safety plans are needed.

Natural Environments Provide Multiple Benefits to Health and Well-Being, Especially in Urban Areas

Nearly 75% of European citizens live in urban areas, and this is expected to increase to 80% by 2020. Under the 6th EAP, the Thematic Strategy on the urban environment highlights the consequences for human health of the environmental challenges facing cities, the quality of life of urban citizens and the performance of cities. It aims to improve the urban environment, to make it more attractive and healthier to live, work and invest in, while trying to reduce the adverse environmental impacts on the wider environment.

The quality of life and health of urban dwellers depends strongly on the quality of the urban environment, functioning in a complex system of interactions with social, economic, and cultural factors. Green urban areas play an important role in this context. A multifunctional network of green urban areas is capable of delivering many environmental, social, and economic benefits: jobs, habitat maintenance; improved local air quality and recreation, to name a few.

The benefits of contacts with wildlife and access to safe green spaces for a child's exploratory, mental and social development have been shown both in urban and rural settings. Health is generally perceived to be better by people living in more natural environments, with agricultural land, forests, grasslands or urban green spaces near the place of residence. Furthermore, the perceived availability of green urban areas has been shown to reduce annoyance due to noise.

A Broader Perspective is Needed to Address Ecosystem and Health Links and Emerging Challenges

Much progress has been achieved through dedicated approaches to improving the quality of the environment and reducing particular burdens on human health – but many threats remain. The predominant drive for material well-being has played a major role in the biological and ecological disturbances witnessed today. Preserving and extending the benefits provided by the environment for human health and well-being will require continuous effort to improve the quality of the environment. Furthermore, these efforts need to be complemented by other measures, including significant changes in lifestyle and human behaviour, as well as consumption patterns.

Meanwhile, new challenges are emerging with a wide range of potential, highly uncertain, ecological and human health implications. In this context, technological advancements may provide new benefits – however, history also offers many examples of adverse health impacts from new technologies.

Nanotechnology, for example, may allow the development of new products and services which are capable of enhancing human health, conserving natural resources or protecting the environment. However, the unique features of nanomaterials also raise concerns about potential environmental, health, occupational and general safety hazards. The understanding of nanotoxicity is in its infancy, as are methods for assessing and managing the risks inherent in the use of some materials.

Given such knowledge gaps and uncertainties, an approach to responsible development new technologies, such as nanotechnologies, could be achieved through 'inclusive governance' based on broad stakeholder involvement and early public intervention in research and development. The European Commission has, for example, consulted experts and the public regarding the benefits, risks, concerns and awareness of nanotechnologies to support the preparation of a new action plan for 2010 to 2015.

The increasing awareness of multi-causality, complexity, and uncertainties also means that the EU Treaty principles of precaution and prevention are even more relevant than before. More recognition of the limits of what we can know, in time to prevent harm, is called for, as is the need to act on sufficient, rather than overwhelming, evidence of the potential harms to health, given the pros and cons of action versus inaction.

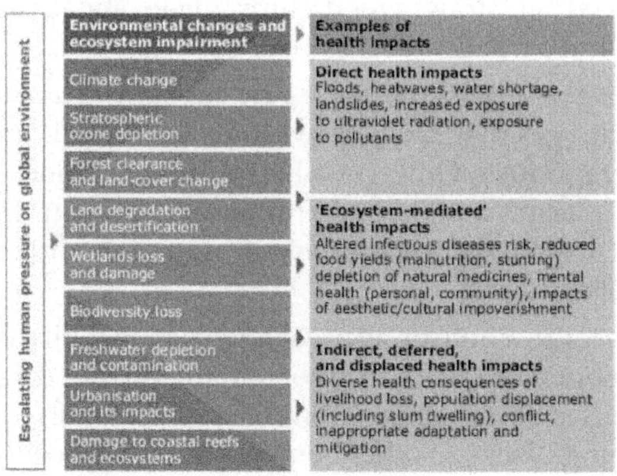

Fig. : Harmful effects of ecosystem change on human health.

Note: Not all ecosystem changes are included. Some changes can have positive effects (food production, for example).

SUSTAINABLE CHEMISTRY 2030

"In 2030, Stenungsund will become a manufacturing hub for sustainable products in the chemical industry. Our business is based on renewable raw materials and energy, and contributes to a sustainable society," says Lars Josefsson, President of INEOS.

The vision of a sustainable society and a thriving chemical industry was developed in 2010 by chemical companies AGA, Akzo Nobel, Borealis, INEOS and Perstorp - all active in the chemical cluster in Stenungsund on the Swedish west coast. Their chemical products are found in tubes, cables, plastics, paints, detergents, fuels, medicines and thousands of other places.

After World War II there was a rapid development of products based on fossil fuels, particularly oil and gas. The term "plastic age" was coined in 1951 when the first plastics exhibition was held in Sweden, "All about plastics."

Plastics are just one example of products manufactured in the chemical industry, and fossil fuels account for more than 90 percent of the raw materials. There are many reasons to try to break dependence on oil and gas, among them access to declining reserves together with the climate effects of burning fossil fuels.

Let's look at how green chemistry, good business and a sustainable future can become a reality.

Chemistry is the Science of Opportunity

In 20 years, large-scale fossil-based chemical industries could be relegated to history's scrap heap - if you believe the visionaries from Stenungsund.

The vision: "Sustainable Chemistry 2030" is based on the interplay between a number of different components, including:

- Transition to renewable resources. This is about biological materials such as rapeseed oil, straw, wood chips, algae, sugar cane, and waste from forestry and agriculture.
- One man's waste is another's raw material. Household waste can be turned into gas that can be used to make the building blocks of plastics and other products. Other possible raw materials include industrial waste, recycled plastics and more. With new technology, it is now possible to separate undesirable constituents from cables and recycle thousands of tons of plastic cable.
- That one takes a holistic approach to sustainability issues and that companies cooperate with each other and the surrounding community. Some companies are processing renewable raw materials and selling the chemical building blocks to other companies thereby contributing to the next step in the chemical process.

This 'Chemistry Cluster' in Stenungsund aims to use biogas and bio-ethanol plants to process renewable raw materials and various waste products. Excess heat from factories is sold back to the district heating system, the trains run on renewable fuels and purchased electricity is produced from renewable sources. The common heating plant is run on renewable fuels and waste.

This is a new kind of industrial ecosystem that is already happening. The infrastructure is mostly in place.

The really big news is that the organic molecules like ethylene, propylene, and natural gas in the future will no longer come from oil and gas but from renewable sources. The chemical processes in the factories are the same as before and the end products do not change.

"We were an integrated cluster that started this process in 2010 and formed a common vision. It required many players to achieve it, including the collaboration with academia, politicians and other industries. Another crucial reason why we achieved our vision was that there was a common and long-term chemistry policy," says Lars Josefsson, President of INEOS, at a seminar about 2030, where various speakers described how the project "Sustainable Chemistry 2030 "became a reality.

The Future

The focus on green chemistry and renewable raw materials has already begun in Stenungsund. Companies participating in the vision have begun the journey towards 2030, and here are some examples:

- Renewable fuel from rapeseed oil – In Stenungsund, Perstorp AB, has Scandinavia's largest plant for the production of biodiesel (RME). It produces most of the renewable fuel components that today is part of Swedish diesel and Perstorp RME stands out in terms of sustainability. Life cycle analysis has shown that more than 60 percent of fossil carbon is replaced by biodiesel, compared with the EU criteria for bio-fuels of 35 percent.
- Durable cables – Borealis is a new high-pressure plant that manufactures durable cable products. It cost SEK 4 billion to build. The company is now focusing more on developing materials for high-quality cables, especially for long-distance electricity transmission, with a lifespan of 50 years. Borealis is a world leader in the development of cable materials and has a research department of about 100 people. Electrical cables for power are in great demand and contribute to solving the world energy supply.
- Environmentally friendly car washing – Every year, Akzo Nobel in Stenungsund spends about SEK 80 million on research and development. The company has developed products such as the water-based degreaser used in car washes. The products have revolutionized the cleaning of hard surfaces and in a paradigm shift have evolved from the use of solvent-based cleaners to aqueous degreasing.

- Renewable paint components - Since 2010, Perstorp has manufactured the world's first renewable penta component (Voxtar) for aqueous alkyd paint. The raw materials consist of biomethanol and bioethanol. Voxtar is used to make about 15 percent of the ink and provides improved drying characteristics and stability.

Then, Now and in the Future

There are many pieces to fall into place for the vision of "Sustainable Chemistry in 2030" to be realized. As usual, it is a combination of factors relating to consumer demands, legislation, policy, education, and investment capital. Let us assume that we are sitting at a seminar in 2030 listening to explanations of why green chemistry created a thriving chemical industry. Perhaps the speakers present the following arguments:

- There were political instruments that were useful in creating sustainable chemistry. Politicians were far sighted, which made it possible for businesses to plan ahead and dare to invest in the conversion.
- Consumers continued to demand green products and actually bought such products to an increasing extent. The same applied to public tenders.
- Companies got wind in their sails, made money and continued to bet on the winning concept. Interest from investors increased and capital was used to invest in the Swedish chemical industry.
- The plastic and chemical industry succeeded in reaching out to the world and focused on changing their raw materials and promoting their benefits.
- Chemistry's reputation grew and chemistry programs at colleges and universities became increasingly attractive. Working at a chemical company was a high status job. The industry's commitment to innovation and collaboration with academia was a benefit to both sides.

Chapter 2

History and Drivers of Sustainability in the Chemical Industry

SUSTAINABLE MANUFACTURING

There is no unique and common definition for sustainable manufacturing but the US Department of Commerce's Sustainable Manufacturing Initiative describe it as: "The creation of manufactured products that use processes that minimize negative environmental impacts, conserve energy and natural resources, are safe for employees, communities, and consumers and are economically sound." We have to remember that sustainable manufacturing is part of a larger concept, sustainable development, which emerged in the early 1980's in response to increased awareness and concern over the environmental impact of economic growth and global expansion of business and trade.

In 1992 during the Earth summit in Rio de Janeiro, sustainable production was introduced and adopted as one of the guiding principles for business and governments in transitioning towards and achieving sustainable development. As sustainability is becoming an expected business practice by companies, large and small, sustainable manufacturing is being defined, developed and implemented by manufacturing companies and their networks of suppliers and customers. Today, emerging regulatory and social trends around sustainability create both pressures and opportunities for chemical companies at global and EU levels. Legislative requirements, stakeholder expectations and companies' own business and Responsible Care strategies are driving the development of more sustainable chemical products and supply chains. Clear trends are already surfacing:

- The introduction of REACH creates new pressures on specific
- Substances; the emergence of eco-design, Green Public Procurement (GPP),

- Ecolabel criteria and waste prevention schemes is creating demand for more sustainable products; rising consumer interest in sustainable goods is giving incentives
- Retailers to develop sustainability measures for their suppliers.

The move towards sustainable products will take many years to progress through legislative and business processes. During this time, retailers, consumers and non-governmental organizations (NGOs) will continue to call for transparency and clear statements about the constituents of the goods they purchase. In this period, doing business with a good environmental practice is an important aspect for investors, regulators, customers and the communities. Sustainable manufacturing can bring to industry an economic profit, a good reputation and can attract investments. These benefits are not just a game 2 for big business, but also new firms and small businesses can also play an exciting role. Start-ups, small and medium-sized enterprises (SMEs) with their flexible business models and less reliance on established ways of working, can also benefit, evolving and innovating quickly to gain advantage over on the competition.

For example the green marketplace value trillions; the global market for lowcarbon products is already estimated to be worth over 5 trillion USA Dollars.

Chemical industry

In May 2012 the European chemical industry published its first sustainability report, outlining the sector's visionto play a key role in ensuring that by 2050 over nine billion people live well, within the resources of the planet. Making this vision a reality means, among other things, that our industry strives to be sustainable in terms of its operations and to be a key enabler of a sustainable society through the excellence of its employees and the benefits of its products. It is important that sound science be used as a critical element to ensuring industry, society and governments make good choices in terms of product selection, technology options and framework conditions. Today, resource efficiency is becoming as important an element of the sustainability of products as chemicals safety. This document is an introduction to how sustainability may apply to chemicals in practice. It explains approaches that industry can use to make science-based choices in product development and assessment. It is particularly aimed at smaller companies, as part of Cefic's programme to help deliver the chemical industry's vision. The chemical industry has a very important role in the global industry. On a global scale it produces an annual turnover of more than € 1,700 billion, generating 9% of international trade. Chemical industry is placed upstream of various sectors such as construction, transport, food, health, personal cleanliness and home, clothing, electronics, etc..; this industry is also able to supply intermediate products for downstream industries and it contributes directly to create materials for the consumer market. Since chemical industry have a strategic position it is fully involved in the question of industrial sustainability outlined above.

Moreover it is often perceived by the public opinion as one of the biggest causes of the environmental degradation. Chemical industry has a structure very various, it includes different areas, for example: petrochemicals, chemistry for the production of industrial intermediates or products, inorganic and organic intermediates, polymers, dyes, medicines and products for agriculture. A reality so varied obviously multiplies the problems posed by sustainability, but at the same time increases the opportunities to resolve them. The issues connected with sustainability can be summarized as follows: pollution caused by chemical processes and products, risks arising from hazardous chemicals, reduction of sources of raw materials.

Chemical Industry, due to Responsible Care Programme improved its performance in all the three issues but still need to increase the performance to contribute to the sustainable development of the planet. In fact a significant proportion of final products of the chemical industry are outspread in the environment, not only by the chemical industry that created them but also by industries that hire them in various application areas and by the final consumers. We must also stress the difference between the negative effects that result from polluting agents and those from dangerous products. Highly toxic, flammable, explosive products can be the cause of the dramatic events that involve people and things but always on a local scale. The pollutants are certainly dangerous to the environment and can also act in insidious way for their long-term effects (effects that are not so easily and unequivocally recognizable) and on a global scale. The prevention of the effects of pollution and the protection of accidents caused by the release of chemicals must be kept distinct. The third problem identified is related to the depletion or exhaustion of the sources of the raw materials used, whether mineral or organic fossil, even for a local area, with consequent changes in the scenario of prices and the economic balance of the production and sales. This will lead to the development and use of new or modified products and processes to enable a shift towards the use of renewable resources as raw materials.

In recent year, also for a bigger awareness and sensitivity for this argument, we have an introduction of new legislations with stricter and stringent procedures for: industrial processes, characteristics of the products used and placed on the market, the quality and quantity of emissions and wastes generated. At the same time, also some managers, as part of Responsible care policy, have adopted voluntarily some technical measures for the reduction, for the waste treatment and, where possible, for the recycling of products at the end of their use. The introduction in Europe of REACH (Registration, Evaluation and Authorization of Chemicals) gave an important contribution to index chemical substances.

Life Cycle of Product

The chemical industry can help people live better, within the limits of the planet; so we need to find ways to provide the likes of housing, health and hygiene in the most resource efficient way, with minimum burdens placed on the environment. It applies not just to produce themselves, but to the whole product life cycle – from cradle-to-grave or even cradle-to-cradle. It includes the impacts of raw material sourcing, manufacture, packaging, transport and distribution, retail, use and then post-use recovery or disposal. Any sustainability assessment of products or services needs to be both integrated and based on a life cycle approach. The business of the Industry may be the manufacture of complete products that are destined for consumers or components and intermediate products that are transported to other businesses and that may become part of the final products for sale or products used to provide services. Whatever they make, their business uses resources and services provided by the natural environment (*e.g.* metals, materials, fossil fuels, soil, water, biodiversity) and discharges "by-products" (*e.g.* wastes, emissions) into the environment. As a consequence, all this actions have an impact on the environment. We can represent the pattern of the life cycle of a product as in figure:

Tab. 1 - The life cycle of a product

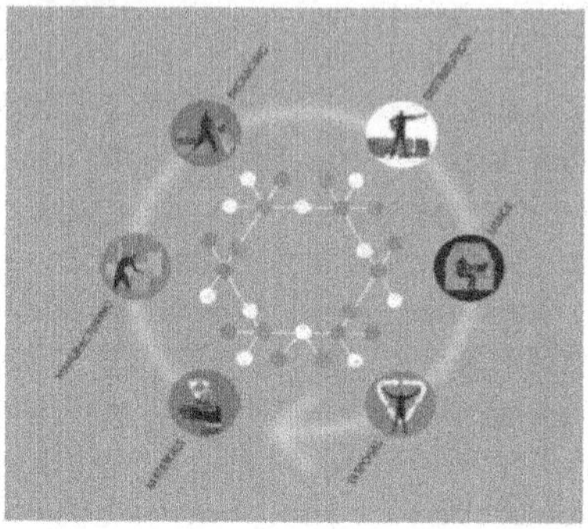

Source: CEFIC

Figure above shows an integrated life cycle; in this cycle we have to consider all three pillars of sustainability: environmental, social and economic.

Considering these aspects together helps identify how to deliver benefits in all areas, or at least to inform decisions if trade-offs between the different pillars are required.

Tab. 2 - Pillars of Sustainable Development

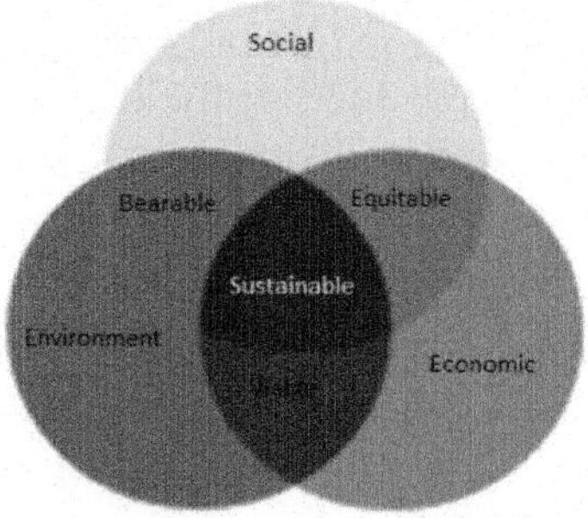

Source: European commission; Institute for Environment and Sustainability

The same applies to decisions within each pillar. For example, improvement in one environmental area - such as air, water, land or biodiversity - may have trade-offs in another area. It is important to integrate all pillars of sustainability in any decision-making process. We can see in the next figure the basic interaction between your facility and the environment and the impact it may have on the environment throughout the lifecycle of the products that it produces. Even though the actual production processes are far more sophisticated, your environmental impacts are principally formed in the following three stages:

Tab. 3 - Environmental impacts stages

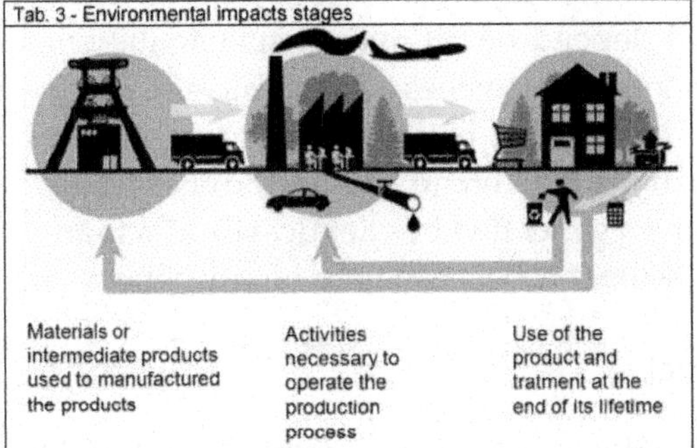

Source: European commission; Institute for Environment and Sustainability

What is a Closed Loop Approach?

Companies are increasingly moving from linear production models to closedloop value chains. This means that a by-product from one process becomes the raw material for another process. Closing the loop can help to improve resource efficiency and security of supply and reduce costs for waste management and disposal. The move towards closed- loop models is thus accelerated by rising commodity prices and increasing scarcity of resources. Moving from a linear to a circular economic model is also promoted by policymakers in some regions. "Circular economy" legislation has been enacted in countries such as Germany and China. We can see in the figure that there are different ways for closing the loop:

Tab. 4 - Ways for closing the loop

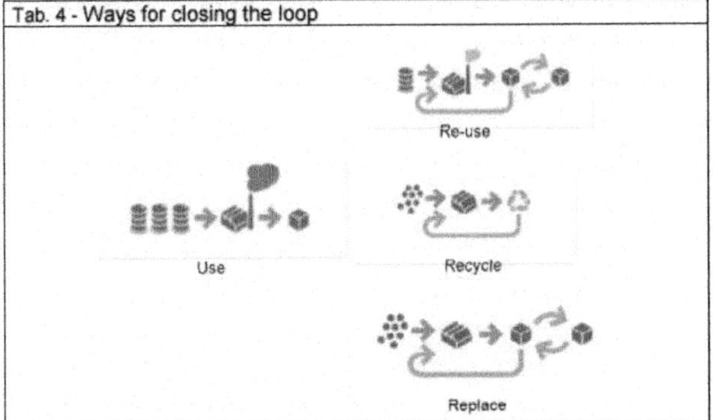

Source: European commission; Institute for Environment and Sustainability

- The simplest closed loop is re-use, where a product is re-used without
- Need for remanufacturing; when recovery and material recycling is needed, then closed loop
- Recycling involves taking the product or material back for use in the same application; a wider loop occurs where used materials are recovered for
- Subsequent use in a different application.

The priority for material use is that the loop is closed, rather than material being lost from the system and disposed to landfill. The choice between the different options will depend on an overall sustainability evaluation, which will include environmental effectiveness, economic efficiency and social acceptance. While closing, the loop makes both economic and environmental sense in many cases, it is important to take a broad view of the energy use, emissions and overall environmental impact generated by chemical products and processes. Tools such as Life Cycle Assessment (LCA) can help identify the most sustainable strategy.

Managing Products

Sustainable products provide environmental, social and economic benefits while respecting public health, welfare, and environment over their full life cycle, from the extraction of raw materials to final disposal. Sound product and process management practices are crucial to make this happen

Product Safety Management

A key aspect of sustainability is ensuring that products - existing and new - are safe for their intended uses. Product stewardship involves the active management of chemical products on site and beyond, and a dynamic engagement with downstream users and suppliers.

Process Management

Companies seek to optimise their processes to reduce energy and other resource use and waste generation. In some instances, this can be achieved by completely rethinking the way the substance is manufactured by using different technologies. For example, a membrane system could be used instead of a solvent to separate two substances. Chemistry based on renewable materials might present a number of opportunities in this area. For example, a fermentation process can be used instead of a traditional petrochemical one to manufacture some alcohols such as butanol. Similar to product safety, it is important that a life cycle approach be used when assessing process safety impacts of manufacturing changes. Improvements in one part of the chain can result in trade-offs in other parts of the chain. 10 Process safety risk assessments should be conducted across the supply chain to understand the full impact.

Product and Chemical Services

Sustainable management of chemicals also includes substance-by-service replacement, also known as chemical leasing. Chemical leasing is defined by the United Nations Industrial Development Organization (UNIDO) as a service-oriented business model that shifts the focus from increasing sales volume of chemicals, towards a value-added approach. The producer mainly sells the functions performed by the chemical, and functional units are the main basis for payment. Functional units can be, for example, the number of pieces cleaned or the amount of area coated. This system is used in some applications, but it is not applicable to all chemicals and their uses due to feasibility or performance issues. Chemical leasing is an option among other substitution possibilities, but its economic, environmental and social benefits should be demonstrated in each case.

EVALUATING PRODUCT SUSTAINABILITY

A range of assessment tools exist to evaluate the sustainability of products and processes, focusing on different aspects (environmental, economic and social) and using different methods.

Life Cycle Assessment (LCA)

Life cycle assessment methodologies were developed in the 1970s after the energy crisis. While life cycle thinking simply involves taking a life cycle approach, Life Cycle Assessment (LCA) is a structured tool for assessing environmental burdens across the whole product life cycle, either to identify improvement areas or to make comparisons with other product or service systems. LCA allows to asses the impact associated with all the phases of a process from cradle-to-grave, *i.e.* from definition till recycling. It provides tools for compiling inventories (energy, material inputs and environmental releases), evaluating the potential impacts associated with identified releases and inputs, and interpreting the results for decision making. A Standard exists today for LCA (ISO 14040 and 14044) and this is built on by a large usage community of practitioners and solution providers. GHG (Greenhouse Gases) emission reduction can be consider as a subset or simplified LCA, as current and future regulations on GHG reduction emission are based on existing LCA methodologies and practises. The LCA goal is to compare the full range of environmental effects assignable to products and services in order to improve processes, support policy and provide a sound basis for informed decisions. The term Life Cycle Scientific refers to the notion that a fair, holistic assessment requires the assessment of raw-material production, manufacture, distribution, use and disposal including all intervening transportation steps necessary or caused by the product's existence. A Life Cycle Assessment is carried out in four distinct phases as illustrated in the figure.

Tab. 5 - Life Cycle Assessment phases

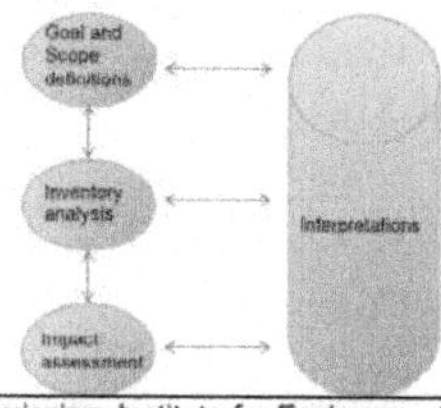

Source: European commission; Institute for Environment and Sustainability

Choosing the Right Parameters

In LCA, the functional unit is the unit of comparison between different products or services. As its name suggests, it is related to the function, or service delivered, by the product. The choice of the most appropriate functional unit is essential in any LCA, since it is the unit against which all of the life cycle impacts are normalized. For instance, in some cases it is appropriate to compare "per tonne of product produced", but for consumer goods it is often more appropriate to consider "per consumer use". Laundry detergents, for example, can be assessed per wash-load. It is also important that the geographical scope and the defined system boundaries of the data reflect the purpose of the study. For example, to inform decisions at an EU level, the study should reflect the product life cycle across the EU, using data with geographical relevance. Real situation data should be provided even if the production falls outside of the geographical scope of the EU.

LCA Tools and Databases

A variety of LCA software tools are commercially available to help conduct assessments. Examples include tools such as SimaPro, TEAM, Umberto, Gabi, and others. In some cases it is necessary to collect life cycle inventory data from the system under study, but there are also a range of sector- and industry-wide databases that can be used to run LCA studies. Examples include databases by Plastics Europe, EcoInvent, the European Reference Life Cycle Database (ELCD), and others. The type of question that is being asked will determine the type of data required, for example whether it should be industry average, best available technology or actual measured data. Using industry standard data will facilitate conducting LCA studies, but the results have to be interpreted with care. In particular, it is important to ensure that comparisons use data of comparable age, geography, quality standard and scope.

Data Analysis

A Life Cycle Analysis is only as valid as its data; therefore, it is crucial that data used for the completion of a life cycle analysis are accurate and current. When comparing different life cycle analyses with one another, it is crucial that equivalent data are available for both products in question. If one product has a much higher availability of data, it cannot be justly compared to another product which has less detailed data. Data validity is an ongoing concern for Life Cycle Analyses. Due to globalization and the rapid steps of research and development, new materials and manufacturing methods are continually being introduced to the market. This makes it both very important and very difficult to use up-to-date information when performing an LCA. If an LCA's conclusions are to be valid, the data must be recent.

Carbon Footprinting

Recent developments have focused on just one of the environmental impacts over the product or service life cycle - that of greenhouse gas emissions, or the "carbon footprint". This reflects the recent focus on climate change and the need to manage and reduce carbon emissions. A carbon footprint has historically been defined as "the total set of greenhouse gas (GHG) emissions caused by an organization, event, product or person." However, as calculating the total carbon footprint is impossible due to the large amount of data required. Carbon footprinting can help identify where significant greenhouse gas emissions arise in a product or service life cycle and help target improvements. But to ensure an overall environmental improvement, other environmental aspects - such as water - also need to be considered, as do social and economic aspects for an overall sustainability assessment. Standard methodologies for calculating a carbon footprint have been developed by several organizations. Examples include the British Standard BSI PAS 2050, ISO 14067 (in preparation) and WRI/WBCSD Greenhouse Gas Protocols.

EU Environmental Footprint Methodology

The EU Product Environmental Footprint is a multi-criteria measure of the environmental performance of a product or service throughout its life cycle. Product Environmental Footprint information is produced for the overarching purpose of seeking to reduce the environmental impacts of goods and services. The methodology is being developed by the Joint Research Centre (JRC). It is intended to become an integral part of the existing regulatory framework through its introduction as a mandatory requirement into specific pieces of legislation, such as the Ecolabel. The Product Environmental Footprint criteria take into consideration the recommendations of similar internationally recognized product environmental accounting methods and guidance documents.

Identification of Hotspots

A valuable aspect of life cycle thinking is identifying where environmental hotspots occur in product life cycles, so that strategies can be devised to address them. While a full LCA can require considerable time and resources, industry now has LCAs for many existing product categories, and these can be used to guide future efforts.

Economic and Social Analysis

Companies use a range of tools to assess and manage the economic aspects of production. Analogously to LCA, Life Cycle Costing (LCC) generates cost figures related to a product using the same system boundaries as in LCA, but focuses on its monetary impacts. There is no general standard which prescribes or describes how to conduct an LCC or ensure comparability between different applications. Therefore, institutions conducting LCA and LCC simultaneously need to address the issue of system boundaries and time scales in order to ensure consistency and comparability. LCA has to be combined with LCC in order to cover both the economic and environmental dimension of a complete sustainability analysis. Companies conducting LCC and LCA have to ensure that different departments, such as accounting, finance and environmental management, work hand in hand in order to avoid inefficient, redundant work and inconsistent data. The combination of environmental and economic values is known as Ecoefficiency, and will be covered by the upcoming ISO standard 14045. EcoEfficiency can be understood as economic-ecological efficiency. It links environmental impacts with economic issues by measuring the environmental impacts per monetary unit earned.

Assessing the social aspects of product sustainability is not an easy task. First, the social dimension of sustainability is a complex issue. Second, in addition to questions on assessment methods, there are very practical obstacles, such as the availability of data and the consensus on the procedure by industry and the public. A basic problem is that many of the social aspects – such as ensuring safe working practices, provision of highskill employment, or absence of child or forced labour in the supply chain – are qualitative aspects covered at the level of the company, rather than quantitative measures that can be attributed to individual products. Data on environmentally relevant inputs and outputs related to one product unit (usually 1 kg or 1 MJ) can be found in life cycle assessment databases, but so far there are no similar databases for social aspects

THE SITUATION NOWADAYS

Traditionally, the end of pipe system is the way to reduce the environmental impacts of industrial production activities. These technological solutions are not essential parts of the manufacturing process and don't change the process itself. Unlike end-of-pipe technologies, the cleaner production (or pollution prevention) is based on precautionary principle: the focus shifts to the cause of pollution, which is the industrial process. The concept of cleaner production also includes the efficient use of resources and reduction of waste. The improvement of environmental performance in accordance with the principles of cleaner production requires changes to processes, products, organizational structures. Although the practice of cleaner production remains within the organizational boundaries of each company, it is the first step towards a more integrated approach to the environment. It is in fact central to an eco-efficient production (doing more with less) and for the transition to the closed-loop production systems.

Applying the concepts of cleaner production at the macro level leads directly to the concept of industrial ecology. In Europe, a limited number of sectors of the economy contribute significantly to environmental pressures: agriculture, industry, electricity generation, transport, construction (buildings and infrastructure) and manufacturing base (refinery products, chemicals, non-metallic minerals, basic metals). The main eco-innovation in the chemical industry regarding waste minimization in chemical processes, replacing conventional chemicals with less toxic and less impact on the environment and the use of renewable raw materials for the production of chemicals and materials.

Open fields for ecoinnovation in the chemical concern heterogeneous catalysts, biocatalysis, the use of alternative solvents, the design of solvent-free chemical processes and the development of innovative technologies such as micro-reactors, reactors rotating disk, the continuous-flow reactors, reactors with microchannels and the catalytic membrane reactors. Much work has been done on alternative techniques to provide energy to chemical processes, in particular the use of radiofrequency and microwave. The use of renewable raw materials for the production of chemicals and materials is certainly the most promising field of eco-innovation, green growth of the chemical industry. The use of biomass in the production of bio-products commercial and industrial provides considerable benefits of environmental and socio-economic development.

One of the objectives of the Working Group was the creation of '"Yearbook on Industrial Research for Sustainable Chemistry". In this chapter it has been made an analysis of the various activities of research and development of some Companies associated with Federchimica. In particular, we examined the activities in the field of research for both incremental and innovative products and processes. The survey also considers the research activities aimed at obtaining a saving in the use of water and a reduction of carbon dioxide emissions into the atmosphere. The table also includes actions that companies play in the field of biotechnology research, treatment of wastewater and chemistry from renewable sources.

These figures show that it is also possible to obtain the economic benefits. Here is an example to some estimates provided by companies participating in the project, in which they were obtained, thanks to the research of new products and processes, sustainable savings for water consumption (about 22% per finished product), reduction of emissions (about 25% for greenhouse gases in 7 years) and waste (about 50% of industrial sites do not send waste to landfill, thanks to research for treatment on site). Finally, it is shown the importance of collaboration between companies and public research institutions, in fact, the exchange of information with Italian or foreign Universities optimizes the results reached in the research.

CHEMICAL INDUSTRY

Fig. Oil refinery - an example of chemical industry.

The chemical industry comprises the companies that produce industrial chemicals. Central to the modern world economy, it converts raw materials (oil, natural gas, air, water, metals, and minerals) into more than 70,000 different products.

History

Although chemicals were made and used throughout history, the birth of the heavy chemical industry (production of chemicals in large quantities for a variety of uses) coincided with the beginnings of the Industrial Revolution in general.

Industrial Revolution

One of the first chemicals to be produced in large amounts through industrial process was sulfuric acid. In 1736, the pharmacist Joshua Ward developed a process for its production that involved heating saltpeter and allowing the sulfur to oxidize and combine with water. It was the first practical production of sulfuric acid on a large scale. John Roebuck and Samuel Garbett were the first to establish a large-scale factory in Prestonpans in 1749, which used leaden condensing chambers for the manufacture of sulfuric acid.

In the early 18th century, cloth was bleached by treating it with stale urine or sour milk and exposing it to sunlight for long periods of time, which created a severe bottleneck in production. Sulfuric acid began to be used as a more efficient agent as well as lime by the middle of the century, but it was the discovery of bleaching powder by Charles Tennant that spurred the creation of the first great chemical industrial enterprise. His powder was made by reacting chlorine with dry slaked lime and proved to be a cheap and successful product. He opened a factory in St Rollox, north of Glasgow and production went from just 52 tons in 1799 to almost 10,000 tons just five years later.

Soda ash was used since ancient times in the production of glass, textile, soap, and paper, and the source of the potash had traditionally been wood ashes in Western Europe. By the 18th century, this source was becoming uneconomical due to deforestation, and the French Academy of Sciences offered a prize of 2400 livres for a method to produce alkali from sea salt (sodium chloride). The Leblanc process was patented in 1791 by Nicolas Leblanc who then built a Leblanc plant at Saint-Denis. He was denied his prize money because of the French Revolution.

However, it was in Britain that the Leblanc process really took off. William Losh built the first soda works in Britain at the Losh, Wilson and Bell works on the River Tyne in 1816, but it remained on a small scale due to large tariffs on salt production until 1824. When these tariffs were repealed, the British soda industry was able to rapidly expand. James Muspratt's chemical works in Liverpool and Charles Tennant's complex near Glasgow became the largest chemical production centres anywhere. By the 1870s, the British soda output of 200,000 tons annually exceeded that of all other nations in the world combined.

These huge factories began to produce a greater diversity of chemicals as the Industrial Revolution matured. Originally, large quantities of alkaline waste were vented into the environment from the production of soda, provoking one of the first pieces of environmental legislation to be passed in 1863. This provided for close inspection of the factories and imposed heavy fines on those exceeding the limits on pollution. Methods were soon devised to make useful byproducts from the alkali.

The Solvay process was developed by the Belgian industrial chemist Ernest Solvay in 1861. In 1864, Solvay and his brother Alfred constructed a plant in the Belgian town of Charleroi and in 1874, they expanded into a larger plant in Nancy, France. The new process proved more economical and less polluting than the Leblanc method, and its use spread. In the same year, Ludwig Mond visited Solvay to acquire the rights to use his process, and he and John Brunner formed the firm of Brunner, Mond & Co., and built a Solvay plant at Winnington, England. Mond was instrumental in making the Solvay process a commercial success; he made several refinements between 1873 and 1880 that removed byproducts that could slow or halt the mass production of sodium carbonate through use of the process.

Expansion and Maturation

The late 19th century saw an explosion in both the quantity of production and the variety of chemicals that were manufactured. Large chemical industries also took shape in Germany and later in the United States.

Fig. : The factories of the German firm BASF, in 1866.

Production of artificial manufactured fertilizer for agriculture was pioneered by Sir John Lawes at his purpose-built Rothamsted Research facility. In the 1840s he established large works near London for the manufacture of superphosphate of lime. Processes for the vulcanization of rubber were patented by Charles Goodyear in the US and Thomas Hancockin England in the 1840s. The first synthetic dye was discovered by William Henry Perkin in London. He partly transformed aniline into a crude mixture which, when extracted with alcohol, produced a substance with an intense purple colour. He also developed the first synthetic perfumes. However, it was German industry that quickly began to dominate the field of synthetic dyes. The three major firms BASF, Bayer and Hoechst produced several hundred different dyes, and by 1913, the German industry produced almost 90 percent of the world supply of dyestuffs and sold about 80 percent of their production abroad.

The petrochemical industry can be traced back to the oil works of James Young in Scotland and Abraham Pineo Gesner in Canada. The first plastic was invented by Alexander Parkes, an English metallurgist. In 1856, he patented Parkesine, a celluloid based on nitrocellulose treated with a variety of solvents. This material, exhibited at the 1862 London International Exhibition, anticipated many of the modern aesthetic and utility uses of plastics. The industrial production of soap from vegetable oils was started by William Lever and his brother James in 1885 in Lancashire based on a modern chemical process invented by William Hough Watson that used glycerin and vegetable oils.

By the 1920s, chemical firms consolidated into large conglomerates; IG Farben in Germany, Rhône-Poulenc in France and Imperial Chemical Industries in Britain. Dupontbe came a major chemicals firm in the early 20th century in America.

Products

Polymers and plastics, especially polyethylene, polypropylene, polyvinyl chloride, polyethylene terephthalate, polystyrene and polycarbonate comprise about 80% of the industry's output worldwide. These materials are often converted to fluoropolymer tubing products and used by the industry to transport highly corrosive materials. Chemicals are used to make a wide variety of consumer goods, as well as thousands of inputs to agriculture, manufacturing, construction, and service industries. The chemical industry itself consumes 26 percent of its own output. Major industrial customers include rubber and plastic products, textiles, apparel, petroleum refining, pulp and paper, and primary metals. Chemicals are nearly a $3 trillion global enterprise, and the EU and U.S. chemical companies are the world's largest producers.

Sales of the chemical business can be divided into a few broad categories, including basic chemicals (about 35 to 37 percent of the dollar output), life sciences (30 percent), specialty chemicals (20 to 25 percent) and consumer products (about 10 percent).

Basic Chemicals & Commodity Chemicals to Polymers and Speciality Chemicals

Basic chemicals, or "commodity chemicals" are a broad chemical category including polymers, bulk petrochemicals and intermediates, other derivatives and basic industrials, inorganic chemicals, and fertilizers. Typical growth rates for basic chemicals are about 0.5 to 0.7 times GDP. Product prices are generally less than fifty cents per pound.

Fig. : New polypropylene plant PP3 in the Slovnaft oil refinery (Bratislava, Slovakia)

Polymers, the largest revenue segment at about 33 percent of the basic chemicals dollar value, includes all categories of plastics and man-made fibers. The major markets for plastics are packaging, followed by home construction, containers, appliances, pipe, transportation, toys, and games.

- The largest-volume polymer product, polyethylene (PE), is used mainly in packaging films and other markets such as milk bottles, containers, and pipe.
- Polyvinyl chloride (PVC), another large-volume product, is principally used to make pipe for construction markets as well as siding and, to a much smaller extent, transportation and packaging materials.
- Polypropylene (PP), similar in volume to PVC, is used in markets ranging from packaging, appliances, and containers to clothing and carpeting.
- Polystyrene (PS), another large-volume plastic, is used principally for appliances and packaging as well as toys and recreation.
- The leading man-made fibers include polyester, nylon, polypropylene, and acrylics, with applications including apparel, home furnishings, and other industrial and consumer use.

The principal raw materials for polymers are bulk petrochemicals.

Chemicals in the bulk petrochemicals and intermediates are primarily made from liquefied petroleum gas (LPG), natural gas, and crude oil. Their sales volume is close to 30 percent of overall basic chemicals. Typical large-volume products include ethylene, propylene, benzene, toluene, xylenes, methanol, vinyl chloride monomer (VCM), styrene, butadiene, and ethylene oxide. These basic or commodity chemicals are the starting materials used to manufacture many polymers and other more complex organic chemicals particularly those that are made for use in the specialty chemicals category.

- Other derivatives and basic industrials include synthetic rubber, surfactants, dyes and pigments, turpentine, resins, carbon black, explosives, and rubber products and contribute about 20 percent of the basic chemicals' external sales.
- Inorganic chemicals (about 12 percent of the revenue output) make up the oldest of the chemical categories. Products include salt, chlorine, caustic soda, soda ash, acids (such as nitric acid, phosphoric acid, and sulfuric acid), titanium dioxide, and hydrogen peroxide.
- Fertilizers are the smallest category (about 6 percent) and include phosphates, ammonia, and potash chemicals.

Life Sciences

Life sciences (about 30 percent of the dollar output of the chemistry business) include differentiated chemical and biological substances, pharmaceuticals, diagnostics, animal health products, vitamins, and pesticides. While much smaller in volume than other chemical sectors, their products tend to have very high prices—over ten dollars per pound—growth rates of 1.5 to 6 times GDP, and research and development spending at 15 to 25 percent of sales. Life science products are usually produced with very high specifications and are closely scrutinized by government agencies such as the Food and Drug Administration. Pesticides, also called "crop protection chemicals", are about 10 percent of this category and include herbicides, insecticides, and fungicides.

Specialty Chemicals

Specialty chemicals are a category of relatively high valued, rapidly growing chemicals with diverse end product markets. Typical growth rates are one to three times GDP with prices over a dollar per pound. They are generally characterized by their innovative aspects. Products are sold for what they can do rather than for what chemicals they contain. Products include electronic chemicals, industrial gases, adhesives and sealants as well as coatings, industrial and institutional cleaning chemicals, and catalysts. In 2012, excluding fine chemicals, the $546 billion global speciality chemical market was 33% Paints, Coating and Surface Treatments, 27% Advanced Polymer, 14% Adhesives and Sealants, 13% additives and 13% pigments and inks.

Speciality chemicals are sold as effect or performance chemicals. Sometimes they are mixtures of formulations, unlike "fine chemicals," which are almost always single-molecule products.

Consumer Products

Consumer products include direct product sale of chemicals such as soaps, detergents, and cosmetics. Typical growth rates are 0.8 to 1.0 times GDP.

Consumers rarely if ever come into contact with basic chemicals but polymers and speciality chemicals are the materials that they will encounter everywhere in their everyday lives, such as in plastics, cleaning materials, cosmetics, paints & coatings, electronic gadgets, automobiles and the materials used to construct their homes. These speciality products are marketed by chemical companies to the downstream manufacturing industries as pesticides, speciality polymers, electronic chemicals, surfactants, construction chemicals, Industrial Cleaners, flavours and fragrances, speciality coatings, printing inks, water-soluble polymers, food additives, paper chemicals, oil field chemicals, plastic adhesives, adhesives and sealants, cosmetic chemicals, water management chemicals, catalysts, textile chemicals. Chemical companies rarely supply these products directly to the consumer.

Every year, the American Chemistry Council tabulates the U.S. production volume of the top 100 basic chemicals. In 2000, the aggregate production volume of the top 100 chemicals totalled 502 million tons, up from 397 million tons in 1990. Inorganic chemicals tend to be the largest volume, though much smaller in dollar revenue terms due to their low prices. The top 11 of the 100 chemicals in 2000 were sulfuric acid (44 million tons), nitrogen (34), ethylene (28), oxygen (27), lime (22), ammonia (17), propylene (16),polyethylene (15), chlorine (13), phosphoric acid (13) and diammonium phosphates (12).

Companies

The largest corporate producers worldwide, each with plants in numerous countries, include BASF, Bayer, Braskem,Celanese/Ticona, Arkema, Dow, DuPont, Eastman Chemical Company, Evonik Industries, ExxonMobil, Formosa Plastics, Givaudan,INEOS, LANXESS, LG Chem, LyondellBasell, Mitsubishi, Monsanto, PPG Industries, SABIC, Shell, Shin-Etsu, Sinopec, Solvay, Sumitomo Chemical and Wanhua along with thousands of smaller firms.

Fig. : Process engineers design, construct and operate plants.

Continents and Countries

In the U.S. there are 170 major chemical companies. They operate internationally with more than 2,800 facilities outside the U.S. and 1,700 foreign subsidiaries or affiliates operating. The U.S. chemical output is $750 billion a year. The U.S. industry records large trade surpluses and employs more than a million people in the United States alone. The chemical industry is also the second largest consumer of energy in manufacturing and spends over $5 billion annually on pollution abatement.

In Europe the chemical, plastics and rubber sectors are among the largest industrial sectors. Together they generate about 3.2 million jobs in more than 60,000 companies. Since 2000 the chemical sector alone has represented 2/3 of the entire manufacturing trade surplus of the EU.

In 2012 The chemical sector accounted for 12% of the EU manufacturing industry's added value. Europe remains world's biggest chemical trading region with 43% of the world's exports and 37%of the world's imports, although the latest data shows that Asia is catching up with 34% of the exports and 37% of imports. Even so, Europe still has a trading surplus with all regions of the world except Japan and China where in 2011 there was a chemical trade balance. Europe's trade surplus with the rest of the world today amounts to 41.7 billion Euros.

Over the 20 years between 1991 and 2011 the European Chemical industry saw its sales increase 295 billion Euros to 539 billion Euros a picture of constant growth. Despite this the European industry's share of the world chemical market has fallen from 36% to 20%. This has resulted from the huge increase production and sales in the emerging markets like India and China. The data suggest that 95% of this impact is from China alone. In 2012 the data from the European Chemical Industry Council (CEFIC) shows that 5 European countries account for 71% of the EU's chemicals sales. These are Germany, France, United Kingdom, Italy and the Netherlands.

The chemical industry has shown rapid growth for more than fifty years. The fastest-growing areas have involved the manufacture of synthetic organic polymers used as plastics, fibres and elastomers. Historically and presently the chemical industry has been concentrated in three areas of the world, Western Europe, North America and Japan (the Triad). The European Community remains the largest producer area followed by the US and Japan.

The traditional dominance of chemical production by the Triad countries is being challenged by changes in feedstock availability and price, labour cost, energy cost, differential rates of economic growth and environmental pressures. Instrumental in the changing structure of the global chemical industry has been the growth in China, India, Korea, the Middle East, South East Asia, Nigeria, and Brazil.

Technology

From the perspective of chemical engineers, the chemical industry involves the use of chemical processes such as chemical reactions and refining methods to produce a wide variety of solid, liquid, and gaseous materials. Most of these products serve to manufacture other items, although a smaller number go directly to consumers. Solvents, pesticides, lye, washing soda, and portland cement provide a few examples of product used by consumers.

Turbine generator systems

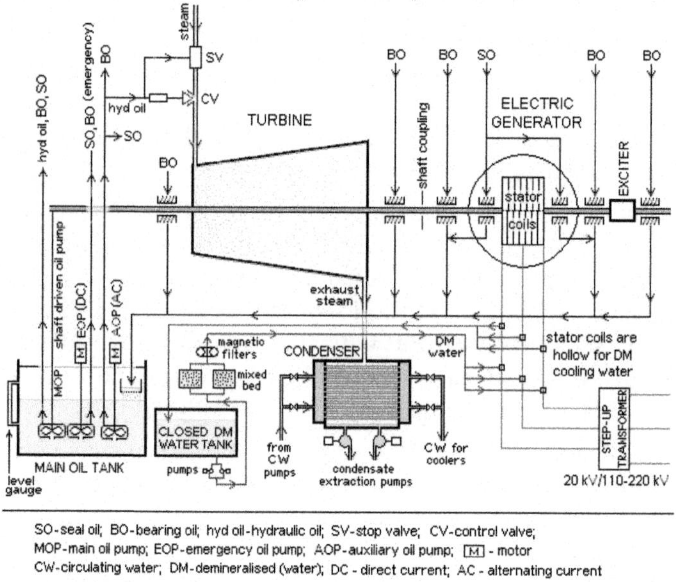

Fig. : This is a process diagram of a turbine generator. Engineers working to produce a sustainable process for use in the chemical industry need to know how to design a sustainable process in which the system can withstand or manipulate process-halting conditions such as heat, friction, pressure, emissions, and contaminants.

The industry includes manufacturers of inorganic- and organic-industrial chemicals, ceramic products, petrochemicals, agrochemicals, polymers and rubber (elastomers), oleochemicals (oils, fats, and waxes), explosives, fragrances and flavors. Examples of these products are shown in the Table below.

Product Type	Examples
inorganic industrial	ammonia, chlorine, sodium hydroxide, sulfuric acid, nitric acid
organic industrial	acrylonitrile, phenol, ethylene oxide, urea
ceramic products	silica brick, frit
petrochemicals	ethylene, propylene, benzene, styrene
agrochemicals	fertilizers, insecticides, herbicides
polymers	polyethylene, Bakelite, polyester
elastomers	polyisoprene, neoprene, polyurethane
oleochemicals	lard, soybean oil, stearic acid
explosives	nitroglycerin, ammonium nitrate, nitrocellulose
fragrances and flavors	benzyl benzoate, coumarin, vanillin
industrial gases	nitrogen, oxygen, acetylene, nitrous oxide

Although the pharmaceutical industry is often considered a chemical industry, it has many different characteristics that puts it in a separate category. Other closely related industries include petroleum, glass, paint, ink, sealant, adhesive, and food processing manufacturers.

Chemical processes such as chemical reactions operate in chemical plants to form new substances in various types of reaction vessels. In many cases the reactions take place in special corrosion-resistant equipment at elevated temperatures and pressures with the use of catalysts. The products of these reactions are separated using a variety of techniques including distillation especially fractional distillation, precipitation, crystallization, adsorption, filtration, sublimation, and drying.

The processes and product or products are usually tested during and after manufacture by dedicated instruments and on-site quality control laboratories to ensure safe operation and to assure that the product will meet required specifications. More organizations within the industry are implementing chemical compliance software to maintain quality products and manufacturing standards. The products are packaged and delivered by many methods, including pipelines, tank-cars, and tank-trucks (for both solids and liquids), cylinders, drums, bottles, and boxes. Chemical companies often have a research-and-development laboratory for developing and testing products and processes. These facilities may include pilot plants, and such research facilities may be located at a site separate from the production plant(s).

Chemical Manufacturing Locations

The scale of chemical manufacturing tends to decline from largest = Petrochemicals and commodity chemicals, then specialty chemicals and smallest = fine chemicals. The commodity and petrochemical manufacturing units are on the whole single product continuous processing plants. Specialty Chemical and Fine Chemical manufacturing are mostly made in discrete batch processes.

Those chemicals made on the largest of scales are made in a few manufacturing locations around the world, for example in Louisiana in the US, on Teesside in the Northeast of England in the UK and in Rotterdam in the Netherlands. Not all of the petrochemical or commodity chemical materials that are produced by the chemical industry are made in one single location but groups of related materials often are to induce industrial symbiosis as well as material, energy and utility efficiency and other economies of scale. Speciality and Fine Chemical Companies are often found in similar locations but in many cases they are to be found in multi sector business parks.

The large scale manufacturing locations often have clusters of manufacturing units that share utilities and large scale infrastructure such as power stations, port facilities, road and rail terminals. In the United Kingdom for example there are 4 main locations for commodity chemical manufacture: near the River Mersey in Northwest England, on the Humber on the East coast of Yorkshire, in Grange mouth near the Firth of Forth in Scotland and on Teesside as part of the Northeast of England Process Industry Cluster (NEPIC). To demonstrate the clustering and integration, mentioned above, some 50% of the UK's petrochemical and commodity chemicals, are produced by the NEPIC industry cluster companies on Teesside.

THE ORIGINS OF ENVIRONMENT SUSTAINABILITY IN CHEMICAL INDUSTRY

This working chapter explores the evolution of environmental strategies in the German chemical industry between the 1950s and the 1980s. Historical research on corporate environmental policies remains skewed. A vibrant environmental history literature has often identified business as a primary agent in environmental degradation, but it has usually not delved deeper into corporate strategies. The business history literature has only recently engaged with the topic of business and the environment,And the history of corporate responsibility more generally. This may reflect a disciplinary bias towards exploring issues related to innovation and growth, or else the failure to achieve them, rather than addressing negative outcomes of such innovation and growth.

It is well-established that the emergence of concerns that industrialization and modernization were damaging the natural environment date back to the nineteenth century. These concerns were driven by early ecologists, philosophers and scientists rather than business leaders, although by the late nineteenth century industrial pollution in major US cities such as Chicago and St. Louis was so evident that it encouraged a handful of business leaders to organize to persuade the business community to voluntarily seek ways to control such pollution. During the era of German industrialization in the late nineteenth and early twentieth century, conservationists also lamented the dangerous impact of chemical effluents on water and air. However after 1914 the era of world wars, virulent nationalism, and economic crisis did not provide a favourable context for the continued growth of environmental consciousness. Environmentalism became, if anything, associated with right-wing political extremism. Historians were distressed the discover that the Nazi regime introduced pioneering environmental legislation in Germany during the 1930s, although given the other policies of the criminal regime claims that the Nazi's were "proto-Greens" are clearly wholly misplaced.

After the end of World War 2, attention was more focused on economic recovery, but the renewal of a more widely based environmental movement was propelled by a growing number of accidents and pollution incidents. There was a new articulation of environmental threats. Rachel Carson's *Silent Spring* had a much-observed impact in raising awareness of the detrimental environmental impact of pesticides. In 1972 the United Nations held the first Conference on the Human Environment in Stockholm, which was followed by a new environmental legislation in Europe.

Rising environmental concerns, and the legislation which it stimulated, had substantial consequences for businesses, and especially for those who were directly associated with activities deemed polluting. In post-war France, Boullet has described the emergence of environmental strategies in heavy industries, including chemicals, in response to regulation, ecological disasters and shifting public opinion. The literature on the United States is larger. Hoffman, in a study of the chemical and petroleum industries, identified an evolution of corporate strategies over time. During the 1960s many US corporations believed that environmental concerns were exaggerated, and if there were problems, that they could handle them. In the wake of the 1970 Environmental Protection Act, there was a somewhat sullen commitment to technical compliance with the law. During the 1980s, as environmental activism rose, US firms began to engage more proactively in establishing environmental rules.

Hoffman's study forms part of a wider literature on the history of environmental management and strategy in the post-war chemical industry, including studies by Colten and Skinner, Markowitz and Rosner, and Ross and Amter. For the most part, these studies are critical of the industry's claims to be following best practices during the post-war decades, and frequently criticise a gap between the public statements of firms and what they actually knew about pollution. On a similar line, there are studies by Kehoe and by McGucken, concerning the pollution caused by the discharge of detergents into lakes and rivers, although these authors also note the scientific uncertainties faced by policy makers and companies. These studies have limited firm-level analysis. One exception is Smith's survey of the environmental policies of DuPont, which is more positive about the company's environmental strategies, and also stresses the uncertain toxicological knowledge on which decisions needed to be made. In other industries and countries, firm-level studies have also pointed towards imperfect information as a constraint, as well as organizational deficiencies within the firms themselves.

This working paper takes a firm and industry level perspectives, and focuses on postwar Germany. The choice of country is deliberate. The German Federal Republic (hereafter West Germany) as a country has often been considered as a European pioneer in green policies. In 1979 the Green political party was founded in West Germany, and became the first European Green party to make a major political impact in 1983, when it won 28 seats in the Bundestag (the West German parliament).

German business has also been seen as a pioneer in addressing environmental issues. A 1995 study asserted that German business had been a top spender "for a cleaner environment... since the early 1970s." However it is not clear how this happened, or whether it is really true, as there has been less historical research on green business strategies in West Germany compared to the United States. Most research has focused on the issue of industrial pollution primarily before 1945, and is heavily focussed on government policy. While business has been identified as a major polluter, firm strategies have not been discussed in detail in recent studies of the evolution of environmental policy, including Uekötter's comparative study of German and American policies on air pollution before 1970. Uekötter mentions Bayer only once and Henkel not at all.

The theory of why some firms become "greener" than others is still being developed. The international political economy literature has sought explanations in the varieties of capitalism model. Mikler, in a comparative study of the automobile industry, has argued that firms in coordinated market economies, such as Germany and Japan, are more inclined to adopt greener strategies than their counterparts in liberal market economies, such as the United States and Britain. It has been suggested that firms in such coordinated market economies operated with a framework of collaborative-consensus setting between business and government, that companies incorporated consumer attitudes rather than price signals in their strategies, and that soft topics counted more in internal corporate strategy.

In the organizational sociology literature, there is also an emergent body of theory seeking to understand why some firms selectively disclose information about their environmental impacts, more colloquially known as greenwashing. This literature builds on earlier work which has explored how an organization's visibility affects compliance with institutional pressures. There is evidence that greater visibility makes organizations more concerned with their legitimacy, although other research has suggested that the more powerful an organization the more it can afford to resist pressures from external stakeholders. Marquis and Toffel have employed a large-scale comparative sample on corporate environmental disclosure to distinguish between "generic visibility," such as just being big and well-known, and "domain-specific visibility," such as being identified specifically with an issue like worker safety, which exposes it to special scrutiny and is more likely to prompt more transparent disclosure.

This working paper will examine the strategies of two prominent chemical companies, Bayer and Henkel to establish how far the assumption that post-war German business was greener than others, is supported. This question is comparative, and comparisons will be made with, especially, the better-researched American case. Bayer and Henkel are not "representative" of the German chemical industry in a formal sense, but they do provide valuable cases to test hypotheses. Both companies have been described, and have self-identified, as "eco-pioneers." They differ in size, product markets and ownership structure, but share a common regional base. Both companies have also provided access to their corporate archives, enabling a more nuanced understanding of corporate motivations and internal organizational issues.

Bayer and Henkel – History and Governance

Bayer and Henkel shared many similarities as long-established members of the top-tier of the German chemical industry. They were founded in 1863 and 1876 respectively. Both were located in the state of North Rhine-Westphalia (NRW) near the river Rhine. Henkel was headquartered in Holthausen, in the south of Düsseldorf, after 1878, while Bayer settled in Leverkusen in 1900, where it made its head office twelve years later. Both companies became closely associated with their home regions. At the turn of the twentieth century Bayer was one of the top three chemical companies in the successful German chemical industry, and

Henkel was the fourth largest. In 2013 Bayer and Henkel were also the second and fourth largest chemical companies in Germany. Yet the companies were not identical twins. While Henkel largely made consumer products, especially cleaning products, Bayer earned many of its profits through producer sales. While both companies were confronted with the environmental impact of their products and processes, Henkel's premier position in detergents exposed it to one of the most publicised issues of industrial pollution after 1945. Finally, while Bayer became a public company as early as 1881, Henkel remained a family firm. In 1975, Henkel changed its legal form to a KGaA, a distinctive German corporate form which was a hybrid between a partnership and a public corporation, and ten years later the company issued preference shares to the public, with no voting rights.

The growth of both companies mirrored the history of modern Germany. Henkel was founded in 1876 by Fritz Henkel and two partners. Henkel's first product was a universal detergent based on silicate. In contrast to similar products that at that time were sold loose, one of Henkel's main innovations was to sell the heavy-duty detergent in handy packets. In 1878, the company launched its first branded detergent, the Henkel's Bleich-Soda (Bleaching Soda). In 1907, Henkel introduced the laundry detergent Persil, a self-acting detergent which made rubbing and bleaching unnecessary. After 1933, although individual members of the Henkel family were critical of the regime, most of the firm's top managers joined the Nazi party and cooperated with the authorities, including using slave labour. After the end of World War 2, five members of the family and another seven members of the Management Board and the Supervisory Board were interned by the Allies. They were rehabilitated in November 1947.

Bayer was founded by dye salesman, Friedrich Bayer (1825-1880), and master dyer Johann Friedrich Weskott (1821-1876) in 1863. The company made synthetic dyestuffs, whose production from coal-tar derivatives had only been invented a few years previously. Dyestuffs were sold primarily to the textile industry, which was growing rapidly during German industrialization. Bayer also diversified, especially into pharmaceuticals. In 1899, it developed and launched Aspirin, its most famous branded pharmaceutical product. By 1914 the company had many international subsidiaries, which were all lost as a result of the War. In 1925 Bayer and the other German dye companies merged to form I.G. Farbenindustrie AG. During the Nazi era I.G. manufactured synthetic rubber and fuel, and infamously was the supplier for the gas chambers in concentration camps. In 1945, the Allies confiscated I.G., and put two dozen of its managers on trial. I.G. was broken up in 12 individual companies, one of these being Bayer.

After 1945 German big business, including Bayer and Henkel, sought to rebuild reputations, promoting themselves as pillars of a "social market economy" which was conceived as an economically successful, but more humane, alternative to American-style capitalism. Family-owned and managed companies such as Henkel were highlighted as exemplars of this German-style capitalism. In fact, an emphasis on the broader societal responsibilities of business was also prevalent in the United States and elsewhere. Spector has traced the roots of the modern

Corporate Social Responsibility movement back to the Dean of the Harvard Business School, Donald K. David, during the late 1940 and 1950s.

As major chemical manufacturers, Henkel and Bayer were important participants in West Germany's post-war "economic miracle." This made the chemical industry, in the language of sociology, highly visible. Table provides data on the growth of the revenues, profits and employment of the two firms between 1950 and 1980.

Table : Revenues, Profits and Employment at Bayer and Henkel, 1950-1980.

	1950	1960	1970	1980	1950-1980 (%)
Bayer					
Revenue (million D-Mark)	599	2,818	9,524	28,825	
CAGR (%)		*16.7*	*12.9*	*11.7*	*13.64*
Profits (million D-Mark)	16 (1952)	132	358	730	
CAGR (%)		*23.49*	*10.49*	*7.39*	*13.44*
Number of employees	30,009	58,500	110,200	181,639	
CAGR (%)		6.90	6.54	5.12	6.12
Henkel					
Revenue (million D-Mark)	248	677	1,820	6,899	
CAGR (%)		*10.5*	*10.4*	*14.25*	*11.60*
Profits (million D-Mark)	8	n/a	54 (1974)	86	
CAGR(%)				*4.76*	
Number of employees	4,961	9,362	14,217	12,736	
CAGR (%)		6.56	4.27	(1.09)	3.19

Sources: Bayer -BA Geschäftsberichte, 1951-1991; Henkel -Feldenkirchen, Menschen, 145 and 186.

This fast growth initially reflected recovery from the wartime nadir rather than profitability. Cassis's comparative study of the profitability of large German, British and

French firms shows Bayer earning only a motley 4% return on shareholder funds between 1953 and 1955. Like all the German companies, it was far behind its American counterparts in commercializing products from petrochemicals. However even Britain's largest chemical company, ICI, only earned a 6.7% return. It is not possible to make a direct comparison with Henkel as the family firm did not disclose profits, but it is known that it badly lagged its American competitors, led by Procter & Gamble (hereafter P & G), in the new technology of synthetic detergents. Both Bayer and Henkel invested heavily to rebuild their competitiveness. The former gained entry into petrochemicals using joint ventures, and invested further in pharmaceuticals. Henkel fought off the entry of P & G into Germany during the 1960s, and by 1968 had almost 50 per cent of the European detergent market.

Between 1970 and 1972 Cassis shows Bayer's return to shareholders as 9.5% and Henkel's at 9.7%, compared to ICI's 9.5%. By this measure the two German companies were in Cassis's ranking of the eight most profitable German companies at that time, although they were significantly less profitable than the automobile companies Daimler-Benz (17.2%) and BMW (14.1%). The oil price rises of 1973 were a more difficult time for chemical companies as indicated by Bayer's slowing revenue and profit growth, and Henkel's falling employment.

The Rise of Pollution as a Political Issue

In the wake of the devastation caused by World War 2, environmental issues were not at the forefront of either political or corporate agendas. Unlike in many American cities, where debates about air and water pollution were underway, German society was struggling with economic reconstruction, the rebuilding of bombed cities, the integration of millions of displaced people, and the ethical collapse of Nazism. The chemical industry was hailed as a pillar of reconstruction, and a champion of Germany's burgeoning export economy. Pollution was accepted as a necessary evil. However, as the economy recovered, environmental problems, especially air and water pollution, emerged in public discourses. Pollution was especially evident in the industrialized region of NRW. A report by independent experts in 1961 showed that there was heavy air and water pollution near the Henkel and Bayer factories. Local residents complained about unpleasant black smoke, carbon black particles, and bad odours. Residents living near the main Henkel factory wrote to the company complaining about increases in carbon black output, gas smells, and soot. During 1962 one resident repeatedly drove his car to the Henkel headquarters showing the black and white particles dirtying his Mercedes.

Residents near the Bayer factory at Leverkusen also complained. In 1958 Bayer had installed a giant illuminated version of its logo, the Bayer Cross, overlooking Leverkusen. It was the largest illuminated advertisement in the world, and might be considered a symbol of "domain-specific visibility." The problem was that it shone over a stinking city. Bayer received growing complaints about foul air, and especially the acidic effluents which stank, destroyed garden plants, caused headaches, and ruined women's nylons. Bayer paid compensation for the nylons, but officially maintained that they must have been of poor quality to be affected by the effluents. In one official report initiated by the city of Düsseldorf covering the years 1958 to 1961, Bayer was directly accused of emitting too much ash and sulphur dioxides into the air, and requested to build taller chimneys with the idea that pushing the air high up in the atmosphere would help to thin out the pollution.

The NRW was no stranger to bad smells. The traditional corporate response to complaints was to stress the importance of the industry. During the early 1950s Henkel's managers regularly stated that odour nuisances could not be further reduced, and that the neighbours would just have to live with them. A decade later Bayer's managers were still asserting in public that the discharged air was not harmful to humans, and that any nuisances should be accepted for the sake of employment and economic development. The pollution just got worse. In the mid-1960s, a phenomena known as "Auto-Pocken" ("car-pox") appeared near Bayer's factories. Effluents caused severe paint damage on 7,000 cars. Bayer paid almost one million D-Mark in compensation (approximately £89,600). The mysterious damage, whose cause could not be determined with certainty, had a distinctly negative impact on Bayer's image in the neighbourhoods affected.

Factory effluents in water were also a source of major concern. Between 1949 and 1952 local authorities repeatedly warned Henkel not to release allegedly dangerous effluents into the public wastewater system. Bayer also faced an alarming number of dead fish close to its factory on the Rhine, which was attributed to a high level of Phenol released by Bayer.

Water impureness became even more obvious in the hot and dry summer of 1959, when the water levels of rivers and lakes were exceptionally low. A significant amount of synthetic colours were detected in the Rhine. While it was not unusual to see a kaleidoscope of colours in the Rhine, the low water level made the effluents even more visible. The water police took samples and sent them to Bayer for further analysis. The long-standing relations between the local authorities and the firm was reflected in the fact that it was left to the company to determine the cause for pollution because, as Bayer managers stated, "they value our analytical competencies."

However it was foam on the Rhine rather than dead fish which really raised public concerns in the NRW. The problem stemmed from the use of alkyl benzene sulphonate derived from cheap petrochemicals benzene and propylene, as the main cleaning agent in synthetic detergents. The surfactants did not bio-degrade, accumulated within treatment plants, and entered rivers and lakes. By the late 1950s, a highly visible foam was found in rivers and lakes across Europe and the United States. In Germany, as elsewhere, the detergents industry was soon identified as the culprit. In 1961, the popular national periodical *Zeit*, issued twice a month, ran a story with the title "The water is sick. What to do with the scum of civilization?" Foamy rivers and car-pox began to erode the high status held by the chemical industry as a pillar of West Germany's economic reconstruction. The industry was well-protected by the post-war legal and political system. The German legal system formally prioritized economic performance over the protection of victims of pollution. The *Gewerbeordnung* (Prussian Industrial Code) of 1869/1900, still in force, stipulated that companies needed a license for new factories, which could be denied if the plant could be expected to cause unbearable risks or nuisances for its neighbours, but for existing factories, the only legal recourse for residents affected by pollution was to engage in private law suits. There was no enthusiasm to disrupt the activities of national champions. In 1959, the Prussian Industrial Code was finally modified to strengthen the rights of residents living around factories to sue for financial compensation. However, it was difficult to prove cause, and local authorities left it to companies to determine the cause of pollution.

Critical voices struggled to get traction. In 1952 Sturm Kegel, the director of the Ruhr Area Federation for Regional Planning, presented a detailed plan to establish *Luftreinhaltegenossenschaften* (air pollution control cooperatives) in NRW, aimed at bringing together companies, local municipalities and regional planners to jointly implement measures against air pollution. Kegel's concept was heavily criticized by public officials and industry representatives and eventually rejected, but it did prompt a debate and helped to raise the profile of air pollution as a political issue in NRW. On the federal level, an SPD member of parliament prepared a proposal for air pollution management for deliberation in the Bundestag in 1955. The professional association *Verein Deutscher Ingenieure* (Association of German Engineers), or VDI, volunteered to draft technical guidelines to monitor air pollution.

Legislators remained cautious to move beyond monitoring. In NRW, the issue of water management was transferred from the Ministry of Economics to the Ministry of Agriculture in 1953 to assure ecological concerns were well-represented. The state government at the time was controlled by the conservative CDU party, which governed NRW between 1950 and 1956 and again between 1958 and 1966, but the business newspaper *Handelsblatt* criticised the move on the grounds that in such a heavily industrialized region the management of a key resource such as water ought to be in the hands of the Ministry of Economics. There were no new restrictions on industry, but a 10-year-plan to improve the water infrastructure and the building of treatment plants emerged. The NRW government made over 75 million D-Mark (£6.75 million) available. The Bundestag also passed a Water Resources Act in 1957, which went into effect in 1960, initiating the building of new water treatment plants and other improvements of the water infrastructure. In 1962 the NRW's Ministry of Agriculture stated that the measures had delivered few visible results, especially in the context of a growing population and rising industrial production. Waste water was still perceived to be a major a problem, and a further State Water Act was passed in 1962. Overall, these water management initiatives imposed few restrictions on industry, and were little more than devices to create structures for studying problems, much as was parallel legislation in the United States. The exception was the detergent industry. The rising public outcry about detergent foam on rivers and lakes led to a Detergents Act being passed in 1961 which legally mandated the use of at least 80 per cent biodegradable detergents from 1964 onwards. During the federal election campaign in 1961, the candidate of the left-wing, Social Democratic Party (SPD), Willy Brandt, put forward a bold demand in a speech at the party convention: "The sky above the Ruhr must become blue again." While this issue was not central to his speech, clean air was emerging as a topic which resonated with voters. Brandt lost the election, but the victorious CDU was alert to public discomfort, and its potential political fallout. The Federal Chancellor's office under Konrad Adenauer had already proposed inquiries into environmental matters in the summer of 1960, "with a view towards the coming elections." The influential weekly *Spiegel* echoed Brandt's phrase in a title story about air pollution in NRW titled "Toward blue skies." In eleven well-illustrated pages, the chapter described dirt-covered and dying plants, sick children, and automobile accidents due to the reduced visibility and roads made slippery by dust particles.

There were no radical changes in either legislation or corporate policies after 1961. However, there was a new willingness on the part of the NRW regional authorities to push back when Bayer, Henkel and the other companies sought to acquire licenses for new factories. In their applications, the companies employed the concept of *Gewerbefreiheit* ("freedom of occupation") which included the freedom to practice any trade or craft and, closely connected to it, the freedom to take one's residence where one wished. Both Bayer and Henkel pleaded for their autonomy on these grounds when it came to locational decisions. Yet an increasingly concerned public informed by critical investigative journalists made their voices heard, and the willingness of local politicians to listen to them was discernible.

CHEMICAL INDUSTRY MOVEMENT TO ACHIEVE AND DEMONSTRATE IMPROVEMENT IN SAFETY

The global chemical industry plays an important role in modern society by contributing to the solution of many human problems: hunger, sickness, sanitation, clothing, housing, transportation, leisure. With 6 billion people on earth, a strong and innovative chemical industry is a condition to sustainable development. However the chemical industry, like any other human activity suffers from time to time of accidents in its plants and during its products transportation, some health problems because of product exposure, problems of waste disposal, air and water pollution, bioaccumulation.

How to get the benefits of the the products while minimizing risk of accidents? Responsible Care is the answer of the industry.

Responsible Care is the commitment by chemical companies to demonstrate continuous improvement of environmental, health and safety performance of their operations and products in a manner which is responsive to the concerns of the public.

Communication is an essential component of Responsible Care. It must address all constituencies which are or might be involved in our operations or with our products. That is our customers and what we call our stakeholders: our suppliers and contractors, local communities, authorities and regulating bodies, the general public and of course our employees.

Communication processes must be both ways so that besides explaining what we do, customers and stakeholders can voice their opinions and concerns. It's these communication processes which should enable this industry to improve its perception by the public and maintain its license to operate. As a US Telecommunication company CEO stated early in this century: All business in a democracy begins with public permission and exists by public approval. The chemical industry is very well aware of that.

Responsible Care is a chemical industry voluntary initiative. It covers all the industry activities from research, process and product development through manufacturing and sales. It aims at an ethical and behavioral change: going away from a regulatory driven approach to a proactive approach taking into account the perceived needs of society all along the product life, from cradle to grave. This approach is much more efficient and demanding because, contrary to end of pipe regulations it calls for improvements at all stages of the product cycle. It's more flexible, more comprehensive and to some extent easier to implement since it is designed by the industry for the industry. It encompasses commitments to communication and experience sharing that would be very difficult to organise through regulation.

Responsible Care is a global, I mean worldwide, initiative. In each country there's an association of chemical companies: the national federation. While companies have their own Responsible Care programmes, it's the federation responsibility to adapt the Responsible Care initiative to the cultural, social and legal environment of the country

In Europe, 21 national federations participate in a Responsible Care Committee, inside CEFIC, the European Chemical Industry Council. CEFIC itself is a member of ICCA, the International Council of Chemicals Associations. There's inside ICCA a Responsible Care Leadership Group whose prime responsibility is to promote Responsible Care while insuring that Responsible Care implementation in various countries is in accordance with its key elements and fundamental features, thus protecting worldwide the integrity of the initiative.

Where and When it Came from and How it Spread Across the World?

Responsible Care started in Canada in 1984. It sprang over to the USA in 1988, then to Western Europe in 1989-1990 and then to the rest of the world. Today, Responsible Care is being implemented in about 40 countries:

21 European countries, including Turkey, none in the former Soviet Union.

In North America: Canada, USA, Mexico.

In South America: Argentina, Chile, Brazil, Colombia, Peru.

In Africa: South Africa, Zimbabwe.

In Pacific/Asia: Australia, New Zealand, Philippines, Japan, India, Taïwan, Hong-Kong, Malaysia, Singapore

The Key Elements of Responsible Care

The involvement and support of a company's Chief Executive Officer or the most senior management person operating in the country is critical to the establishment of overall responsible care ethic within an organization.

Thus the first requirement for a company to be recognised as a participating member in the Responsible Care initiative and its first fundamental feature is the signature by its CEO of guiding principles. These principles are broad statements defining the fundamental ethical responsibilities and operational philosophies of the company, consistent with Responsible Care.

At this time it is not required that all fundamental features of Responsible Care, as described later, are fully implemented for a company to be recognised as following the Responsible Care approach. However the company must present an acceptable implementation plan of these features.

It's chemical companies which manufacture, sell, distribute chemical products. All the field action is at company level : Responsible Care implementation builds on companies' programmes.

The national federation role is to promote the initiative so that most chemical companies inside the country join Responsible Care. National federation also check that the guiding principles are consistent with Responsible Care and that the fundamental features implementation plan is credible. The national federations monitor their members progress, help sharing experience, edit local codes, guidelines, guidance documents and may set minimum requirements. The national federations own and protect the Responsible Care title and logo which signal that Responsible Care implementation meets participation requirements. Thus Responsible Care implementation is a joint effort of companies and federations.

The Fundamental Features of Responsible Care

The fundamental features that companies participating in the initiative are committed to implement are:

A formal commitment on behalf of each company to a set of guiding principles signed by the CEO that I already mentioned.

A series of codes, guidance books, check lists that I will describe next.

The progressive development of indicators against which improvements in performance can be measured.

An ongoing process of communication on health, safety and environmental matters with interested parties inside and outside the industry.

Provision of fora in which companies can share views and exchange experiences on implementation of the commitment.

Adoption of a title and logo which clearly identify national programmes as being consistent with the concepts of Responsible Care.

Considerations of how best to encourage all federation members companies to commit and participate in Responsible Care.

Systematic procedures to verify the implementation of Responsible Care by the member companies.

The titles of the next 4 codes are self explanatory.

Product Stewardship is Responsible Care applied to products: how to ensure that they are designed with due consideration to their potential environmental effects throughout their lifecycle and also how to ensure that they will be handled, used, disposed of safely all along what we call the supply chain from producer to final user and disposal.

Management systems verification was recently added in response to stakeholders concerns about the credibility of the Responsible Care Initiative. It checks that codes implementation is supported by some kind of Deming's wheel management system: Plan (define organisation, provide ressources, set targets and schedule to meet objectives), Do (implement), Check (monitor, review), Act (improve policy and objectives).

The Current Status of Responsible Care

Like codes, programmes are at very different stages, depending on countries and when the Responsible Care initiative was started.

A few countries already developed and implemented all their codes while achieving significant results.

Accidents reduction: the chemical industry is in most countries amongst those with the lowest fatalities and lost time accidents rates for its personnel.

We also achieved significant waste and effluent reduction. As an example, in Canada under Responsible Care, members companies from the Canadian Chemical Producers' Association have since 1992 reduced their aggregate emissions of all pollutants to water by 99.2 % (140 000 tonnes down to 1 100 tonnes), largely through the conversion of former wastes such as sulfuric acid to saleable products such as gypsum.

We have also published details on many initiatives on improving safety of chemicals transportation and distribution and shared a lot of experience through workshops and seminars and much more.

For countries with a few years of experience with Responsible Care, there are two areas of communication which are already working well. One is communication with local communities through local citizen advisory panels and we see locally that public interest is shifting from plant safety which is of less concern to product safety. By the way, CEFIC is organising a workshop on Product Stewardship in Lyon, next April, to encourage more business people to engage in Product Stewardship programmes. A second area is environmental reporting: most major chemical companies now publish local and global environmental reports, sometimes certified by external auditors, displaying their commitment to Responsible Care and supplying a lot of datas on health, safety and environment to demonstrate their continual improvement.

Collecting aggregate performance data is complex at ICCA level or even in Europe, at CEFIC level because of different definitions. We are presently working on that and our plans for Europe call for publication of aggregated performance data on 16 core parameters, implemented from now till year 2000. These core parameters are listed on next slide. Now that programme development is significant in many countries, we must better assess our progress and demonstrate that we have real programmes and not paper programmes, in other words we must improve verification. First step of verification is self assessment. We, at CEFIC are presently developing a Self Assessment Questionnaire on Responsible Care implementation. The US CMA (Chemicals Manufacturers Association) recently added a Management System Verification code. This verification requires people external to the company, including the local community, to be included in the plants auditing teams. Next step for CEFIC to consider is third party auditing. Some countries are already there like Austria. They claim to have a third party verification system consistent with ISO 9000, ISO 14000, and EMAS (the European Eco Management and Audit Scheme) in place.

At the ICCA Responsible Care Leadership group, we review annually a progress report which each industry federation sends to us on their programme. We are preparing more and more detailed questionnaires to help assess that real progress is made. An ICCA Responsible Care Status Report was published last year for the first time. It describes Responsible Care implementation status in 40 countries. This is it and I will be pleased to provide a copy to those of you who are interested.

At European level, CEFIC has been publishing a Responsible Care status report for 3 years now. Again, I will be glad to forward it to those interested.

Many companies are already reporting all these parameters and much more. Our problem is to get agreement in definitions so that we can aggregate data and report as federations or association of federations. There are discrepancies even on what could be seen as simple data such as Lost Time Accident Rate. They are sometimes reported for 200 000 hours, sometimes in number of occurrences per workers. A lost time accident is 3 days absence from work in some countries, 1 day in others, work related illness may be included and so on. Obviously we need more time before we can publish aggregate data on more complex issues like occupational illness or selected substances emissions.

Concerns

With its communication and experience sharing, the Responsible Care initiative is much more than a Public Relation exercise although we of course highly value an improved public opinion about our industry.

Nothing more than regulations?: It's also much more ambitious than just meeting regulations as it aims at an ethical, behavioural and cultural change inside the industry.

No performance improvement?: Achievements and performances improvement which companies report and which we will try to aggregate at CEFIC and ICCA levels in the previously mentioned status reports, are evidence of real, significant performance improvements although we are not yet able to aggregate data.

No public perception improvement?: We do have some evidence of improvement of public perception, but this improvement is seen close to plants which have local citizen advisory panels. Surveys show that the general public still has a poor perception of the chemical industry.

No stakeholder involvement?: We have been busy a long time developing codes, agreeing on initiatives.

We must now live up to our commitment to improve the relationship with all our stakeholders which obviously include our workers and their organisations.

A lot of Responsible Care implementation work is actually done by big companies who have experienced and knowledgeable people to do it and then this experience can be shared with SMEs mainly through national federations.

We already do have action in developing countries. For instance the industry is participating in the UNEP-IFCS (United Nations - Environmental Protection and Intergovernmental Forum on Chemical Safety) seminars throughout the world on POPs (Persistent Organic Pollutants) awareness raising and there are many training programmes on the use of agricultural products.

We are willing to extend our Responsible Care activities, using multinational companies as a network while taking into account local sensitivities.

THE ENVIRONMENTAL MOVEMENT

The Magnitude of Concern about the Environment

With the revelations of glasnost, citizens became acutely aware that environmental conditions in their neighborhoods were far from favorable. Though crude by Western standards, local public opinion polling has revealed the magnitude of concern about the environment. In a 1989 USSR Goskomstat survey, 1 in 10 people surveyed said the environment was the country's most serious problem. Of twelve major problems enumerated, cleaning up the environment was listed fourth, behind food supply problems, poor housing conditions, and inflation. Surprisingly, it surpassed such pressing and visible issues as ethnic tensions, social injustice, poor healthcare, and crime. In 1990 and 1991, official surveys reported by USSR Goskomstat revealed that almost one-half of the urban population polled considered environmental conditions in their neighborhoods to be "unsatisfactory." In July 1991, the Russian state statistical agency reported that of those polled in another study, three-quarters considered environmental conditions in their hometown "intolerable." In a survey conducted in the Moscow region in spring 1990, environmental degradation was ranked as the most important social problem. Of those polled, 98.1 percent rated the issue "important" or "very important." Less pressing issues, by comparison, were crime (94.7 percent), food shortages (94.4 percent), and consumer goods shortages

(93.4 percent). In a 1989 survey conducted in Ukraine, environmental problems were "the main concern" of 26 percent of the population, behind economic problems (44 percent) but well ahead of political and cultural issues.

Of all environmental issues, citizens are most strongly antinuclear; so powerful is their aversion, that the mood aptly has been labeled "radiophobia." Two nonbinding, unofficial referendums conducted in 1990 illustrate the level of sentiment against nuclear power. In February, voters in the small Ural Mountains city of Neftekamsk went to the polls to decide the fate of the Bashkir Atomic Energy Station, located 30 kilometers away. The referendum, organized by the Neftekamsk (population 109,000) city government along the lines of a regular election, attracted an 80 percent turnout of voters—a stunning 99 percent of whom voted to halt construction of the plant immediately. Three months later in the Russian city of Voronezh (population 895,000), a local group calling itself Ecological Initiative prodded the city government into holding a referendum on the fate of a nuclear-powered citywide heating system. Of the 82 percent of the voting-age population who went to the polls, 96 percent turned down the scheme. The environmental movement has garnered great respect from the public, in part as a result of its efforts to discover the truth about ecological conditions. In the spring 1990 survey of Muscovites previously mentioned, the Green movement had earned the trust of over 54 percent of the population. Only the Russian Orthodox Church (64 percent) and the military (56 percent) scored better. Far down on the list was the CPSU (39 percent), the official trade unions (37 percent), and the government (28 percent). Anatolii Panov, vice-president of the Zelenyi Svit (Green World) environmental association, claimed his organization enjoyed the highest trust rating of any group in Ukraine in 1991.

The Response of the Government to the Environmental Movement

The best way to gauge the strength of the environmental movement is to examine its impact on the government policy process. Responding to the public's concern, many politicians make a point of showing their awareness of environmental problems and their desire to resolve them. Officials and politicians frequently visit ecological hot spots to render some measure of political first aid. For example, in August 1990, Boris Yeltsin, recently elected chairman of the Supreme Soviet in the Russian Federation, took advantage of this traditionally quiet period in Soviet politics to make a three-week tour of Siberia and the Far East; he wanted to assess environmental conditions. "I received a very strong sense of colossal problems on my trip," he told Radio Moscow. One year later, campaigning for the Russian presidency, Yeltsin again made a point of visiting ecological hot spots such as Chelyabinsk oblast and Sakhalin Island. Obsessed with bolstering its public image, even the KGB took pains to demonstrate its environmental awareness. In October 1990, the Soviet news agency TASS noted that the KGB had prepared a report to challenge the military's plans to use a nuclear device to create an underground storage facility for high-level radioactive wastes near the Ural Mountains city of Chelyabinsk. On the occasion of the agency's seventieth anniversary two months later, Soviet television screened a program illustrating the KGB's environmental consciousness. The agency was so bold as to claim partial responsibility for the government's decision in 1986 to cancel the Siberian rivers diversion project. Although conceding the impact of scientific and cultural figures working in opposition to the scheme, Major-General Eduard N. Yakovlev of the KGB's analytical department added: "We spoke from an objective, unbiased position."

Local as well as national officials often turn to environmental groups for expertise and input. Sometimes the relationship goes beyond this; many environmental officials consider the public a helpful ally in bureaucratic battles against industry. Thus, although Ukrainian industries were required to submit an environmental impact statement (EIS) for development projects, they also were able to pressure the Ukrainian environment ministry to waive its standards and to accept their plans. According to one official at the environment agency, independent EISs conducted by environmental groups along with public pressure were welcomed as an aid to the beleaguered agency in enforcing its regulations. A deputy of the Latvian environmental agency told a local paper that his agency "must make a major effort to consolidate the various movements and organizations of the Greens." The general director of Moldova's environmental agency went much further: "We will utterly and completely support any movement to protect the environment, including through rallies, strikes and picket lines." Said I. I. Deyu, formerly a professor at Moldova State University, "We are trying to do it in a way that people trust us."

Evidence of a growing alliance between government environmental officials and the environmental movement also can be seen in the plethora of ecology-oriented newspapers that have sprung up. By jointly publishing newspapers, environmental groups can gain access to state publishing facilities, and officials seek to tap into the popularity and respect earned by these organizations. In 1991, the Moldovan environmental agency announced that it was publishing an ecological newspaper, *Abe natura,* jointly with the Moldovan Green movement. The Kurgan oblast Committee for the Protection of Nature joined with the local branch of the All-Russian Society for the Protection of Nature to publish *Ekologicheskaya gazeta* (The Ecological Newspaper). The RSFSR environmental agency teamed up with the Ekopress information and publishing association to put out *Zelenyi mir* (Green World).

Environmentalism Strikes the Economy

Rising protests in the 1980s had a strong impact on the region's economy, as environmentalists began to demand a rapid solution to pollution problems in their neighborhoods. In all, over 1,000 production units either were closed or had their output scaled back in 1989 for "violation of environmental protection laws," USSR Goskomstat reported. In the city of Moscow alone, 72 plants and production lines were closed. A trade union leader pointed out that over 100 plants in the chemical industry were idled around the country in mid-1990 because of demands from the communities in which they were located to be "dechemicalized." In 1990, output sacrificed was projected to total 5.2 million tons of fertilizer, 951,000 tons of soda, 400,000 tons of cellulose, 387,000 tons of methanol, 500,000 tons of polymers, and over 250,000 tons of synthetic rubber, among other goods, and leading economists were worrying that the ecological movement would deal a fatal blow to the already failing economy. Indeed, in his report on the performance of the Soviet economy during the first quarter of 1990, USSR Goskomstat chair Vadim Kirichenko blamed a sharp downturn in economic performance, among other causes, on the closing of factories for not meeting "basic ecological demands." Most of these plants were shut down only temporarily as a form of sanction, it must be pointed out, but in many cases, the closure was intended to be permanent.

Energy is one sector in which the public virtually has dictated policy. Media reports suggest that not one nuclear power plant has been safe from opposition as the public has been gripped by radiophobia. In the fallout from the Chernobyl accident, public pressure had forced the abandonment of 60 projects by January 1991, including the much-disputed Crimea, Rostov, Tatariya, and Bashkiriya atomic energy stations (AESs), and expansions at the Smolensk, Khmelnitski, Tver (Kalinin), and Zaporozhye stations. The generating capacity of these plants was projected to total 160 million kilowatts. The fates of numerous others remain uncertain as officials have been forced to undertake a complete reappraisal of nuclear energy development plans. As if to beat a dying horse, one week after its inauguration in the summer of 1990, the popularly elected Supreme Soviet of the Russian Federation voted to issue a moratorium on nuclear projects beginning in 1991. One month later, the Supreme Soviet of Ukraine passed a five-year moratorium on all nuclear projects in that republic, and after a turbine fire at Chernobyl's power unit No. 2 in October 1991, the Ukrainian parliament voted to decommission immediately the crippled reactor and the two other units (1 and 3) still operating on the site by 1993. With less nuclear power on which to rely in the future, the region will have to turn more to oil, coal, gas, and hydroelectric resources. These options are limited as well, however, as environmental pressure mounts against the diversion of water and the submersion of land for hydroelectric projects and against the continued use of aging and dirty coal-fired power plants. In one of its first actions, the first democratically elected Moscow city council voted in May 1990 to scrap construction of the Severnaya station—planned to be the largest gas-fired heat and power plant in Europe. Opponents, led by the Moscow Green Party, collected 300,000 signatures against the plant, arguing that it was not necessary. Instead, they pointed out that massive heat losses in the distribution network, estimated to average 25–40 percent, could be reduced with the improved insulation of heating pipes and the construction of small, local heat and power cogeneration facilities. On the other hand, Vasilii Selyunin, a prominent radical economist and government critic, warned: "It is not possible now to stop even one, even the tiniest, power station because of the shortage of electricity."

Unexpectedly, environmental activism has also had an effect in the area of pharmaceuticals. Much of the region's drugs are produced by the metallurgical and chemical industries—two sectors of ill ecological repute. Environmental protests halted production of pharmaceuticals at the infamous Azot (nitrogen) production enterprise in Kemerovo, at Yerevan's Nairit, and at the Kirovgrad copper smelter. In some cases, Radio Moscow reported, these plants were the only ones in the USSR producing certain essential medicines. According to USSR Minister of Health Igor Denisov, the domestic pharmaceuticals industry was able to meet 39 percent of the Soviet Union's demand for drugs in 1990, down from 52 percent five years earlier. Denisov predicted that domestic production would fall to 30 percent of demand in 1991.

In preparation for the Second USSR Congress of People's Deputies in December 1989, the USSR Supreme Soviet Committee on the Protection of Public Health also looked into the medicine crisis; the head of the USSR Ministry of the Medical and Microbiological Industry (Minmedprom) reported that local opposition had prevented the siting of new plants to produce pharmaceuticals, including those to produce disposable syringes, desperately needed to avoid the spread of AIDS. Construction had been halted in Arkhangelsk, Kursk, and Saratov oblasts and in Novosibirsk, the Mari Autonomous Republic, Latvia, and Uzbekistan. Of the 36 Minmedprom pharmaceutical projects slated for the 1986–1990 plan, 21 had yet to be initiated because local forces had refused to allocate any land to the ministry. Public opposition stemmed from Minmedprom's reputation as an inveterate polluter, an example being its petroleum-based livestock supplement operations at Kirishi, outside St. Petersburg. The labor daily *Trud* noted pessimistically that the pharmaceuticals industry was "falling apart with even greater speed" than the rest of the Soviet economy. "Until we convince the population that our production units can be safe—and there are already some that are—we will not move the sector forward," gravely testified Minister Valerii Bykov. "'Green' extremism will not let the pharmaceuticals industry take a breath." "We have already seen cigarette shortages," concluded *Trud*. "Should we prepare for aspirin shortages?"

The Diversity of the Environmental Movement

Despite major successes like that at Semipalatinsk, most proenvironment groups are loosely organized, small, and often short-lived. In April 1990, the co-chairman of the Leningrad Green Party counted approximately 60 "ecological organizations" in that city, most of which had fewer than a dozen members. Environmental groups usually focus on a single issue in their own locality. As elsewhere in the world, the rallying cries are numerous, and groups span the ideological spectrum, ranging from fiercely nationalist organizations to apolitical bird-watching societies. Sovetskaya Rossiya, the staunchly conservative Russian newspaper, sponsored the Committee to Save the Volga, whereas the Committee to Save the Ob is based in the scientific research center of Novosibirsk. Reflecting its constituency, the latter group announced plans to conduct research and environmental impact studies on all developments threatening the Ob and Irtysh river basins. Founded in October 1989, the Chernobyl Union, with member chapters throughout the former Soviet Union, aims to serve as a support and legal advocacy group for victims of the nuclear accident as well as the 600,000 people it estimates have participated in the cleanup operation, 50,000 of whom reportedly have become unable to work.

Although each group has its own eclectic agenda, many express similar principles and objectives—ecological glasnost, comprehensive environmental monitoring and assessment, public education, grassroots cleanups, and direct political action. For example, environmental NGOs often spend a large share of their resources to establish themselves as alternative and credible sources of information. Although government agencies have made major strides in bringing environmental information into the public domain, much remains hidden or unknown. "A lot of interesting information is out there," says Sheryl Belcher, coordinator of Greenpeace International's "Children of Chernobyl" project based in Kiev: "[Environmental groups] need to access it." Thus, environmentalists persistently lobby for full disclosure of government data. Paradoxically, however, they and the public at large remain highly suspicious of much official data that are released, arguing that accurate data would indicate problems so severe as to be too risky politically to be made public. Such a situation lies in stark contrast to that in the United States, where environmental groups rely heavily on access to official government data through the Freedom of Information Act, for example, to pursue their objectives, including challenges to the government.

Where reliable data are either lacking or unavailable, groups have sought to acquire their own monitoring equipment and to conduct independent studies. Using a battery of radiation monitoring devices, Vladimir Mikheev of Krasnoyarsk's Green World association, conducted several informal surveys revealing high radiation levels in the Yenisei River north of the Krasnoyarsk-26 nuclear facility. Often environmentalists turn to their counterparts in the West for monitoring equipment as well as information that is considered independent and credible. Greenpeace's "Children of Chernobyl" project, for example, imported a special truck outfitted with diagnostic equipment to travel around Ukraine and sample environmental conditions. Because membership in NGOs frequently includes highly qualified scientists, groups often aim to serve as an independent source of expertise for policymakers and the public. The Chelyabinsk Ecological Fund, for instance, retains several medical and scientific specialists to assess and to inform the public about environmental conditions.

Similarly, many environmental groups place a strong emphasis on educating the public, especially the youth, about ecology and the environment. Activists often describe the general public as being "ecologically illiterate"—a consequence of the Soviet government's low priority on raising environmental consciousness and teaching ecology in public schools, its suppression of information, and its hostility to independent thought. Anatolii Grebenyuk, a parliamentarian from Kyrgyzstan, stressed the legacy of Soviet environmental education thus: "We have been taught since kindergarten that we have no ecological problems, and that we can live for millions of years with no concerns. Today we see this is not so. It is very difficult to change people's opinions, to make them realize the severity of our situation." To correct this attitude, many groups have opened environmental libraries (with donations often coming from abroad) and educate the public through their own publications, the mass media, and the schools.

The Tactics of Environmental Groups

Despite the liberalization of politics in much of the former Soviet Union, democratic politics remain in a nascent and tentative stage. Democratic institutions and instruments such as referendums, public opinion polling, parties, lobbyists, public hearings, fundraising and advertising are still in the early stages of development. The lawsuit, the mainstay of environmentalists in the United States, is not an effective or realistic option in the newly independent republics, given that environmental laws remain weak and often unenforceable. This problem is compounded by an often unsupportive judiciary. As a result, organizing traditional mass demonstrations—with attendance frequently numbering in the tens of thousands—has been the most visible and popular tactic of environmental groups and has served to send the most potent messages to the authorities about the legitimacy of their public's demands. Environmentally concerned individuals need not form organized groups to be effective participants in the policy process. Scientists, for example, with their high prestige in society and privileged access to information, have formed effective lobbies, as was the case with the relatively unorganized yet successful campaign against the Siberian rivers diversion plan. Physicists and writers formed a strategic alliance to oppose successfully the development of nuclear power, most notably in Ukraine: The physicists were able to disseminate their authoritative information and forebodings about the government's plans, exploiting writers' access to such media as the influential newspapers Literaturnaya Ukraina, Komsomol'skaya pravda, and Moscow News.

Analysis of the political agendas of various ecological groups makes clear that not all "green" organizations are what they appear to be. The Russian ultranationalist group Pamyat' has been charged with creating the innocuous-sounding All-Union Ecological Society in an attempt to use environmental issues to add a veneer of legitimacy to its less benevolent pursuits (*e.g.*, its harassment of Jews and non-Russians). Leaders of the Social-Ecological Union (SEU) have leveled similar charges against the Committee to Save the Volga. In addition to political opportunists, the environmental movement also attracts people and institutions motivated by material impulses, and their actions may not bode well for nature. Environmentalists have accused Communist Party apparatchiks of founding or supporting front organizations, such as the Green Party of Leningrad, to draw support away from real opposition groups. In Chelyabinsk oblast, conservative interests put forth the Green Party, while the "real Greens" in the region are said to be represented by the Democratic Green Party. Similarly, leaders of the SEU have accused the Green movement of being a tool of the Communist Party and agroindustry: "The one who pays the piper calls the tune," they write.

Some ministries also have formed their own environmental front organizations much in the way U.S. industries form political action committees with ecologically correct-sounding names. The Ecological Foundation, a group that announced its intention to establish a bank that would finance purchases of environmental technology, reportedly is controlled by the oil and chemical industries' ministries. Acquisitive individuals are also seeking to profit from widespread concern about the environment. It is relatively easy for an organization to open a bank account and solicit donations by publicizing the account number, but there is no mechanism to ensure that the money collected is spent on environmental improvement.

The Politics of the Green Movements

Practically all Soviet successor states have some form of Green party or political organization. This process of creating Green movements has been most advanced in the Baltic states of Estonia, Lithuania, and Latvia; the first Green Party was formed in Estonia in spring 1988. The Baltic region's Greens scored remarkable successes in early elective politics: Juhan Aare of Estonia and Vaidotas Antanaitis of Lithuania, running as official Green candidates, won election to the USSR Supreme Soviet in March 1989. Both served on the USSR Supreme Soviet's Ecology Committee before withdrawing from the parliament in 1990 in recognition of their republics' political sovereignty. In the 1990 republican elections, the newly created Green Party of Lithuania won three seats in the state's Supreme Council.

Georgia's Greens movement is described as a "public-political organization" and bases its action on the following principles: ecological safety, democracy, and nonviolence. Leaders of the Armenian Union of Greens conceive of their movement as one of human rights, such as fighting for the right to breathe clean air. As a result, they decided not to create a Green party because they hoped to count a large number of people and political parties in their movement without making a bid for power.

By far the most politically significant Green movement to have developed in the former Soviet region is Ukraine's Zelenyi Svit (Green World). In the late 1980s, Communist officials at the regional and local levels, sensing the strength of environmentalism and Zelenyi Svit, cooperated with environmentalists to oppose the construction of potentially harmful enterprises. After democratic elections held in 1990, Zelenyi Svit counted 7 of its adherents among the 105 members of the Ukrainian parliament. In June 1989, members of Zelenyi Svit created the Ukrainian Green Party. The success and popularity of the Ukrainian environmental movement can be attributed in part to the high visibility and close proximity of environmental threats: from Chernobyl and the numerous other nuclear power plants in the republic, from wide-scale devastation in the Donets Basin, and from the ecological decline of the beloved Crimea and the Black Sea. Success may also be attributed to the symbiotic relationship between the movement and its leader Yurii Shcherbak: The popular writer, doctor, and scientist brought a great measure of visibility and respect to Zelenyi Svit and to the Ukrainian Green Party; the movement, in turn, provided him with a strong group of supporters and a vehicle with which to further his political ambitions and, ultimately, to propel him into the government as minister of the environment in 1991.

The Environment and Labor

One of the most significant changes as a result of perestroika during the Soviet era was the reemergence of an independent and insurgent labor movement after decades of being smothered by an official labor bureaucracy that was ultimately loyal to the Communist Party and government instead of the workers. Strikes and other labor actions became commonplace, if not ubiquitous, attesting to a high level of worker dissatisfaction. Protests centered around traditional economic issues such as pay and benefits, work rules, and autonomy of their enterprises. Many strikes were organized with purely political goals or as part of nationalist and ethnic protests. Quality-of-life issues such as food supply, housing, and the environment also figured high on strikers' agendas. According to official reports, almost two-thirds of all strikes in 1990 centered around economic demands, 15 percent had a political nature, and about 4 percent were called to protest environmental conditions.

Improved working conditions have been one demand of labor. This is understandable because the workplace is exceedingly dangerous: In the late 1980s, about 14,000 workers were killed while on the job, 20,000 were maimed, and an estimated 10,000 suffered from work-related maladies every year. More than 9 million people were forced to work under harmful conditions, such as excessive noise, vibrations, and dust. According to one state study, 20 to 38 percent of all illnesses in the Soviet Union were believed to have been "connected with conditions in the workplace." Conditions have only deteriorated under the rigors of economic reform, as new constraints have prompted managers to divert funds away from occupational safety. According to one specialist on the issue, such economizing was "a very widespread problem." Commented Dr. E. Petrosyants: "The following is occurring at many enterprises: Having received the right to distribute a portion of their profits [managers are] releasing funds to boost wages and bonuses. Simultaneously, they reduce investment in occupational safety down to nothing." Thus, managers pay their workers more to put up with deteriorating conditions. Asked Rabochaya tribuna, "Is it worth the price?" Labor's support for environmental cleanup was strongest in the late 1980s and early 1990s. Workers at the Volga Automobile Works, the former Soviet Union's largest car factory, for example, attempted a strike in September 1989; in addition to pay increases and more vacation time, the workers demanded improved working conditions and a cleanup of their city of Tolyatti, which also has been polluted by chemical factories. In the Bashkir capital of Ufa, workers held a three-hour warning strike on May 21, 1990, paralyzing the city. They were protesting a chemical spill at a local chemical plant that left 600,000 residents—over half the city's population—without safe running water for a fortnight. In Karaganda, Kazakhstan, 12,000 coal miners joined with Nevada-Semipalatinsk to call for a halt to nuclear testing at Semipalatinsk, 350 kilometers east of their city.

The strength and anger of labor were most clearly demonstrated in the wildcat strikes that paralyzed the coal industry and sent the Soviet government into a state of crisis for over two weeks in July 1989. After an initial walkout at one mine in the Kuznetsk Basin (Kuzbass) in central Siberia, strikes spread to other mines in the region as well as in Kazakhstan and the Donets Basin in Ukraine, the largest mining region. Throughout decades of centralized planning, the government packed these areas with large heavy industrial complexes to make use of the local coal. As a result, all of these areas still rank high as being environmentally distressed; Kemerovo oblast in the Kuzbass is arguably the worst. "I was in the Kemerovo region where the strikes took place," says Aleksei Yablokov. "People were walking around with masks because you couldn't breathe the air." After the strike ended, Pravda examined conditions there. The state of people's health was "critical," concluded the newspaper: One-half of all workers in the oblast suffered from chronic ailments, and 87 percent of all children were born with "mental and physical anomalies." The problem stems in part from the poor working conditions of women. Health workers in the city of Kemerovo report that there are "practically no healthy women" working in industrial enterprises, and 99 percent of all expectant mothers have been classified as "at risk." "If these trends continue," noted Argumenty i fakty, "by 2000 not one single infant will be born healthy." Said Vladislav Stergilov, a local environmental activist: "Only with the strike did people really become aware of the environmental crisis."

In its economic platform, the Kuzbass Union of Workers, an independent union set up by the coal miners, placed ecology at the top of its list of social concerns:

Several industrial centers in the Kuzbass are located on the verge of ecological disaster. Illnesses, caused by a polluted environment, victimize oblast residents not only at work but also at rest. The ministries, oblivious to the region's interests, commit ecological crimes. The Kuzbass Union of Workers declares as its objective to work for legislation which will protect oblast residents from environmental degradation and which will create such an economic mechanism as to make dangerous production unviable. We reproach the oblast's previous management structure for allowing an unbalanced and rapacious approach towards natural resources and the interests of the Kuzbass population; we will demand a program of ecological recovery for the oblast, environmental assessments of construction projects, for the provision of the population with reliable instruments for the monitoring of chemical and other forms of pollution of the environment and foodstuffs. Still, for most blue-collar workers, the environment is a nonissue. Most are willing to discount their future in favor of immediate material gains, especially if that means holding onto a job—a growing concern as economic reform threatens tens of millions of workers in the 1990s. Conditions in the Arctic mining and metal refining region of Norilsk (population 250,000) provide a case in point. On most days, a sickly yellow-grey pall of sulfur dioxide, nitrogen oxides, carbon monoxide, phenol, and chlorine from the region's light metal smelters settles over the region. One can literally taste the sulfur in the air. Air and water pollution has poisoned local lakes and scorched the fragile tundra: "It's like Hell," summed Andrei Ivanov-Smolenskii of the SEU. Most residents do not live past the age of 50.

After decades of rapid immigration and growth, the city's population decreased by 8,000 people between 1987 and 1990—attesting to increased dissatisfaction with the quality of life in the far north. By 1991, however, outmigration slowed and many workers began to return to Norilsk in search of their old jobs; in the south, economic reform had boosted the cost of living sharply and good-paying jobs had disappeared. In contrast, miners' wages in Norilsk stood at 40,000–50,000 rubles per month in the summer of 1992, while the local minimum wage (including a premium for hardship conditions in the far north) was approximately 10,000 rubles. By comparison, the average monthly wage in Moscow stood at about 6,000 rubles.

Vladimir Shishkov, a manager at the Volga Automobile Plant in Tolyatti, related the following story to Moscow News about a trip to Chelyabinsk to negotiate supply contracts for his firm: "At the shop producing lead-containing products, I saw men almost naked to the waist and without respirators. I asked how I could help them: with working clothes or protective means? They answered: if we could just have at least 50 cars to sell to our employees each year." The Odessa city government, under heavy pressure from environmental groups, decided in the summer of 1990 to shut down a local chemical plant. The city's leaders, in turn, quickly found themselves in conflict with the plant's employees, who challenged the measure in order to save their jobs. Higher-ups in the chemicals ministry ignored the city's order. The city resorted to a public referendum on the plant "as the last means" to resolving the impasse. The result was that 83 percent of voters were in favor of the plant's closing.

In the Soviet era, industrial ministries perfunctorily absorbed the cost of fines for pollution from their enterprises as a routine business matter. With the transition to free-market relations and the rise of private and collective ownership, enterprises now are forced to pay these fines out of their own revenues and profits, hence impinging on their social benefits and wage funds. The result has been predictable: Miners striking on Sakhalin Island demanded, among other things, that fines assessed against their enterprise for polluting the environment be lifted. In October 1990, oil and gas workers in the Tyumen region threatened to shut down their pipelines unless the government lifted pollution fines assessed on employees of the firm; accidents had garnered the west Siberian concern 26 lawsuits for violation of environmental regulations. Moscow News commented:

The oil workers found themselves unprepared for the demands of the oblast procuracy and committee for the protection of nature recently created in the oblast. For decades, oil poured onto the ground from the active pipelines; for decades, an indulgent system meticulously protected the guilty parties. And suddenly... popular concern, new people in the soviets, million-ruble suits against enterprises, criminal cases, fines.

The Tyumen workers were dismayed at being held responsible for the pipelines, which were laid through swamps and were shoddily built by other firms. In addition to the removal of the fines, the workers demanded that their enterprise be permitted to keep more of the hard-currency earnings it produced. Radio Moscow pointed out that the workers were not intending to pocket all of the money; instead, they planned to use much of it to purchase new drilling equipment and to renovate the pipelines, which were in "a dangerous condition."

CHEMICAL INDUSTRY'S TRADE ASSOCIATION UNDERMINES THE POLICIES

The American Chemistry Council (ACC) may not be a household name, but its influence on chemical policies has affected millions of Americans. Many chemicals in your household are there because ACC lobbying and advertising has prevented or delayed federal regulation of them.

Of course not all chemicals found in consumer products are dangerous, and "chemophobia" needs to be tempered with science. But there is plenty of evidence that many common chemicals do pose significant health risks to consumers and workers, and we need more effective regulation to address those risks.

The ACC has too often worked to obstruct such regulation, following a pattern modeled by the tobacco industry: deny the science, bring in its own experts to counter the evidence, launch misleading advertising campaigns, and pressure decision makers to abandon restrictions on the chemical's use.

A Major Lobbying Force

Founded in 1872 as the Manufacturing Chemists' Association, the ACC today boasts an annual budget of more than $100 million. Its board members include chemical and petrochemical industry heavyweights such as Dow, DuPont, Marathon Petroleum, and ExxonMobil. In 2013-14, the ACC spent $23 million on federal lobbying, and it is active in state-level lobbying, both directly and through the American Legislative Exchange Council (ALEC).

The ACC also has a sizable budget for political contributions to members of Congress. Several members sitting on key committees overseeing chemical policy have received sizable donations from the ACC and its member companies in recent years. And the proverbial "revolving door" between government and industry has been spinning hard for the ACC and its member companies: in the 2013-14 election cycle, nearly 70 percent of ACC's registered federal lobbyists had previously held jobs in Congress or in the executive branch.

Pushing against Chemical Safety Reform

ACC influence has been a factor in several recent chemical policy battles:

TSCA Reform

The Toxic Substances Control Act of 1976 (TSCA, pronounced "Tosca"), though well intentioned, has been remarkably ineffective. In four decades, the EPA has completed reviews and issued bans or restrictions on just nine of the 84,000 chemicals currently registered for commercial use in the U.S.

As a result, many chemicals long known to pose serious health risks, such as formaldehyde and asbestos, are not adequately regulated at the federal level. To fill this gap, many states have issued regulations of their own.

As of our report's release in July 2015, two bills addressing TSCA reform were being considered by Congress. The ACC has endorsed and lobbied for these bills, both of which would pre-empt state regulations, though the House version would give states more freedom to regulate before EPA takes action.

Whatever TSCA reform bill does pass, chemical regulation is still likely to be a slow process: under current proposals, it would take the EPA 50 years or more to assess even 1,000 of the most toxic chemicals in commerce.

Safety at Chemical Manufacturing Facilities

In recent decades there have been about 30,000 documented chemical facility accidents per year, resulting in more than 1,000 deaths. Recent catastrophes such as the plant explosion in West, Texas or the spill at Elk River in West Virginia have demonstrated how devastating such accidents can be for unprepared communities.

Despite the seriousness of this problem, the ACC has worked to limit regulatory oversight of chemical plant safety. As a result, companies are not required to implement technologies that can reduce risk. And chemical facilities can foist responsibility for emergency response onto local communities without ensuring that local authorities have the information or resources they need to respond to a disaster. Compounding the problem, access to information about risks at specific chemical facilities is tightly restricted—and the industry, citing security concerns, has lobbied to keep it that way.

Trade Negotiations

The ACC hasn't confined its lobbying efforts to domestic chemical policy. It has also worked to influence the outcome of international trade negotiations such as the Trans-Atlantic Trade and Investment Parternship (TTIP) agreement.

For instance, ACC has joined other industry groups in targeting EU regulation of pesticides containing endocrine disruptors (EDCs), which have been shown to cause a variety of health problems. So far, this effort is paying off: in May 2015, the EU announced that it would delay implementation of a ban on 31 pesticides containing EDCs. A leaked document obtained by the Center for International Environmental Law reveals that this is part of a larger strategy to slow regulatory developments at all levels.

How ACC Undermines Chemical Safety Science: Five Examples

Here are some common hazardous chemicals that the ACC has helped to keep unregulated or under-regulated:

Bisphenol-A (BPA)

- What it is: A synthetic compound used as an additive in the production of polycarbonate plastics and epoxy resins. BPA is commonly found in many products, including beverage containers, food packaging, and compact discs.
- Health effects: BPA is an endocrine disrupter that mimics estrogen. It is a reproductive, developmental, and systemic toxicant, especially harmful to children. Animal studies suggest that BPA exposure could lead to reproductive disorders, diabetes, and cardiovascular disease.
- Regulatory status: Largely unregulated in the US, but banned or limited in many other countries, including the European Union and Canada.

- ACC's role: The ACC has lobbied against legislation to regulate BPA and has launched a PR campaign featuring the slogan "Listen to the Science." The campaign's website refers to several flawed studies and uses misleading quotations to lend seeming scientific authority to the group's own pronouncements.

Flame Retardants

- What they are: A group of chemicals added to furniture, clothing, electronics, and building materials in order to meet flammability standards. (Recent evidence suggests that flame retardants may not in fact be very effective in delaying the spread of fire.)
- Health effects: Studies have linked flame retardant exposure to lower IQ in children, early puberty in girls, endocrine disruption, birth defects, and cancer.
- Regulatory status: No federal rule governs the use of flame retardants; strict state regulations, especially in California, have had some impact on the industry.
- ACC's role: The ACC has responded to California's strong regulation push with a misleading website, relentless lobbying, "astroturf" activism, and industry-sponsored studies, oe of which was published without a required conflict-of-interest disclaimer and has been strongly criticized by leading fire scientists.

Formaldehyde

- What it is: A colorless, flammable organic compound used in building materials, medicinal and personal care products, and furnishings.
- Health effects: Formaldehyde's status as a human carcinogen has been confirmed by the World Health Organization's International Agency for Research on Cancer (2004) and the U.S. Department of Health and Human Services (2011). In addition to the long-term cancer risk, short-term formaldehyde exposure causes nausea, headache, and eye, ear, nose, throat and skin irritation.
- Regulatory status: Formaldehyde was one of 62,000 chemicals grandfathered in when the TSCA was first passed in 1976, so testing to establish its safety has never been required, even as evidence of its carcinogenic effects has continued to accumulate.
- ACC's role: As with BPA and flame retardants, the ACC has posted a website with misleading information about formaldehyde's health impacts. ACC lobbying has won a five-year delay in EPA implementation of 2010 legislation setting limits on formaldehyde emissions from composite wood products.

Silica

- What it is: Silica (also known as quartz or silicon dioxide) is a very common mineral, used in a variety of building materials such as concrete, bricks, and glass.
- Health effects: Silica dust is created when cutting, grinding, drilling, or mining materials containing silica. When this dust is inhaled, it can cause silicosis, a progressive, irreversible, sometimes fatal lung disease.
- Regulatory status: Worker exposure is regulated by OSHA, which defines Permissible Exposure Limits (PELs) for silica dust. However, current silica PELs are more than 40 years old, and in recent years OSHA has proposed updating the standards to better protect workers.
- ACC's role: In 2013, a group of OSHA opponents, including the ACC, contributed $151,266 to a group of 16 senators—who three months later submitted a letter to OSHA requesting an extension of the hearing process. As of 2015, the silica rule is still tied up in the rule-making process.

Spray Polyurethane Foam

- What it is: A building insulation material that can contain harmful chemicals, including diisocyanates.
- Health effects: Diisocyanates are respiratory, skin, and mucus membrane toxicants that can cause asthma or trigger severe asthma attacks in sensitive populations.
- Regulatory status: As part of its Green Chemistry Initiative, California has added SPF products to a list of priority products that companies are encouraged to avoid in favor of safer alternatives.
- ACC's role: The ACC has argued publicly against the inclusion of SPFs on the priority products list, despite scientific evidence of their harmful effects.

Recommendations

Our chemical policies should protect the public, not the chemical industry's profits. The chemical industry and its trade association should be held accountable for their work to influence decision makers and undermine science. The public has the right to know both about the chemicals in the products around us and about industry efforts to influence elected officials and regulators.

Chemical Policies Should Protect the Public

- Congress should pass legislation to strengthen TSCA so that more chemicals are reviewed and regulated.

- The United States should reject any international trade agreement that compromises science-based public health and safety protections concerning chemicals.
- The EPA should revise the Risk Management Program (RMP) to prioritize disaster preparedness, prevention, transparency, access to information, and overall industry accountability. Specifically, the agency should
    - o Fully utilize web tools and social media to ensure timely, accessible, and public access to RMP information;
    - o Require chemical facilities to evaluate, document, and use safer chemicals and processes wherever possible;
    - o Require chemical companies to take greater responsibility for accident response, rather than shifting costs to local governments;
    - o Require comprehensive reporting and investigation of all incidents.

The Political Activities of the ACC and its Members Need to be More Transparent

- The Securities and Exchange Commission should issue a rule requiring publicly traded companies to disclose both their direct and indirect political activities. 1.2 million people have already signed a petition asking the SEC to do this.
- Congress should approve legislation, such as the DISCLOSE Act, enhancing disclosure of indirect political contributions.
- Investors and their representatives should pressures companies to disclose all direct and indirect political spending, disclose whether they agree with the scientific and policy positions of their trade and business associations, and influence the scientific and policy positions of their trade groups , or leave groups whose positions do not align with the company's.
- Companies should insist that their associations accept the best available science on chemicals and their impacts and adopt policy positions that reflect this acceptance; where their position differs from their trade group's , they should publicly state such differences, attempt the influence the trade group's position from the inside, or leave the group if differences are irreconcilable.
- Consumers should hold their congressional leaders and companies accountable by demanding access to information on the effects of chemicals they may be exposed to; where there is strong evidence that chemicals in commercial use have harmful health effects, they should pressure companies to shift to safer ones.

TECHNOLOGY DEVELOPMENT IN CHEMICAL INDUSTRY

The Indian chemical industry is among the established traditional sectors of the country, playing an integral role in the country's economic development. This sector, forming part of the basic goods industry, is a critical input for industrial and agricultural development. The industry has a weight of 14% in the Index of Industrial Production (Base year 1993-94 = 100), giving an indication of the importance the sector holds in the country's industrial growth. A robust chemical industry is a harbinger of significant economic and strategic benefits to the nation.

The chemical industry is among the most diversified industrial sectors, including basic chemicals and its products, petrochemicals, fertilisers, paints, gases, pharmaceuticals, dyes, etc. The sector covers over 70,000 commercial products, and provides the building block for many downstream industries, such as finished drugs, dyestuffs, paper, synthetic rubber, plastics, polyester, paints, pesticides, fertilisers and detergents.

The industry includes a wide variety of products, from basic chemicals to research-driven specialised products, at different levels across the industry supply chain. The fundamental nature and diversity of the industry is best understood from the fact that the industry itself is the largest consumer of its products, accounting for around 33% of total consumption.

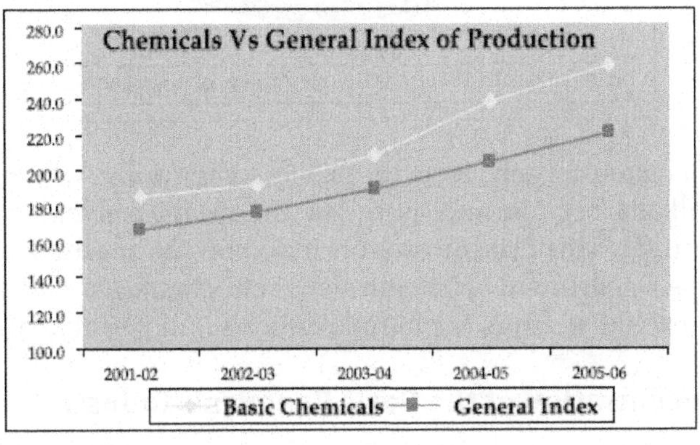

As stated in the Annual Report of 2005-06 of the Department of Chemicals & Petrochemicals, GoI, the domestic chemical industry contributes about 17.6% to the total output in the manufacturing sector, 13-14% to total exports and 8-9% of total imports into the country. The sector has a share of 3% to the country's total GDP. Its contribution to the revenue kitty of the Government is around 18-20%.

The domestic industry's turnover is estimated to have crossed US$ 30 bn (Rs 1,300 bn), which is slightly over 1% of the global production. In world ranking, India stands 12th in terms of production.

Industry Structure

The chemical industry can be broadly classified into two segments - organic and inorganic chemicals. Organic chemicals cover over half of all known chemical compounds, and includes petrochemicals, drugs, cosmetics, agrochemicals, etc. Inorganic chemicals comprise of alkalis, dyes & dyestuffs.

Based on a more functional classification, chemicals may be divided into basic, specialty and fine chemicals.

The basic chemicals industry forms the largest part of the chemical industry and is characterised by capital intensive, high volume, low margin products. Specialty and fine chemicals are low volume, high margin in nature. It is estimated that nearly 70% of fine chemicals produced in India are used by the pharmaceutical and agrochemical industries. Specialty chemicals include adhesives, additives, antioxidants, biocides, corrosion inhibitors, cutting fluids, dyes, lubricants, pigments, etc..

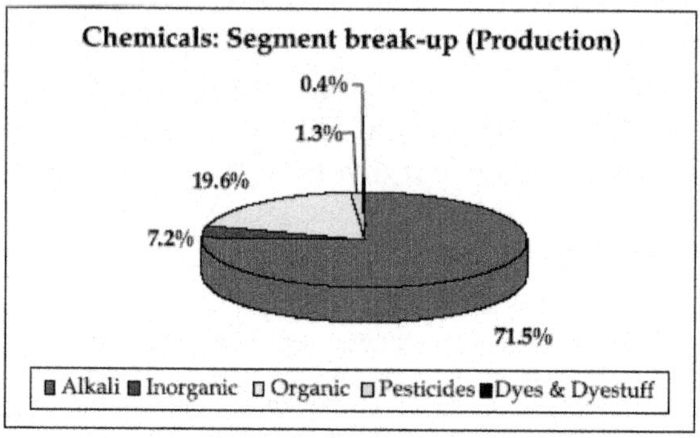

This publication largely focuses on basic chemicals, which can be further divided into alkalis, organic and inorganic chemicals, pesticides and fungicides, dyes and dyestuffs. This classification is based on the product categorisation as provided by the Department of Chemicals & Petrochemicals, GoI, for the chemical industry, and excludes drugs & pharmaceuticals and petrochemicals.

Regional Concentration of the Basic Chemicals Industry

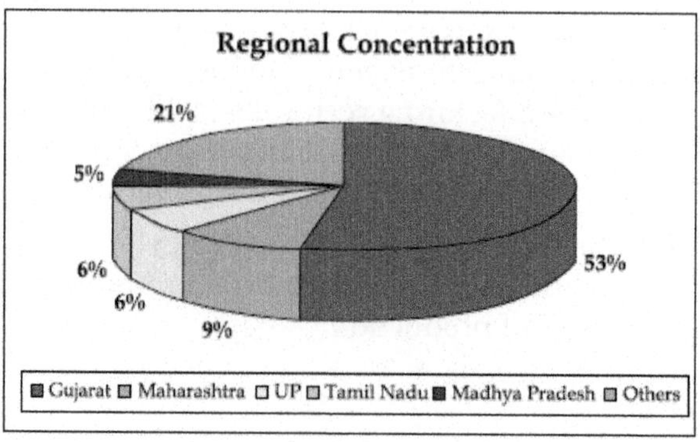

Though the chemical industry is spread across the country, there is relatively a high concentration along the west-coast, largely due to the proximity to raw materials and ports. Gujarat alone is estimated to contribute around 53% to the total production in the country, followed by Maharashtra, which contributes 9%.

The other major producing states include UP, TN, MP and Punjab. On the other hand, in the case of heavy chemicals segment, especially inorganic chemicals, fuel availability is a determining factor, and hence there is a concentration of these companies around power plants. Due to the regional concentration of chemical companies in certain pockets, logistics costs for the industry have tended to become a significant position of total costs.

Industry Sub-segments

The annual production of basic chemicals between FY02-FY05 has been growing at 7% per year, while consumption has been rising at 5% per annum. Imports and exports have also been rising at 7.25% and 37.6% respectively, implying that India has been a net exporter of chemicals. As per the Department of Chemicals, on the largest imports, in value terms have been for organic chemicals followed by dyes & dyestuffs, while the largest export item in value terms is also organic chemicals and chlor-alkali chemicals.

Table : State-wise Capacity & Production of Major Chemicals.

State	CAPACITY			PRODUCTION			% SHARE IN PRODUCTION		
	Mar. 04	Mar. 05	Mar. 06	2003-04	2004-05	2005-06(P)	2003-04	2004-05	2005-06
Gujarat	4,331	4,449	4,474	3,744	3,940	4,055	53.0	53.4	53.1
Maharashtra	1,004	1,014	1,011	708	725	698	10.0	9.8	9.1
Uttar Pradesh	486	527	637	423	440	486	6.0	6.0	6.4
Tamilnadu	509	509	525	468	475	468	6.6	6.4	6.1
Madhya Pradesh	475	477	463	318	331	341	4.5	4.5	4.5
Punjab	342	364	364	279	288	312	3.9	3.9	4.1
Andhra Pradesh	312	312	416	228	229	256	3.2	3.1	3.3
Rajasthan	270	278	410	158	159	248	2.2	2.2	3.2
Kerala	311	261	280	201	215	185	2.8	2.9	2.4
West Bengal	248	256	255	151	160	163	2.1	2.2	2.1
Pondicherry	110	110	110	100	124	141	1.4	1.7	1.9
Assam	122	122	124	106	95	100	1.5	1.3	1.3
Jharkhand	92	92	108	84	90	81	1.2	1.2	1.1
Karnataka	83	83	83	54	61	62	0.8	0.8	0.8
Orissa	38	38	38	37	34	37	0.5	0.5	0.5
Haryana	30	30	30	6	7	6	0.1	0.1	0.1
Bihar	10	10	10	3	2	0	0.0	0.0	0.0
Total	8772	8932	9337	7067	7375	7640	100.0	100.0	100.0

Chlor Alkali

Chlor alkalis, the largest segment of basic chemicals produced in India, is a volume driven, low margins industry and accounted for around 72% of the total production in FY06. Caustic soda, soda ash and chlorine are products of this industry, forming the basic building block for the chemical processing industry. End-users of this segment include aluminium, dyes, pharmaceuticals, glass, newsprint, paper & boards, soaps & detergents, viscose, textiles, water treatment, pesticides industries, to name a few. The scale of margins for alkalis, to a large extent, depends upon the levels of industrial activity in sectors such as metals, textiles and pharmaceuticals.

Being a capital intensive sector, the sector is largely dominated by big players. Hence, energy costs form a key determinant of the profitability of the industry. Chlor alkali companies are largely concentrated along the west coast due to the availability of salts, a key raw material.

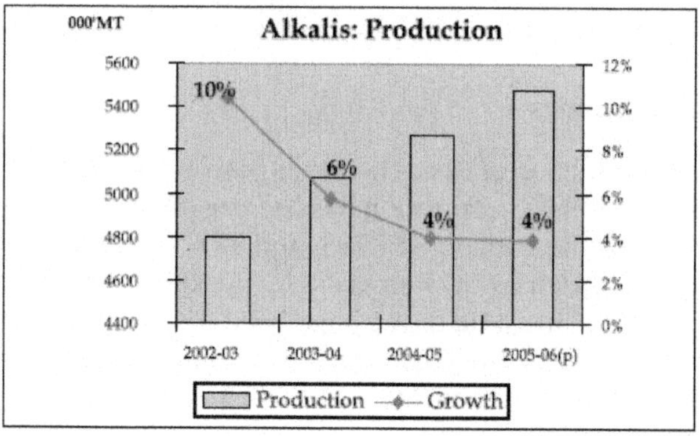

The Chlor alkali segment has been witnessing a robust 6% growth in production since FY02, with production in FY06 touching an estimated 5.5 mn tonnes. This is one of the few industries where supply exceeds demand, thus having tremendous potential for exports. Since FY03, the sector has been a net exporter. Exports of chlor alkali have been growing at 52% annually since 2001, while imports are rising at close to 6% per year.

Technology has played a key role in this segment in adopting better production techniques. Around 60% of mercury-based caustic-chlorine plants in the country have shifted from mercury cell technology to membrane cell technology that has been recommended as a viable production alternative.

Table : Performance of Chlor Alkali Chemicals.

	2001-02	2002-03	2003-04	2004-05	CAGR
Capacity	6,057	6,256	6,393	6,422	1.9
Production	4,342	4,792	5,070	5,272	6.7
Imports	203	220	201	239	5.6
Exports	107	239	217	380	52.6
Consumption @	4,438	4,773	5,054	5,131	5.0

Organic Chemicals

Organic chemicals form the second largest segment of the chemical industry. This segment has the largest number of products classified under it, most of which are knowledge driven. As a result, R&D forms a considerable part of the manufacturer's costs. Acetic acid, phenol, methanol, formaldehyde, nitrobenzene, citric acid, etc are part of this segment, most of which are used in drugs, pesticides, etc. This segment accounts for 20% of total chemical production in the country.

This segment produces a large number of products and combinations thereof, and is dominated by medium and small players, with specialised focus in particular product segment. Being largely a technology driven segment, these manufacturers have exhibited the competence to produce chemicals of high quality standards. This segment too is concentrated in Western India.

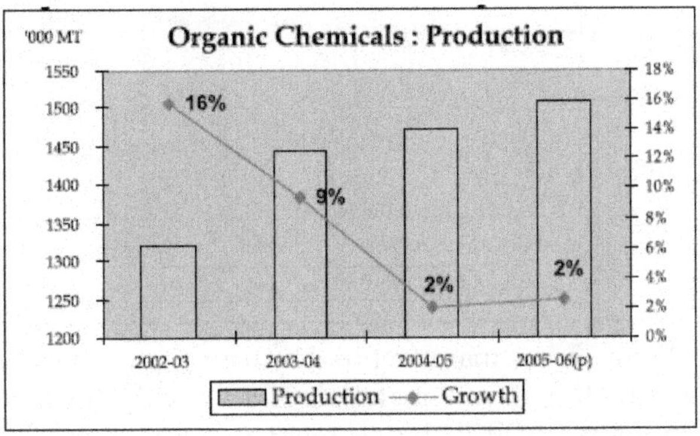

During FY06, production of organic chemicals is estimated to have increased by 2% to 1.5 mn tonnes. This sluggish rate of growth has been persisting since FY05, while between FY02-FY05 production grew at an impressive 7%. Exports of organic chemicals have been impressive, both in terms of value and volume. Between FY02-FY05, exports grew by 55% while imports went up by 6% annually in quantity terms. This rise in exports can be attributed to the significant value-addition to the Indian product list, especially in the drugs and pharmaceutical segments.

During FY05, exports of organic chemicals rose by 36% in volume terms to 63 thousand tonnes, with a 66% jump in value terms to Rs 5.7 bn. Imports, although dropped by 6% to 561 thousand tonnes in volume in FY05, saw a rise of 26% in value to Rs 12.3 bn during this period. The domestic market forms a large part of the consumer base for the XVI industry. However, due to price competitiveness, products like methanol are mainly imported into India.

Table : Performance of Organic Chemicals.

	2001-02	2002-03	2003-04	2004-05	CAGR (%)
Capacity	1,281	1,403	1,601	1,669	9.2
Production	1,140	1,318	1,445	1,473	8.9
Imports	477	395	598	561	5.6
Exports	17	23	46	63	54.8
Consumption @	1,600	1,690	1,997	1,971	7.2

Inorganic Chemicals

This segment comprises of products such as aluminum fluoride, calcium carbide, carbon black, potassium chlorate, sodium chlorate, titanium dioxide and red phosphorous, etc. End-users of these chemicals include manufacturers of soaps & detergents, glass, fertilisers and alkalis.

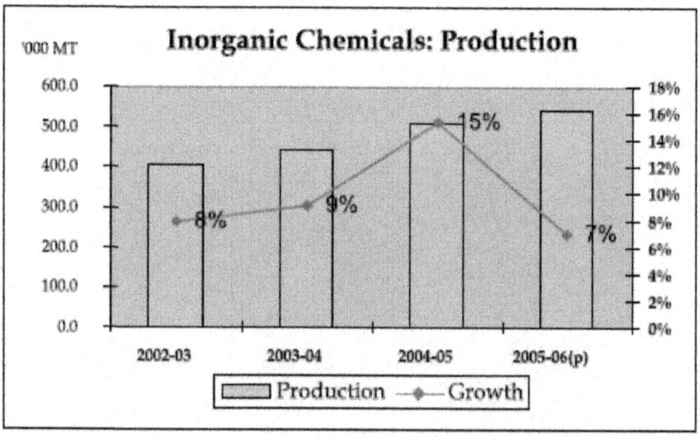

Inorganic chemicals are manufactured by using naturally occurring minerals, and therefore availability of raw materials is the key determinant for the development of the industry. In 2005, this segment accounted for 7% of total chemical production in India. There are large variations among products in the segment owing to the required skills and technology used, and the value addition to the products.

The inorganic chemicals sector is the fastest growing sub-segment, having recorded an average annual growth rate of around 11% between FY02-FY05. During FY06 production is estimated to have risen by 7% to 544 thousand tonnes. Exports of inorganic chemicals too have been rising by over 50% annually since FY02. Imports have been erratic and on the decline in volume terms, though the value of imports has been high, implying the high-value imports by domestic players.

Table: Performance of Inorganic Chemicals.

	2001-02	2002-03	2003-04	2004-05	CAGR (%)
Capacity	487	487	568	642	9.6
Production	374	404	441	508	10.7
Imports	53	99	116	97	22.3
Exports	40	34	64	146	54.0
Consumption @	387	469	493	459	5.9

Pesticides

Pesticides are one of the most important constituent of the agro-chemicals sub-segment, which has played an almost revolutionary role in the Indian agricultural sector. According to estimates, India is the second largest manufacturer of agrochemicals in the world with 165 pesticides registered in the country.

India has one of the most dynamic generic pesticide manufacturing base, with more than 60 technical grade pesticides being manufactured indigenously by 125 producers consisting of large and medium scale enterprises, including about 10 multinational companies, and more than 500 pesticide formulators spread all over the country. India is also a dominant producer of Isoproturon, a weedicide, accounting for nearly 25% of the world's production.

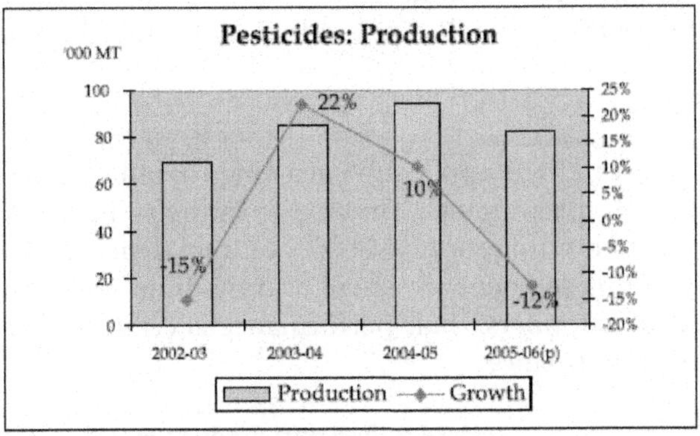

This segment is also a knowledge driven segment and R&D plays an important role. It has witnessed consolidation over a period of time and the presence of MNCs has been expanding through increased acquisitions of local players. Nonetheless, a large number of small and medium players are associated with the production, distribution and marketing of agrochemicals.

Use and production of pesticides is directly related to the crop situation, and indirectly to monsoons. Cash crop producers are the major consumers of pesticides. During FY06, production of pesticides dropped by 12% to 82,000 tonnes mainly due to lower intake of pesticides in the cotton crop, which typically consumes the highest proportion of pesticides.

However, exports have been one of the growth enablers for this sector, which also helps to hedge the risk of weather conditions. Although most of the exports have been undertaken by MNCs through their established distribution channels across the globe, Indian players have also shown competence in select product segments.

During FY05, exports of pesticides grew by 9% to 22,000 tonnes. However, unit realization remained under pressure and fell by 9% thus leading to a decline in the export value by 1.6% to Rs 5,180 mn. Some key export destinations for India include US, UK, France, Netherlands, Belgium, Spain, South Africa, Bangladesh, Malaysia and Singapore. India meets most of the domestic requirement of pesticides and imports are limited to only few innovative products.

Table : Performance of Pesticides.

	2001-02	2002-03	2003-04	2004-05	CAGR (%)
Capacity	138	139	134	146	1.9
Production	82	72	84	94	4.7
Imports	1.0	0.6	2.0	2.0	26.0
Exports	13	17	20	22	19.2
Consumption @	70	55.6	66	74	2.0

India has lately emerged as a global base for generic agrochemicals, sales of a significant proportion which are to traders and not to end-users. The reason for this is that apart from lack of necessary sales and distribution infrastructure, companies have yet to obtain relevant product registrations to enable direct sales.

Dyes and Dyestuff

Dyes and dyestuff finds application in a range of industries, including paints, leather, textiles, ink, plastics, etc. The textile industry is the largest consumer for this segment, accounting for nearly 80% of total demand. A large number of manufacturers in this segment are small and medium scale players. However, being intermediate suppliers, their performance is critically linked with that of their end-users' performance.

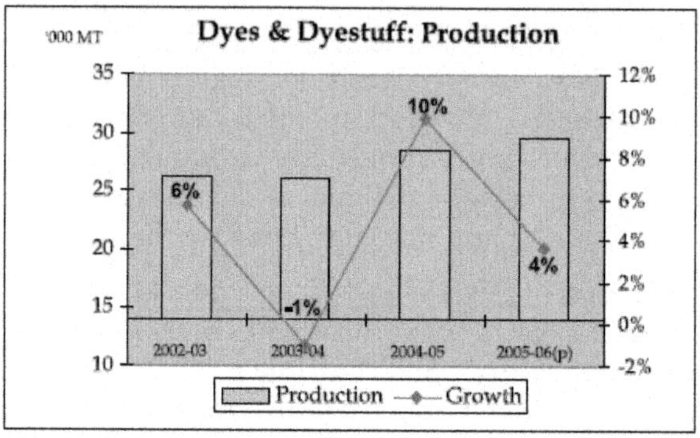

From being an importer and distributor during the 1950s, the dyes and dyestuff industry has come a long way. India today accounts for 6% of total world production of dyes, and Indian dyes are exported to the East Asia, Africa, EU, etc. During FY06, production of dyes grew by 4% to 30,000 tonnes. As per data available from the Department of Chemicals & Petrochemicals, exports of dyes and dyestuff witnessed a fall of 7% during FY05 to 112,000 tonnes in volume and a drop of 6% in value to Rs 2.27 bn. In contrast, imports went up by 21% to 15,000 tonnes in volume and by 20% to Rs 3.56 bn in value during this period.

Table : Performance of Dyes & Dyestuffs.

	2001-02	2002-03	2003-04	2004-05	CAGR (%)
Capacity	49	49	54	54	3.29
Production	25	26	26	28	3.85
Imports	9	10	13	15	18.56
Exports	100	122	131	112	3.85
Consumption @	-66	-86	-92	-69	1.49

Key Strengths & Drivers

The chemical industry forms the backbone of the Indian manufacturing base. Some key strengths of the sector that can drive growth for the industry include low cost manpower, large domestic market, strong forward and backward linkages and conducive policy environment.

Considering the vastness of this sector, some of the common growth drivers that could be identified for the sector include:

Macroeconomic Factors

Being largely an intermediate product, a strong economic growth is an important factor for sustaining demand for the chemical industry. In fact, the per capita consumption of most of the finished products under this sector is far below the world average, giving a hint to the potential growth for the industry.

Integration Along the Value Chain

The industry participates in different stages of the value chain by producing intermediates and finished goods. This makes integration of processes easier. In fact, growing competition in select chemical segments has forced the industry to scale up production, which, inter-alia, requires backward or forward integration in some cases. Higher consolidation and capacity building has driven growth.

Focus on R&D

Specialty and fine chemicals are essentially a knowledge-based industry, which requires sustained investment in R&D. During the past few years, the chemical industry has witnessed a rise in R&D and technology up-gradation. This has led to many new products being introduced in the market, thus boosting demand.

Outsourcing and Contract Manufacturing

On the strength of low-cost production and world-class technology, India is being looked upon as a preferred destination for outsourcing and contract manufacturing. This has led to higher utilisation of capacity and revenue generation for the participants.

The Indian chemical industry today is emerging from a protected environment into highly competitive global market, and at the same time the domestic market is already reaching a mature level where demand potential for chemical end-products is on the rise. In these changing circumstances, the industry faces some key challenges.

Power Costs

Chemicals, especially heavy chemicals, are power intensive sectors, and sustained supply of power is imperative. Volatility in power supply and prices of crude oil has been impacting the margins of the chemical companies.

Technology

Although India has shown remarkable improvement in technological innovation, it still lags behind international standards. Chemical sectors are one, where technological changes are rapid and needs continuous up-gradation and innovation.

Infrastructure

Poor infrastructure, like roads, rail and ports are other detrimental factors. Regional concentration of the chemical industry requires better infrastructure and logistics to reach across the country.

Dumping

Growing international competition and low customs duty on some of the chemicals have led the dumping of certain chemicals in domestic industry. Notably, most of the cases related to anti-dumping duty in India relates to chemical sector.

ADVANCES IN SCIENCE AND TECHNOLOGY AND THE CHEMICAL WEAPONS CONVENTION

With the second review conference of the Chemical Weapons Convention (CWC) approaching in April, a raft of studies have appeared making clear that fundamental changes in science and technology are affecting the implementation of the treaty and that it must be adapted to take account of them.

The most significant development is the revolution in the life sciences and related technologies, including a growing overlap between chemistry and biology. There is a vastly increased understanding of the functioning of biological systems as a result of the mapping of the human and other genomes as well as of advances in structural biology and the study of proteins (proteomics). Information technology and engineering principles are increasingly integrated into biology. The intersection between chemistry and biology has further expanded thanks in part to the automation of synthesis and screening of chemical compounds enabling laboratories to assess vast numbers of new chemical structures and a much-enhanced understanding of how certain "chemicals of biological origin" act. Technological advances supplement these trends, for example, providing for more efficient means of delivering biologically active chemicals to target populations or targeting organs and receptors within an organism.

These developments are expected to bring many benefits, including new medical treatments and methods of pest control. At the same time, the capacity to discover or design new chemical structures that may have utility as chemical warfare agents has also increased significantly. Novel agents can be created far more quickly than ever before. In addition, advances in manufacturing technology have shortened other time requirements, enabling shortcuts in the progression from research and development to full-scale manufacturing. Changes in the chemical industry have dispersed technology and facilities, complicating verification and traditional nonproliferation strategies.

As a result, the time and effort needed to field a new chemical weapon has shrunk, particularly in the early stages, while the capability to detect such actions has not grown significantly. These trends and a recently increased interest in the use of incapacitants for law enforcement purposes raise at least the threat that states could skirt or quickly break out of the CWC prohibitions on developing and acquiring chemical weapons. It has also enlarged the overlap between the two otherwise quite separate treaties governing chemical and biological weapons, the CWC and the Biological Weapons Convention (BWC). States-parties need to adapt the implementation of the CWC to account for these changes or risk diminishing confidence in its effectiveness and endangering its viability.

Advances in Technology and Industry

To be sure, many traditional obstacles remain to the development of chemical weapons in a state-level program. Most importantly, any potential agents must meet tough requirements before they can be fielded. These include the possibility of industrial-scale production, tactical mixtures that can be effectively disseminated and are sufficiently stable for long-term storage, effective dissemination equipment/devices, and adequate means of ensuring that one's own forces are protected.

These constraints do not apply to the threats of terrorist chemical weapons use. The threat that terrorists may use toxic chemicals, however, correlates more closely with the accessibility of toxic materials than with the evolving scientific capability to develop novel agents in the laboratory. That is not to dismiss the concern, but the problem is less one of enforcing an international norm such as the CWC than of how states, companies, and research institutions can control access to chemical facilities and materials.

Technological change can also bring significant benefits to the fight against chemical weapons. For example, advances in nanotechnology are expected to help in developing more effective protections against agents, such as new detection devices (faster, cheaper, more sensitive, and more selective sensors), and improved filtration materials, means of decontamination, and medical countermeasures.

Still, these scientific and technological gains undoubtedly come with new concerns. Nanotechnology offers the possibility to engineer "smart" materials that respond to specific stimuli. It also promises a more efficient and targeted drug delivery *via* the respiratory system and other pathways. For example, it could facilitate the entry of toxic chemicals into the body or specific organs, in particular the brain, for selective reaction with specific gene patterns or proteins or for overcoming the immune reaction of the target organism. These developments may have significant applications as new medicines and treatments. They could, however, also be exploited for the development of new chemical warfare agents or the fine-tuning of existing ones. Any offensive chemical weapons program begun today would surely take advantage of these new methods and concepts.

Technological advances continue to change the manufacturing processes in the chemical industry. New processes are being introduced to increase efficiency and yield, including a range of new catalytic processes. The use of multipurpose equipment and the adoption of on-demand principles has become a common feature. Manufacturers adjust to changing market demands and increasingly use technologies and equipment that allow them to switch production output on short notice. Although allowing chemical plants to be converted to the production of new products, this also means that there is a risk of standby capabilities appearing that could easily and quickly be switched to supplying an offensive chemical weapons program.

A more recent development is the use of microreactors, which have begun moving from the laboratory to (limited) industrial production. One of the driving factors is safety: chemicals that are otherwise hazardous to manufacture, handle, or store can be produced safely on-site when needed. In addition, capital costs for such facilities are low, and many chemical reactions show improved reactivity, product yield, and selectivity when performed in microreactors. Microreactors allow companies to scale up a chemical process from laboratory to industrial scale more quickly and easily. Production output can be increased by combining multiple reactors in batteries, a process also known as "numbering-up."

Of course, these advantages could also apply to new chemical warfare agents. The use of microreactors can significantly shorten the time required to synthesize new toxic chemicals for testing and development purposes. Microreactors can apply combinatorial principles to synthesize a series of related compounds for test purposes, or they can be used to make small quantities of toxic chemicals easily and quickly during the development of a new agent for weaponization. Worryingly, if states were to produce chemical weapons in such reactors, there would be fewer clues that might indicate to outsiders that such production was taking place. Traditional industrial-scale chemical weapons production facilities require heavy-duty ventilation systems, scrubbers, and high stacks. Because processing highly toxic or corrosive materials in microreactors produces fewer waste streams and reaction yields are much higher, there is less need for telltale pollution abatement systems.

In addition, structural change in the chemical industry could also pose risks to CWC implementation. Driven by market forces, the industry is moving from its traditional production locations (Japan, the United States, and western Europe) to new places in Asia, eastern Europe, Latin America, and the Middle East. Some of the countries involved in setting up new chemical operations have limited experience in regulating chemicals or weak implementation systems for the CWC. At the same time, international trade in chemicals is on the increase. These are challenges to the CWC's verification system as well as to traditional nonproliferation measures in the chemical field.

Recommendations to the Review Conference

What, then, are the issues that the upcoming review conference ought to consider when it assesses the impact of advances in science and technology on the operation of the CWC?

A first is whether the CWC's schedules, which attempt to characterize agents and precursors by their risk based on past weaponization and other factors, need to be amended. Currently, the Organization for the Prohibition of Chemical Weapons (OPCW), the international body charged with implementing and verifying the CWC, and states-parties tightly control high-risk chemicals (Schedule 1). This is feasible because these agents and key precursors, as a rule, have very few legitimate uses. States and the OPCW less stringently control Schedule 2 and 3 chemicals, which have legitimate uses ranging from smaller-scale specialty chemicals to mass-produced chemicals and thus have much higher thresholds before such production triggers declarations and inspections. Should the growing overlap between chemistry and biology and the emergence of new biologically active compounds necessitate the inclusion of new chemicals into the schedules or are the schedules unfit to deal with these emerging risks? Including new chemicals with potential chemical weapons utility in Schedule 1 would severely hamper their legitimate uses. Listing them in Schedules 2 and 3 might be meaningless given the relatively high thresholds for declaration and inspection; many would simply fall through the net.

If the schedules were to remain as they are, there is a risk that the CWC verification system may get stuck in the past. One way to avoid this fate would be to step up the frequency and effectiveness of verification at other chemical production facilities (OCPFs) producing unscheduled discrete organic chemicals. This relates to a number of issues: the overall number of OCPF inspections that is desirable, the amount of information available on these facilities to the OPCW Technical Secretariat, an improved site selection mechanism for inspections, the level of expertise at the OPCW to ensure that inspectors can adequately assess facility capabilities during an inspection, and the ability to use inspection methods such as sampling and on-site analysis. There is, however, a degree of reluctance to respond to the trends in chemicals manufacturing with a shift in verification focus. Some developing countries see suggestions to shift emphasis from verifying scheduled chemicals to OCPFs as an attempt to exercise control over the use of chemical technology and to shift the burden of verification to the developing world. They are therefore reluctant to accept that the OCPF verification regime needs to be further enhanced in response to trends in the chemical industry.

In a broader context, to adapt the CWC to the new challenges emanating from advances in the life sciences will require consideration of how best to reinforce the "general purpose criterion." This includes national implementation (legislation, regulations, and enforcement); the recognition that the schedules must not limit the scope of the CWC; and the need to ensure effective (self-)governance of the life sciences as well as industry. One example of this is the Responsible Care program, a global voluntary chemical industry initiative to improve health, safety, and environmental performance; communicate with stakeholders; and apply self-regulatory measures to ensure compliance with regulations, including the CWC.

A second issue relates to how verification can make use of new opportunities created by science and technology, such as new or improved verification equipment and methods. The OPCW has identified gaps in its tool box, for example, with regard to analyzing biomedical samples. This capability gap affects the OPCW's ability to investigate allegations of chemical weapons use and should be closed swiftly. There are also efforts under way to improve further the OPCW's capability to conduct environmental sampling and analysis on-site and to use other inspection methods. The review conference should encourage the OPCW Technical Secretariat to make best use of technological advances so as to maintain a high standard with regard to its verification methods and equipment.

A third issue is whether technological and scientific advances might aid the destruction of chemical weapons, particularly old and abandoned chemical weapons and those dumped in the sea. This is an area where the review conference could encourage further cooperation between states-parties and the sharing of assessments, experience, and technological know-how.

A fourth issue is how advances in science and technology will help improve protection against chemical weapons. Many of these advances can help upgrade the protection against chemical agents, enhance decontamination capabilities, or lead to new medical countermeasures. This is important as states harden their structures against the menace of chemical terrorism. It will also have a deterrent effect against the use of chemical weapons by states that still remain outside the regime. The review conference should recognize the need to improve the protection against chemical weapons further, encourage cooperation and exchanges between the states-parties in this field, and call on the Technical Secretariat to help states-parties develop and improve their protective capabilities. For example, this could include exercises to simulate national and international response mechanisms to chemical incidents.

A fifth issue relates to the CWC objective to enhance international cooperation in the peaceful uses of chemistry. Recent scientific and technological advances will create many opportunities in this respect, but they must be pursued in recognition of the fact that international cooperation must be fully consistent with the CWC's disarmament and nonproliferation obligations. There is therefore good reason to maintain a strong link between international cooperation programs and OPCW efforts to promote national implementation and ensure effective verification. The review conference should recognize that the OPCW is not a development agency but that its efforts with regard to industry verification as well as helping states-parties adopt national regulations and controls in the chemical field will facilitate trade and investment into the emerging chemical sectors of developing countries.

Finally, the arms control community has over the past four decades kept a clear demarcation between chemical and biological weapons. The CWC and the BWC have taken different directions with regard to a number of implementation issues, most prominently verification. In the real world of research and, increasingly, industry, the borders between the two fields are getting blurred. What the implications will be for the two regimes has yet to be fully understood. At the national level, there are signs of what one might call "regime conversion," with some countries combining their national implementing agencies and mechanisms for the two conventions. There also is an overlap of efforts within the scientific and industrial communities to adopt governance mechanisms and ethical codes to prevent the misuse of chemical and biological sciences for hostile purposes.

The situation at the international level, however, is more complicated, particularly with regard to verification. It is difficult to understand how the CWC can successfully address the verification dimension of the increasing convergence between chemistry and biology without running into some of the same difficulties that prevented the adoption of a BWC verification protocol in 2001-2002. There is thus a risk that the CWC review conference might set itself up for failure if it aimed too high. At the same time, it cannot ignore the advances in the life sciences and their effect on the CWC. The challenge will be to find the right balance in further developing the treaty's verification system while enhancing national implementation, developing self-governance mechanisms, and involving all stakeholders in the implementation process.

CORPORATE SUSTAINABILITY STRATEGIES IN CHEMICAL INDUSTRY

The chemical industry continues to face widespread public mistrust despite being an enabler of advances that are key to solving global challenges as well as efforts to improve product and process safety. Increased transparency and stakeholder dialogue about industry's considerable corporate social responsibility efforts should help reshape public perception.

Despite developing many of the technologies key to solving the global challenges of climate change, population growth, and sustainable development, the chemical industry is ranked among the least-trusted industries in a recent survey of the general population. Industry executives say corporate social responsibility (CSR) efforts, including a focus on triple bottom line factors, measure social, environmental, and economic impacts—also known as the three Ps of people, planet, and profits—and increased dialogue and transparency with stakeholders are key to addressing this mistrust.

"While most of the public recognizes that chemicals are essential and improve the quality of life, the public also often distrusts the chemical industry," says Godefroy Motte, chief sustainability officer at Eastman Chemical. "We as an industry recognize the need to work with all stakeholders in addressing this mistrust." Industry needs to be more transparent about its sustainability efforts to date and where progress still needs to be made to improve its image, Motte says.

Thorsten Pinkepank, director corporate sustainability relations at BASF, says that making the "enabling role of chemistry" for sustainability solutions more transparent to consumers has been a challenge. "Over the past 20 or 30 years, the chemical industry, including BASF, has learned that transparency is the foundation for building public trust. We therefore share information about us and what we do; how and why we do it. We report our progress annually."

BASF engages with its stakeholders through dialogue, projects, and partnerships, Pinkepank says. "Industry initiatives such as Responsible Care and the Global Product Strategy are effective approaches. As part of our Responsible Care management system, we have set ourselves ambitious goals for environmental and health protection, safety, and security. We assess our progress on a yearly basis and publicly disclose the status in our externally audited, integrated report."

Debra Phillips, managing director for Responsible Care at ACC, says CSR is no longer just about measuring and reporting what happens within a company's fence line. CSR has evolved to incorporate a more holistic perspective across the supply chain. "The chemical manufacturing industry is evolving alongside [CSR] by assessing our industry's overall impacts and contributions to the entire supply chain, including innovations that contribute positively to life-cycle impacts of products that use chemistry," she adds.

CSR efforts provide benefits beyond improving industry's image, Pinkepank says. "The strategic and organizational implementation of sustainability helps us to identify risks in an early stage while simultaneously opening up new business areas for BASF," he adds. "In order to minimize risks, we set globally uniform environmental, safety, security, health protection, product stewardship, and compliance standards, as well as labor and social standards. We ensure that sustainability is integrated into the development and implementation of our business units' strategies and research projects."

Regaining Public Trust

Producers acknowledge that industry faces an uphill climb. Past accidents have made the public skeptical of industry's efforts, while misinformation and lack of effective chemicals management legislation in the United States have caused consumers to question whether toxic chemicals are in everyday items. Healthy Child Healthy World released a report in January entitled The Toxic Takeover of Baby Nurseries: Chemicals of Concern Found in Almost Every Common Product and Furnishing.

As detection science has become more advanced, even the most minute amounts of a chemical raise the potential for concern. A Campaign for Safe Cosmetics report in 2009 found trace amounts 1,4-dioxane, classified by EPA as a probable carcinogen, in a number of leading baby shampoos. Johnson & Johnson (J&J) notes that 1,4-dioxane is a byproduct of the process that makes cleansers mild and nonirritating; is found naturally in some foods, including tomatoes and coffee; and is only measured in the parts-per-million range in its products.

Nonetheless, the company decided to take action to limit even those trace amounts. "Numerous regulatory agencies around the world have studied 1,4-dioxane and determined that it poses no harm at the trace levels found in personal care products," J&J says. "Still, many people have expressed their concern, and we want you to have complete confidence that the levels in our beauty and baby-care products are extremely low. So, we're in the process of reducing them even further."

Pressure is being felt along the entire supply chain. The Campaign for Safe Cosmetics issued a report late last year ranking retailers on their commitment to cosmetics safety. "Retailers that sell personal-care products are the gatekeepers of safety for their customers," says Janet Nudelman of the Campaign for Safe Cosmetics and the Breast Cancer Fund in a statement. "If the nation's biggest retailers commit to stop selling cosmetics with toxic chemicals linked to disease, manufacturers who want to keep selling to those retailers will comply. There is a rich history of retailers using their purchasing power to effect positive market change. When retailers said no to [bisphenol A (BPA)] in baby bottles or to old-growth lumber, the market responded."

Retailers are already responding, with a push for more sustainable packaging and the elimination of certain chemicals from finished products. Walmart in 2006 announced that it would ban three substances: propoxur and permethrin, both used in household insect-control products; and nonylphenol ethoxylates, an ingredient in some cleaning products. Walmart later announced it would not sell certain children's products containing BPA and launched an initiative to reduce phosphates and increase use of more sustainable packaging.

The Consumer Specialty Products Association (CSPA) says under its product stewardship initiative Product Care, member companies voluntarily agree to follow a code of management best practices and adhere to a set of principles that include an emphasis on protecting health, promotion of safety and environmental consideration through product design, and anticipation of product disposal needs. The program is a platform for downstream specialty product manufacturers and marketers to promote the production and distribution of safe and effective formulated products. The program interacts upstream and downstream with other recognized initiatives such as Responsible Care, Responsible Distribution, and the Retail Sustainability Initiative, CSPA says.

Phillips says the chemical industry has a role to play in making sure chemicals are used safely and responsibly along the supply chain. ACC recently formed a value chain outreach committee to reach relevant value-chain representatives and provide downstream users with data and background to make science-based procurement and policy decisions informed by sustainability and safety considerations, she says. "We are reaching out further down the value chain stream than we have done in the past to provide our value chain partners with information not only about the safety of our products but also about the energy efficiency of our plants and manufacturing processes that contribute to a more sustainable overall footprint," she says.

ACC and its member companies also recently developed a new Responsible Care Product Safety Code, a comprehensive set of commitments that each ACC member will make with respect to how it develops information and assesses, manages, and communicates the safety of its chemical products, Phillips says. "The code champions transparency, accountability, and science-based product development. It goes above and beyond what is required by law [and] adopts industry best practices, and its implementation will be validated by independent, third-party auditors."

Motte says social media and the rapid sharing of information have boosted visibility and knowledge sharing, a phenomenon that creates both opportunities and challenges. "This is why transparency with all of these groups is so important. Many chemical companies and associations have worked to improve their communications and engagement with stakeholders to build trust, but it has been and will continue to be an ongoing journey," Motte adds.

Greater engagement with stakeholders is one of the reasons Eastman joined the World Business Council for Sustainable Development (WBCSD), an association of 200-plus international companies advocating for corporate sustainability and cooperation for a more sustainable future for business, society, and the environment, Motte says. "We also regularly seek input and openly communicate with citizens and community leaders in the communities in which we operate," he adds. "The objective of our community advisory panels (CAPs) is to provide citizens living in plant communities with the opportunity for open dialogue with company representatives."

BASF has set up 78 CAPs worldwide, mostly at its larger production sites, Pinkepank says. A CAP is "a forum for open and honest dialogue between citizens and plant management," Pinkepank says. "By encouraging a two-way flow of information, we hope to enhance communication with the communities in which we operate."

Triple Bottom Line

No matter the industry, CSR is supposed to be a win-win by committing to operating a business with a focus on improving performance on triple bottom line–factors that measure impacts on people, the planet, and profits.

For the chemical industry, the easiest link to make between CSR and the bottom line are savings that come through reduced energy and resource use. Companies are quick to note, however, that the most important benefit comes from the innovation cycle: integrating CSR into business practices fosters innovative ideas and practices and leads to improved competitiveness.

"Ensuring safety, minimizing our environmental impact, and complying with all applicable laws and regulations are the only paths to earning and maintaining public trust," Pinkepank says. "That is why for BASF sustainable development means the combination of long-term-oriented economic success with environmental protection and social responsibility. The strategic and organizational implementation of sustainability helps us to identify risks in an early stage while simultaneously opening up new business areas for BASF. In the future, sustainability will increasingly become a starting point for new business opportunities."

At Eastman, innovation and sustainability work together to drive business results. "Growing and maintaining a pipeline of sustainably advantaged products is a core element of our growth strategy," Motte says. "Eastman's business philosophy embraces the triple bottom line—economic growth, environmental stewardship, and social responsibility. In fact, we leverage the symbiotic relationship between innovation and sustainability to drive growth and operational excellence. We make our operations more efficient, minimize risk across our business, deepen our customer relationships, and realize new growth opportunities."

NEW ISSUES AND REGULATIONS

Because the chemical industry is highly regulated, helping member companies to achieve full regulatory compliance in the most cost-effective way is a key priority for CBA.

CBA therefore engages with the legislative and regulatory process in Whitehall, Westminster, and Brussels. Its expert resources are focused in three main areas: advocacy, implementation, and information and advice.

Advocacy

CBA is consulted at an early stage in the legislative and regulatory process and aims to influence new proposals to ensure they do not undermine the industry's competitiveness and profitability. CBA member companies benefit from the excellent working relationships which have been developed over the years by CBA personnel with key players in the legislative and regulatory process.

CBA's specialist committees examine new legislative and regulatory proposals and frame a response on behalf the Association. Following the approval of CBA Council, this response is submitted to the government department or regulatory authority. This process is supported by CBA personnel in face-to-face meetings with the department or authority concerned.

On major pieces of legislation, such as the REACH legislation, CBA's campaign of advocacy on behalf of its members has to be sustained over several years. In other cases, advocacy is required over a shorter period. This process can involve intensive media campaigns as well as high-level confidential meetings with decision makers. CBA tailors its advocacy approach to the issue involved in order to best serve and protect the interest of its member companies.

In many cases, advocacy is a continuous process. Because CBA is recognised as the authoritative voice of the chemical supply chain, it engages with a wide range of official bodies and industry organisations in the UK, Europe, and elsewhere - as illustrated here.

Implementation

CBA also assists in the implementation of new legislation and regulations through its comprehensive programme of seminars and workshops - a service which is becoming increasingly important for CBA member companies. In 2012, the number of events organised by CBA increased by 10% - and the number of delegates attending rose by more than 15% on a like-for-like basis.

CBA has made an industry-leading commitment to deliver 'compliance training on demand' to its member companies. Subject only to two conditions: that each training event is commercially viable and can be scheduled into its existing event calendar - CBA's in-house experts will deliver the type of compliance training required by its member companies. Examples of recent workshop and seminar topics are:

- Emergency Response
- Carriage of Dangerous Goods
- REACH
- Environmental Compliance
- Classification, Labelling, and Packaging
- Health & Safety Compliance for smaller firms
- Safety Performance Leading Indicators
- Control of Major Accident Hazards (COMAH)
- Competition Law
- Combustible Liquids in IBCs
- Trade Controls for Chemicals
- Dangerous Substances Explosive Atmospheres
- Regulations (DSEAR)
- Safety Leadership for CEOs

Information and Advice

A key element CBA's Compliance Service consists of day-to-day support for member companies through regulatory information and advice. In 2012, the CBA team handled over 300 individual enquiries from member companies. The wide range of topics is illustrated below:

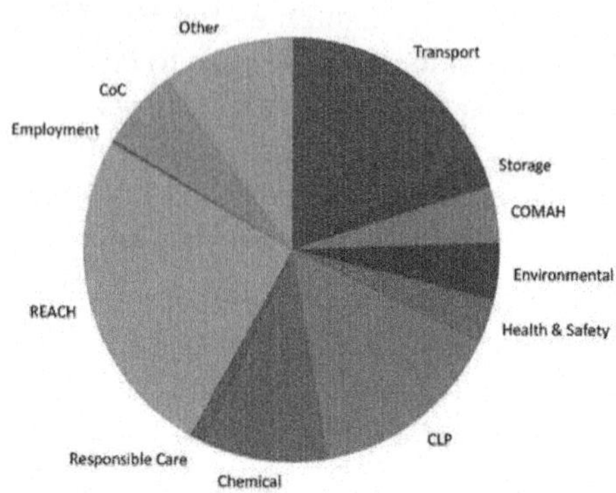

Key :

CoC = Code of Conduct
COMAH - Control of Major Hazards
CLP - Classification and Labelling of Package & Safety Data sheets.

REGULATORY ISSUES IN THE INDIAN PHARMACEUTICAL INDUSTRY

This chapter undertakes a review and assessment of regulatory issues in the Indian pharmaceutical industry. Understanding the regulatory scenario in this sector is extremely crucial not only due to the rapid and ongoing changes at the global level, largely with reference to good manufacturing practices (GMP), good clinical practices (GCP) and good laboratory practices (GLP) but also due to the onus on the regulatory bodies to ensure a healthy supply of quality drugs at affordable prices to the Indian masses.

The present chapter begins with a brief description of the major regulatory bodies monitoring the Indian pharmaceutical sector. It then undertakes a review of the prevailing mechanisms for drug regulation and temporal progression of some predominant policy measures and Acts. The chapter subsequently provides a comprehensive account of the status and key guidelines pertaining to the dimensions of drug pricing, patent related issues, GMP and clinical trials, in addition to a brief review of standards for medical devices and biotech products. It concludes with an assessment of the deficiencies of present regulatory regime and some new initiatives by the State to ensure the production and marketing of safe and efficacious drugs at affordable prices in the domestic sphere and to sustain current growth prospects in the global markets.

Major Bodies Regulating Drugs and Pharmaceuticals

The principal regulatory bodies entrusted with the responsibility of ensuring the approval, production and marketing of quality drugs in India at reasonable prices are:

The Central Drug Standards and Control Organization (CDSCO), located under the aegis of the Ministry of Health and Family Welfare. The CDSCO prescribes standards and measures for ensuring the safety, efficacy and quality of drugs, cosmetics, diagnostics and devices in the country; regulates the market authorization of new drugs and clinical trials standards; supervises drug imports and approves licences to manufacture the above-mentioned products;

The National Pharmaceutical Pricing Authority (NPPA), which was instituted in 1997 under the Department of Chemicals and Petrochemicals, which fixes or revises the prices of decontrolled bulk drugs and formulations at judicious intervals; periodically updates the list under price control through inclusion and exclusion of drugs in accordance with established guidelines; maintains data on production, exports and imports and market share of pharmaceutical firms; and enforces and monitors the availability of medicines in addition to imparting inputs to Parliament in issues pertaining to drug pricing.

The Department of Chemicals and Petrochemicals also oversees policy, planning, development and regulatory activities pertaining to the chemicals, petrochemicals and pharmaceutical sector. The responsibilities assumed by this body are relatively broader and varied in comparison to the other two bodies. The main aspects of pharmaceutical regulation are thus divided between the above two ministries. The Ministry of Health and Family Welfare examines pharmaceutical issues within the larger context of public health while the focus of the Ministry of Chemicals and Fertilizers is on industrial policy. However, other ministries also play a role in the regulation process. These include the Ministry of Environment and Forests, Ministry of Finance, Ministry of Commerce and Industry and the Ministry of Science and Technology. The process for drug approval entails the coordination of different departments, in addition to the DCGI, depending on whether the application in question is for a biological drug or one based on recombinant DNA technology. Issues related to industrial policy such as the regulation of patents, drug exports and government support to the industry are governed by the Department of Industrial Policy and Promotion and Directorate General of Foreign Trade, both under the aegis of Ministry of Commerce and Industry and the Ministry of Chemicals and Fertilizers. With respect to licencing and quality control issues, market authorization is regulated by the Central Drug Controller, Ministry of Health and Family Welfare, Department of Biotechnology, Ministry of Science and Technology (DST) and Department of Environment, Ministry of Environment and Forests. State drug controllers have the authority to issue licences for the manufacture of approved drugs and monitor quality control, along with the Central Drug Standards Control Organization (CDSCO).

Prevailing Mechanisms

This sub-section primarily focuses on major regulatory policies and mechanisms in relation to drug pricing and development of standards for ensuring safety and efficacy.

In India, drug manufacturing, quality and marketing is regulated in accordance with the Drugs and Cosmetics Act of 1940 and Rules 1945. This act has witnessed several amendments over the last few decades. The Drugs Controller General of India (DCGI), who heads the Central Drugs Standards Control Organization (CDSCO), assumes responsibility for the amendments to the Acts and Rules. Other major related Acts and Rules include the Pharmacy Act of 1948, The Drugs and Magic Remedies Act of 1954 and Drug Prices Control Order (DPCO) 1995 and various other policies instituted by the Department of Chemicals and Petrochemicals.

Some of the important schedules of the Drugs and Cosmetic Acts include: Schedule D: dealing with exemption in drug imports, Schedule M: which, deals with Good Manufacturing Practices involving premises and plants and Schedule Y: which, specifies guidelines for clinical trials, import and manufacture of new drugs

In accordance with the Act of 1940, there exists a system of dual regulatory control or control at both Central and State government levels. The central regulatory authority undertakes approval of new drugs, clinical trials, standards setting, control over imported drugs and coordination of state bodies' activities. State authorities assume responsibility for issuing licenses and monitoring manufacture, distribution and sale of drugs and other related products.

Regulatory control of Pharmaceutical sector

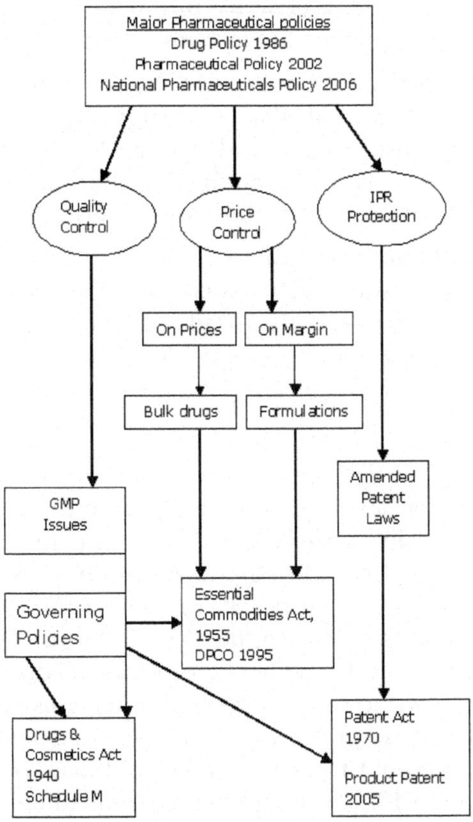

Temporal Progression of Drug Policies & Acts

The Patents Act of 1970, Drug Price Control Order 1970 and Foreign Exchange Regulation Act 1973 played a significant role in terms of the building of indigenous capability with regard to manufacture of drugs. The New Drug Policy of 1978 provided an added thrust to indigenous self-reliance and availability of quality drugs at low prices.

DPCO 1987 heralded the increasing liberalization in the industry. One of the important features of this act was the reduction of the number of drugs under price control to 143.

The major objective of DPCO 1995 was to decrease monopoly in any given market segment, further decrease the number of drugs under price control to 74 and the inclusion of products manufactured by small scale producers under price control list.

In 1997, the National Pharmaceutical Pricing Authority was constituted in order to administer DPCO and deal with issues related to price revision.

The Pharmaceutical Policy 2002 carried forward earlier governmental initiatives in terms of ensuring quality drugs at reasonable prices, strengthening of indigenous capability for cost-effective production, reducing trade barriers and providing active encouragement to in-house R&D efforts of domestic firms.

In 2003, the Mashelkar Committee undertook a comprehensive examination of the problem of spurious and sub-standard drugs in the country and recommended a series of stringent measures at Central and state levels. The regulatory body came in for censure with the committee noting that there were only 17 quality-testing laboratories, of which only seven laboratories were fully functional.

The National Pharmaceuticals Policy 2006, among other initiatives, has proposed a slew of measures such as increasing the number of bulk drugs under regulation from 74 to 354, regulating trade margins and instituting a new framework for drug price negotiations in a move to make drugs more affordable for the Indian masses.

Drug Pricing

As mentioned earlier, pricing policy and industry regulation constitutes one of the key responsibilities of the NPPA. Price control on medicines was first introduced in India in 1962 and has subsequently persisted through the Drug Price Control Order (DPCO). As per the directive of NPPA, the criterion for price regulation is based on the nature of the drug in terms of whether it enjoys mass consumption and in terms of whether there is lack of adequate competition for the drug. The year 1978 witnessed selective price controls based on disease burden and prevalence. The list of prices under DPCO subsequently witnessed a gradual decrease over a period of time. Around 80% of the market, with 342 drugs, was under price control in 1979. The number of drugs under DPCO decreased from 142 drugs in 1987 to 74 in 1995.

Drugs with high sales and a market share of more than 50% are subjected to price regulation. These drugs are referred to as scheduled drugs. The NPPA also regulates the prices of bulk drugs. The MRP excise on medicines was levied by the Finance ministry in 2005. The objective was to increase revenue and lower prices of medicines by using fiscal deterrent on MRP. This change may have had some impact in terms of magnifying the advantage to industries located in the excise free zones. This also succeeded in attracting some small pharmaceutical firms to these zones.

As the report by NIPER, submitted to the Ministry of Chemicals and Fertilizers in 2007 points out, this may have led to tax disparities among firms located in tax exempt zones and tax non exempt areas. This has also led to small firms in non exempt areas requesting for tax subsidies from the government.

For drugs not under price control, firms can set the Minimum Retail Price (MRP). The NPPA only intervenes in cases where drugs have significant sales and where the annual price increases by 10%. This is a recent development, which came into effect in 2007, as in the past the NPPA would intervene only if the annual price increases were more than 20%. This development indicates the heightened sensitivity of the government towards consumer access to medicines at reasonable prices and keeping a check on profit mongering by the industry.

Fixed dose Combinations and Prevalence of Counterfeit and Spurious Drugs

Recently, 294 fixed dose combinations were withdrawn by the Central Drug Control Authority on grounds that these drugs were therapeutically irrational. The order was subsequently stayed by the Madras High court. The issue of the definition of counterfeit drugs is relevant in the context of different drug quality standards prevailing in the Indian market. While exported drugs were of a higher quality (WHO/FDA/EMEA/TGA), to meet the required standards in the country of export, in the case of the domestic market, adherence to local quality standards, fixed by the regulatory body was sufficient. Also absence of transparency in licensing procedures has resulted in the market being flooded with counterfeit and substandard drugs. In this context, the Mashelkar Committee report has referred to a WHO study, which declared that nearly 30% of the Indian market was flooded with spurious, substandard or counterfeit drugs. The government's own estimates have been in the range of 8-10% for substandard drugs and 0.2-0.5% for spurious drugs.

Patents and Data Protection Related Issues

The Indian Patent Act, 1970 was amended through the Patents Amendment Act (2005). A technical expert group was constituted under the chairmanship of Dr R.A. Mashelkar, then Director General of the CSIR. The Committee decision was that it would be TRIPS incompatible to exclude microorganisms from patents and to limit the grant of the patent for pharmaceutical substance to a new chemical entity or a new medical entity involving one or more inventive steps. The committee also opposed the granting of frivolous patents and evergreening and recommended the formulation of detailed guidelines to ensure that only those patents proving 'substantial human intervention' and 'utility' were granted.

As per the provisions of the TRIPS Agreement, member countries have to provide protection to regulatory data submitted for market approval of pharmaceutical products under specific circumstances. The government of India constituted an expert committee under the chairmanship of Mr Satwant Reddy to formulate adequate steps to deal with the issue of data protection. The Reddy Committee report, brought out in 2007, stated that in the context of pharmaceuticals, the present legal regime was inadequate to address the issues related to data protection with respect to provisions. It also underscored the need for more clear and stringent mechanisms within the Drugs and Cosmetics Act to ensure that undisclosed test data was not put to unfair commercial use in India.

Good Manufacturing Practices

Good Manufacturing Practices (GMP) constitute an international set of guidelines for the manufacture of drugs and medical devices in order to ensure the production of quality products. In recent years, GMP protocols are being adopted and followed in over 100 countries, either in the form of regulations (Japan, Korea and United States), or Directives (European Union) or Guides (United Kingdom) or Codes (Australia).

The objective of GMPs is to minimize risks with reference to the manufacturing, packaging, testing, labeling, distributing and importing of drugs, cosmetics, medical devices, blood and blood products, food items etc. These protocols are largely concerned with parameters such as drug quality, safety, efficacy and potency.

WHO GMP Protocols: World Health Organization GMP guidelines were instituted in 1975 in order to assist regulatory authorities in different countries to ensure consistency in quality, safety and efficacy standards while importing and exporting drugs and related products. India is one of the signatories to the certification scheme. The WHO-GMP certification, which possesses two-year validity, may be granted both by CDSCO and state regulatory authorities after a thorough inspection of the manufacturing premises.

Schedule M Compliance: The requirements specified under the upgraded Schedule 'M' for GMP have become mandatory for pharmaceutical units in India w.e.f. July 1, 2005. Schedule M classifies the various statutory requirements mandatory for drugs, medical devices and other categories of products as per the current Good Manufacturing Practices (cGMP). Schedule M protocols have been revised to harmonize it along the lines of WHO and US-FDA protocols. These revised protocols include detailed specifications on infrastructure and premises, environmental safety and health measures, production and operation controls, quality control and assurance and stability and validation studies. Problems related to Schedule M compliance are mostly confined to small-scale pharmaceutical units as large-scale firms have shown greater willingness to comply with the revised norms in order to increase their competitiveness in the global arena. The Central Drugs Standards Control Organization has, however, yet to compile data on the extent of Schedule M compliance by the firms. The Najma Heptullah Committee on Subordinate Legislation, which tabled its report in Parliament recently, is scheduled to compile data on extent of Schedule M compliance shortly. However, according to state regulatory sources, units in states like Gujarat, Karnataka, Maharashtra and Andhra Pradesh have achieved a high percentage of Schedule M compliance in comparison to units in other states.

International regulatory certification for Indian manufacturing units: A principal issue relating to good manufacturing practices is that WHO-GMP is no longer sufficient, particularly for exporting of drugs and related products to developed countries. Regulators from these countries visit Indian firms to carry out a thorough inspection of their manufacturing units before registering the concerned product. A large number of domestic players are seeking international regulatory approvals from agencies like US-FDA, MHRA UK, TGA Australia and MCC South Africa in order to export their products, mostly generics, in these markets. A large number of Indian firms are increasingly seeking at least WHO GMP approval in order to compete for exports to CIS countries and other Asian markets. India has the distinction of having the largest number of US-FDA approved manufacturing units, totaling 100, mainly for production of Active Pharmaceutical Ingredient (API), outside of the United States.

Table : State-wise distribution of manufacturing units with WHO-GMP certification.

State	No of units
Assam	1
Kerala	14
Gujarat	307
Maharashtra	136
Haryana	19
Uttar Pradesh	21
Andhra Pradesh	138
Karnataka	52
Goa	42
Daman	26
Pondicherry	10
Tamil Nadu	5
Rajasthan	4
Delhi	12
Uttarakhand	1
Punjab	7
Madhya Pradesh	24
Orissa	1
Bihar	5
West Bengal	6
Total	814

The Ranbaxy Episode

In the context of the acquisition of international regulatory approvals, the Ranbaxy episode is significant since it provides an indication of the regulatory hurdles likely to be faced by Indian firms competing for the generics market in developed countries, notably United States and Europe. On July 3, 2008, the US Food and Drug Administration (FDA) filed a motion against Ranbaxy in a Maryland court, alleging that the firm had falsified documents submitted to it. This was subsequent to the FDA's raiding of Ranbaxy's offices in New Jersey in February 2007. The FDA officials' allegations were to the effect that Ranbaxy had sold fake or adulterated versions of an HIV drug to patients in Africa. The regulatory body also made allegations about the quality and safety of the generic drugs the firm had sold in the United States and that the firm had refused to hand over documents relating to an audit of its manufacturing unit. The audit had been carried out by Parexel International, a US based pharmaceutical services firm. Ranbaxy subsequently denied the allegations, maintaining that it had made changes recommended by the audits and also referred to the attorney-client privilege, which it claimed, protects the audits. On August 3, 2008, the firm turned over some documents to the FDA. The legal investigation is a first of sorts for an Indian firm and has caused concern among other major domestic Indian players that they would be accorded the same treatment as Chinese manufacturers after FDA officials reported 81 deaths due to production defects in Chinese made ingredients in *Heparin*. If the verdict goes against Ranbaxy, the onus lies on the firm to prove that the manufacturing process deficiencies relate to the plants in question rather than a system wide problem. The FDA's order apply to supplies from the plants located in Ponta Saheb and Dewas.

Clinical Trials

In recent years, India has positioned itself as one of the major players in the clinical trials arena. The recognition for India as a centre for clinical trials has mainly arisen through the providing of contract services to the international pharmaceutical industry in the form of clinical development services.

Clinical trials to establish the safety and efficacy of drugs constitute nearly 70% of research and development costs and the total time taken for drug development constitutes nearly 7-10 years. Well-designed clinical trials provide the requisite data pertaining to safety and efficacy of drugs and impart meaningful results about a given therapeutic intervention in human beings. According to latest estimates made by the Tufts Centre for the Study of Drug Development, while total research costs have increased by 7.4% per year, the costs of clinical trials on human beings has risen by over 12 per cent. Considering the relatively low costs of R&D in India, several MNC pharmaceutical companies, as well as global clinical research organizations are increasingly making India a clinical research and development hub.

The clinical market in India is expected to grow at a consistent rate of 20-25 percent. The recent regulatory revisions in the pharmaceutical industry and stricter patent laws have made it easier to conduct trials, making it the fourth largest market in terms of volume.

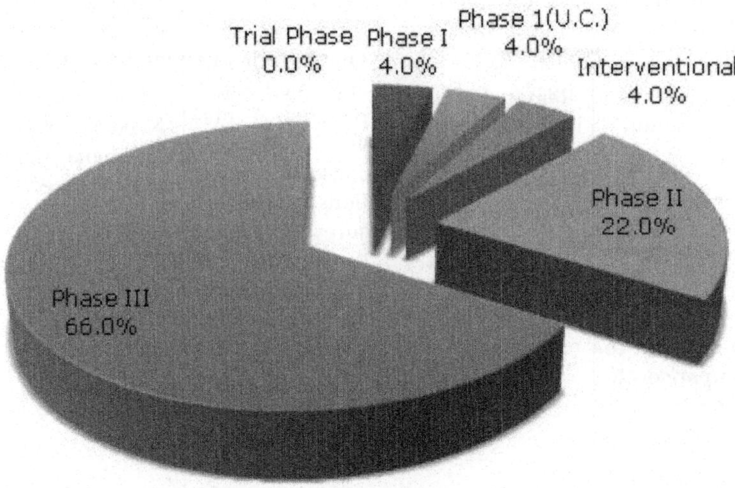

Fig : Phase-wise break-up of clinical trials carried out in India.

A phase-wise break up of clinical trials carried out in India. Phase I trials are essentially carried out to establish pharmacological indications and safety of the drug and are essentially exploratory in nature. Phase II trials provide information related to the efficacy and safety of the new drug in patients. Phase III trials are essentially multi-centric confirmatory trials carried out in larger groups of patients and healthy volunteers, while Phase IV trials involve post-marketing surveillance. The chart clearly indicates that the majority of trials carried out in India fall under the Phase III category.

India's clinical development sector has witnessed a tremendous growth in recent times. In 2005, the revenues from contract R&D for international sponsors totaled $100 million and the sector enjoys an annual growth rate of about 40 per cent. Several global CROs have entered the Indian market in the last few years. Some of these have also entered into alliances with local CROs.

Table : Clinical Trial Cost Differences in India & U.S.

	United States	India
Phase 1	$ 20 mn.	< $10mn.
Phase 2	$ 50 mn.	< $30mn.
Phase 3	$ 100mn.	< $60mn.

Table : Clincial Trials in India: A Swot Analysis.

Strengths	Opportunities	Weaknesses	Threats / Challenges
-Resource pool of well-trained, qualified, English speaking man-power -Diverse genetic pool and disease variation -Numerous government funded and private medical and pharmaceutical institutions with state of art facilities -Cost efficiency (up to 60%) in comparison to USA/ Europe -Fast recruitment of large number of patients -Revamped regulatory regime -Establishment of Clinical Trial Registry	-Relaxation of duties on import of clinical trial samples -Removal of phase lag and permission to conduct Phase I trials concurrently in India along with rest of the world	- Lack of adequate mechanisms to safeguard illiterate and vulnerable patients, prevent informed consent violations and ensure proper functioning of institutional ethics committees	- Need for a strong centralized regulatory regime to effectively monitor GCP guidelines -Need for expertise on data management related specialized services

Policies Relating to Clinical Trials

In this context, it would also be useful to review prominent changes in policies related to clinical trials in the last few decades. Till about a decade ago, regulatory and ethics based environment for the conduct of quality clinical trials in India were conspicuous by their absence. The Central Drugs Standards Control Organization (CDSCO) has played a critical role towards this end. The progression towards Good Clinical Practice (GCP) has largely been a gradual and slow process.

It was in 1988 that local clinical trials for new drug introductions were first made mandatory in India. There was also a phase lag as permissions for trials were granted for one phase behind the rest of the world. Thus, Phase II and Phase III trials were permitted only after these had been carried out elsewhere in the world. The period before 2000 witnessed several incidents of ethical violations related to informed consent and conduct of trials by multinational firms and domestic players as well. In 2000, due to the proactive initiatives of regulators, the Central Ethics Committee on Human Research (CECHR) and Indian Council of Medical Research (ICMR) conceptualized and issued Ethical Guidelines for Biomedical Research on Human Subjects. In 2001, a Central Expert Committee was set up by

Central Drugs Standards Control Organization (CDSCO) to develop Good Clinical Practice (GCP) guidelines in line with the latest WHO and ICH guidelines.

Subsequently, the requirements of data submission on animal testing for permission to undertake Phase I, Phase II and Phase III clinical trials were laid down in the revised Schedule Y of the Drugs and Cosmetics rules.

As per these revisions, the relevant data submitted to the Drugs Control General of India (DCGI), is evaluated with the assistance of expert clinicians & scientists.

Similarly, for registration and approval of new drugs, which have already been registered and used in the country of origin, Phase II trials in about 100 patients is usually insisted upon by DCGI before allowing such products to be marketed in India. Normally, new drug approval is usually granted for a period of about two years. The trials are conducted only after clearances are obtained from the Institutional Ethics Committees. Consent of patients for participation in such trials is an integral part of the regulatory framework.

In 2005, Drugs Technical Advisory Board (DTAB) made GLP practices mandatory for all laboratories and in-house units of pharmaceutical firms and Contract Research Organizations (CROs). In 2007, norms pertaining to the Phase lag have also been revised and Schedule Y now permits Phase I trials to be carried out concurrently in India along with the rest of the world.

For an efficient and ethical growth of the clinical trials industry, the appropriate mechanisms to be adopted include the presence of a strong centralized regulatory regime to effectively monitor GCP guidelines and ensure transparency in the functioning of institutional ethics committees (IECs).

Medical Devices

In June 2007, the DCGI formulated a new set of guidelines for the import and manufacture of medical devices in the country. The guidelines were the aftermath of the JJ Hospital controversy, involving the use of unapproved and untested stents on 60 patients and the subsequent recommendations made by the Mashelkar Committee in 2004.

The immediate outcome of the JJ Hospital controversy was that the Department of Medical Education and Research (DMER) banned the use of unapproved stents and stressed on regulatory approvals from the country of manufacture or US-FDA approval for medical devices.

The Mashelkar Committee subsequently recommended the creation of a specific medical devices division within the CDSCO in order to address the management, approval, certification and quality assurance of all medical devices. This essentially consisted in alteration of the status of sterile medical devices, intended for internal or external use to medical drugs and creation of suitable provisions and amendments to the Drugs and Cosmetics Act of 1940.

The Drugs Consultative Committee approved these recommendations in 2005, ensuring that in future all devices would be licensed for manufacture, distributed and sold by the CDSCO, with special evaluation committees in order to ensure that the concerned manufacturing units complied with the requisite GMP requirements.

The Principal Provisions of these Guidelines are as Follows

Ten categories of sterile devices: cardiac and drug eluting stents, catheters, bone cement, heart valves, scalp vein sets, orthopedic implants, internal prosthetic replacements, IV cannulae and intraocular lenses; would be considered as drugs and consequently regulated.

Importers would have to submit US-FDA clearance, the EU medical device directive or similar approvals from other countries as proof of adherence to quality standards. Expert committees would be set up for evaluation and granting of licences to locally manufactured devices, in the absence of international quality certification.

The approval of the committees would be verified by both Central and State licensing committees.

Some of the problems associated with compliance to these regulations include lack of awareness among smaller firms, high registration fees, delays in granting of licences, restrictions in the entry of new players in the sector and lack of preparation by the firms with respect to documentation requirements.

TTK-Sree Chithra Tirunal Collaboration on Medical Devices

TTK Healthcare Ltd and Sree Chithra Tirunal Institute for Medical Sciences and Technology (SCTIMST) are jointly developing a tissue heart value or bioprosthetic for older patients. The device will be ready in around three years and will subsequently undergo clinical trials. These valves are used in patients above the age of 40 in India and patients above the age of 50 in the United States. The Chitra-TTKvalves mechanical valves are sold at a price of Rs 25,000 each, about a quarter cheaper than the cost of similar imported valves. The tie-up is a significant move towards generating affordable indigenous technology in the wake of high registration fees of imported devices. The bioprosthetic device would have a shorter lifespan of around 15 years in comparison to the existing mechanical valve but the patient would need to take medication only for a period of 6 months in comparison to a longer duration of medication for patients fitted with mechanical valves.

Biotech Products

The Ministry of Environment and Forests under the Environment (Protection) Act of 1986 have notified the rules for the manufacture, use, import, export and storage of hazardous microorganisms or genetically engineered organisms or cells. As per these rules, biological materials are regulated from the R&D stage to their release in the environment. The Institutional BioSafety Committee (IBSC), Review Committee on Genetic Manipulation (RCGM) and the Genetic Engineering Approval Committee (GEAC) to monitor rDNA research, product development and commercialization. The ISBC functions as the nodal point for interaction within the institution for the implementation of the rDNA Biosafety guidelines. The RCGM essentially monitors the safety related aspects of activities involving genetically engineering organisms or hazardous microorganisms. The GEAC undertakes the responsibility of approval of activities involving large-scale use of genetically modified/ hazardous microorganisms and products thereof in research and industrial production and their safety in terms of environmental protection. In addition, the DCGI and state drug controllers as per the Drugs and Cosmetics Act 1945 and its subsequent amendments regulate biologicals.

Deficiencies and Limitations of the current regulatory regime:

- Proliferation of spurious and substandard drugs in the Indian market
- Dual licencing mechanism acts as a deterrent to uniform implementation of regulatory procedures
- Lack of transparency in licencing procedures
- Inadequate regulatory expertise and testing facilities to implement uniform standards
- Need for greater thrust on institutional support to small scale firms to enable speedy implementation of Schedule M upgradation and standardization of drug quality
- Need for greater clarity on patentability of pharmaceutical substances and conditions under which firms can apply for compulsory licences to prevent legal battles between local firms, MNCs and civil rights groups.
- Need for greater coordination, accountability and transparency in functioning among different ministries concerned with drug regulation.

Recent regulatory initiatives:

- Move to establish an integrated regulatory system through the constitution of a National Drug Authority so that quality regulation and price control is performed by the same agency
- Establishment of pharmacovigilance centres at national, zonal and regional levels to monitor adverse drug reactions
- Move to bring nearly 374 bulk drugs under price control and regulate trade margins
- Capability strengthening to monitor clinical trials, including the setting up of the Clinical Trials Registry of India (CTRI)

Chapter 3

INDUSTRIAL TO SUSTAINABLE CHEMISTRY

INTEGRATED POLLUTION PREVENTION AND CONTROL

The Environmental Protection Agency (EPA) is the authority responsible for licensing certain large scale activities in the industrial and agricultural sector. This process has been in effect since 1994 and the licences were originally known as Integrated Pollution Control (IPC) licences. Since then the Protection of the Environment Act, 2003 gave effect to Directive 96/61/EC concerning Integrated Pollution Prevention and Control (IPPC) and this is now the required licensing process.

The aim of an IPPC licence is to prevent or reduce emissions to air, water and land, to reduce waste, and use energy and resources efficiently. Hence, the IPPC process covers all emissions from the activity along with its environmental management. An IPPC licence must be obtained prior to commencement of an activity and certain categories of industry are subject to thresholds. If an activity is operating below a threshold and it is expected that the threshold will be exceeded then it will be necessary to obtain an IPPC licence prior to exceeding the threshold.

The EPA is responsible for monitoring emissions and dealing with any infringements on licences. All emissions must be within set limits which must not be contravened. Offences under the Environmental Protection Agency Act can result in court action by the EPA. The court can impose fines and prison sentences and the EPA can revoke a licence. The IPPC licensing process is based on the Best Available Techniques (BAT) principle. In order for an industry to determine what constitutes BAT a series of sector specific guidance documents have been developed. These are referred to as BAT reference documents or BREFs and provide guidance in terms of complying with an IPPC licence.

Effluent Discharge to Sewer

Where an activity is required to obtain a licence but has an effluent discharge to a local authority sewer, section 97 of the EPA Act requires the EPA to "obtain the consent of the sanitary authority in which the sewer is vested or by which it is controlled". In addition the EPA must include any conditions that the sanitary authority "considers appropriate". The EPA cannot issue a licence unless the sanitary authority is satisfied with the arrangements for effluent discharge to sewer.

Change of Operations

Where a change to a scheduled activity is being considered which could affect environmental emissions, the EPA must be notified of that proposed change. The EPA will then decide whether the change is likely to be significant enough to warrant an application for an IPPC licence (or an IPPC licence review in the case of an existing licence holder).

Main Obligations

The person who carries on an activity:

- Must be aware of the need to apply for an IPPC licence. This involves referring to:
  - (a) Environmental Protection Agency Act, 1992
  - (b) Protection of the Environment Act, 2003
  - (c) Environmental Protection Agency (Licensing) Regulations 1994 to 2004
  - (d) Other various Statutory Instruments and seek appropriate guidance on the interpretation of the above.
- Should consult with the EPA before commencing preparation of the IPPC licence application. This should minimise delays and ensure all relevant information will be submitted.
- Must notify in writing the relevant Local Planning Authority and any other 'prescribed persons' of the intention to apply for an IPPC licence.
- Must publish a notice in a local newspaper within a period of two weeks before an IPPC licence application is made. A site notice must also be erected before lodging the application.
- Must submit copies of the Environmental Impact Statement (EIS) to the EPA where a development "comprising or for the purposes of the activity is required to be accompanied by an environmental impact statement", as part of the licence application. The EIS should be produced in accordance with the EPA's "Guidelines on the Information to be contained in Environmental Impact Statement" which is available on the EPA website.
- Shall not effect any "alteration or reconstruction" which would "materially change or increase emissions", and must "give notice in writing to the Agency" if proposing such changes.

Content of an IPPC Licence Application

The mandatory contents of a licence application are set out in the relevant Statutory Instruments. Some of the particulars include:

- A listing of the raw and auxiliary materials, other substances and the energy used in or generated by the installation.
- A description of the installation and its activities.
- The nature and quantities of foreseeable emissions from the installation into each medium as well as an identification of the significant effects of the emissions on the environment.
- A description of existing and proposed monitoring procedures for all emissions.
- An explanation of how all emissions to the environment will be controlled in particular through application of the Best Available Techniques (BAT).
- The appropriate fee must accompany the application.
- Identify whether the activity is an establishment to which the EC (Control of Major Accident Hazards involving Dangerous Substances) Regulations (S.I. No. 74 of 2006) applies. This can be clarified through contacting the Health and Safety Authority.
- A description of measures to protect the environment after the activity ceases operation along with the measures necessary to return the site of operation to a satisfactory state.

The IPPC licence Application Form and Application Guidance Notes are available from the EPA.

Processing of an Application

Having considered the application, the EPA will issue a 'Final Determination' to :

(a) the applicant/licensee

(b) the relevant public bodies specified in the regulations,

(c) any person who made a written submission or valid objection

The 'Final Determination' will indicate what decision the EPA proposes to make and, where it is proposed to grant a licence, the "conditions, if any, to be attached and the reasons therefor and where a copy of the proposed licence may be obtained". Any person can apply to the High Court within eight weeks of a decision to refuse or grant a licence and seek a judicial review of the validity of the decision.

The timescale for the overall process might be:

(i) For a 'Proposed Determination': A maximum of eight weeks if no additional information is required by the EPA. (If additional information is required the notification may be delayed until two months after this information has been supplied).

(ii) For an Objection:

(a) A period of twenty-eight days is allowed after the date of the 'Proposed Determination' for third parties to lodge an objection. The objection is made to the EPA.

(b) A period of one month is allowed for "each other party to the objection" to "make submissions in writing to the Agency".

(c) An objection should normally be considered by the EPA within four months of expiry of the periods mentioned in (a) and (b) above, but it has the right to specify a longer time duration in exceptional cases.

In summary, most applications should be processed within three to seven months.

Licence Review

The EPA may notify an existing licence holder of its intention to carry out a review of the licence provided that the licence has been in force for at least three years. However, a licence can be reviewed at any time with the consent of (or on the application of) the licence holder or if important new information comes to light (*e.g.* the characteristics or potential impacts of a substance are shown to be greater than originally envisaged).

The procedures in this situation are generally the same as for a first-time application, except that it is the Agency and not the licence holder, which must publish the newspaper notice and notify the relevant planning authority and any other prescribed persons.

ENVIRONMENTAL POLICY FOR INDUSTRIAL EMISSIONS

Industrial activities play an important role in the economic well-being of Europecontributing to sustainable growth but can also have a significant impact on the environment.

The largest industrial installations account for a considerable share of total emissions of key atmospheric pollutants and also have other important environmental impacts, including emissions to water and soil, generation of waste and the use of energy.

The most obvious form of pollution associated with plastic packaging is wasted plastic sent to landfills. Plastics are very stable and therefore stay in the environment a long time after they are discarded, especially if they are shielded from direct sunlight by being buried in landfills. Decomposition rates are further decreased by anti-oxidants that manufacturers commonly add to enhance a container's resistance to attack by acidic contents.

Plastics also put a big chemical burden on the environment. The Oakland Recycling Association commissioned an analysis of the toxic chemical burden that relied heavily on information from EPA data, especially the Toxics Release Inventory. These data were limited because manufacturers within the "miscellaneous plastics sector" did not file reports. Nevertheless, the information available showed that most toxic releases went into the air, and the plastics industry contributed 14% of the national total. Of the top ten manufacturers ranked by total releases, seven made plastic foam products. Significant releases of toxic chemicals included :

- trichloroethane
- acetone
- methylene chloride
- methyl ethyl ketone
- styrene
- toluene
- benzene
- 1,1,1 trichloroethane

Other major emissions from plastic production processes include sulfur oxides, nitrous oxides, methanol, ethylene oxide, and volatile organic compounds.

Less visible but very serious is the pollution generated by producing plastic resin. As ethylene is polymerized, the reactive mixture is scrubbed with dilute aqueous caustic solutions that become high-volume pollutants. The refining process uses waste-minimization methods, but point-source air emissions are still high because of inherent difficulties in handling large flows of pressurized gases. Manufacturing PET resin generates more toxic emissions (nickel, ethylbenzene, ethylene oxide, benzene) than manufacturing glass. Producing a 16 oz. PET bottle generates more than 100 times the toxic emissions to air and water than making the same size bottle out of glass.

Producing plastics can be hazardous to workers, too. Serious accidents have included explosions, chemical fires, chemical spills, and clouds of toxic vapor. These kinds of occurrences have caused deaths, injuries, evacuations and major property damage. A review of the US EPA's data base of 10,000 accidents and spills from 1980-87 shows that nearly 1,600 (16%) of industrial accidents were associated with producing plastics or plastic constituents.

Negative Health Effects – Toxic Additives, Migration into Food

In addition to creating safety problems during production, many chemical additives that give plastic products desirable performance properties also have negative environmental and human health effects. These effects include direct toxicity, as in the cases of lead, cadmium, and mercury; or carcinogens, as in the case of diethyl hexylphosphate (DEHP). Problem chemicals are used as plasticizers, antioxidants, colorants, flame retardants, heat stabilizers, and barrier resins.

A single resin type might be mixed with many such additives, adding complexity to the chemical composition and possibly generating new classes of incompatible resins within the grossly simplified SPI resin code. An example of internal incompatibility is resin type 2, noted earlier as a recycling problem because the blow-mold resin grades and injection-mold grades must be separated for most primary recycling applications.

People are exposed to these chemicals not only during manufacturing, but also by using plastic packages, because some chemicals diffuse (migrate) from the packaging polymer to the foods they contain. Migration potential exists for traces of monomers, oligomers, additives, stabilizers, plasticizers, lubricants, anti-static nucleating agents, and reaction products of the polymer or its additives. Such substances may be toxic.

Diverting Solid Waste

Curbside plastic collection programs are driven in part by a wish to minimize municipal solid waste. Indeed, most legislation dealing with discarded containers has focused on creating mechanisms that would divert municipal plastic waste from incinerators or landfills. These initiatives include container deposit laws and landfill use-reduction laws. Although important, such measures do not solve the problem of over-packaging or reduce the production of plastic packaging. Only source reduction can do this. Germany recognized the need to implement aggressive source reduction and passed a law requiring all manufacturers to arrange to pick up discarded packaging, usually at the point of sale, such as the supermarket. This requirement has encouraged industry to reduce packaging.

In addition, as noted earlier, providing recycling for plastic containers may actually encourage their use and increased production. But as plastic packages become less standardized and morecomplex, recycling becomes more difficult. Already, many plastic packages that consumers have expected to be recycled must be landfilled. Therefore the diversion may not be as effective as proponents intend.

Greenhouse Gas Emissions

The Commonwealth Scientific and Industrial Research Organisation (CSIRO), the Bureau of Meteorology, and Academies of Science from around the world have advised that the world is warming and high levels of carbon pollution risk environmental and economic damage.

The Intergovernmental Panel on Climate Change (IPCC) define climate change as: 'a statistically significant variation in the mean state of the climate or its variability, persisting for an extended period (typically decades or longer)'. Climate change is caused by increases in the total stock of greenhouse gases in the atmosphere. In 2009, the six greenhouse gases measured by the accounting rules of the Kyoto Protocol reached a global level of 439 parts per million (ppm) CO_2-equivalent, an increase of 160 ppm compared to pre-industrial levels.

Australia has adopted a range of responses to climate change. The first pillar of Australia's response is to reduce Australia's GHG emissions and, in order to meet this objective, the Australian Government is developing and putting in place relevant policies through its Clean Energy Future program.

Australia's Clean Energy Future policies

The objectives of the Clean Energy Future program are to "...support Australian businesses and households reduce their carbon pollution, to create the new green-collar jobs of the future and to transform our economy."

The Clean Energy Future policies aim to achieve this through:

- Introducing a carbon pricing mechanism;
- Promoting innovation and investment in renewable energy;
- Encouraging energy efficiency; and
- Creating opportunities in the land sector to cut pollution.

Greenhouse Gas Emissions and the Australian Economy

A carbon pricing mechanism is the first element of the Government's plan for a clean energy future, and is designed to trigger a broad transformation of the economy by breaking the link between emissions and economic growth.

The total greenhouse gases (carbon dioxide, methane, nitrous oxides and fluorinated gases) in Australia, excluding changes due to land use and land use change and forestry (LULUCF), increased by 24% between 1989-90 and 2009-10. During the same period, economic activity as measured by GDP increased by 88%.

The phenomenon where the economy grows at a rate faster than the related pollution is known as decoupling. This can be caused either by structural change in the economy (for instance, where the generally lower-emitting service industries have grown more strongly than higher emitting industries) or by the adoption of technological innovations by businesses or by a combination of both.

In the case of Australia, the decoupling is relative, as GHG emissions are increasing but at a lower rate than economic activity (as measured by GDP).

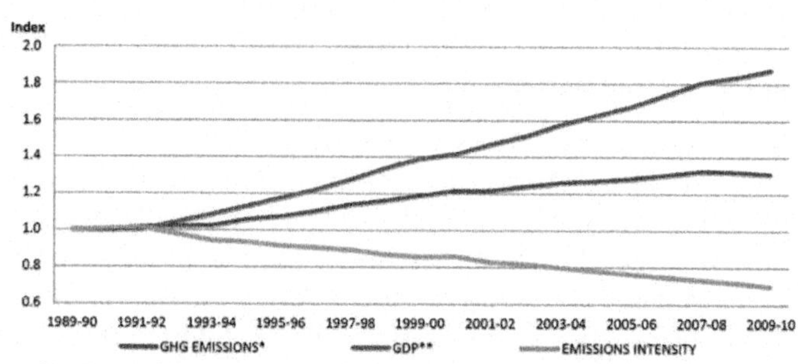

Fig. : Total direct GHG emissions and GDP, 1989-90 to 2009-10.

Figure below presents data on GHG emissions, labour force and economic production by industry.

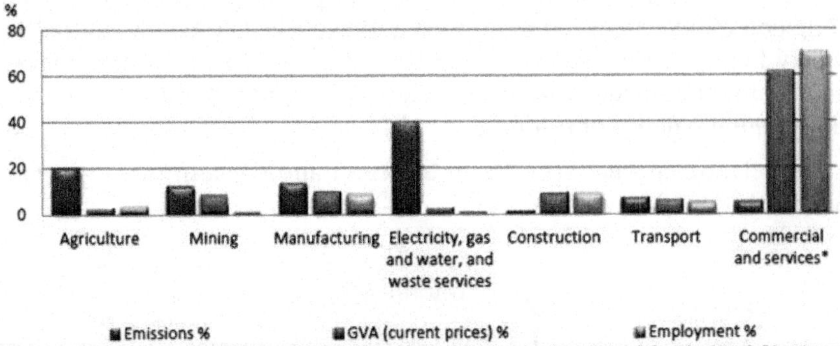

*Commercial Services covers ANZSIC Divisions: F. Wholesale and G. Retail Trade; H. Accommodation; P. Food Services; Q. Education and Health Services; J. Information Media and Telecommunications; K. Finance and Insurance Services; and L. Rental, Hiring and Real Estate Services

Sources: Department of Climate Change and Energy Efficiency; Australian System of National Accounts (ABS cat.no. 5204.0); *Labour Statistics in Brief, Australia, 2012* (ABS cat. no. 6104.0)

Fig. : GHG emissions, GVA and employment for selected industries, percentage of all industries, 2009-10.

In 2009-10, the electricity, gas, water and waste services industry had the highest GHG emissions of any industry, followed by the agriculture, forestry and fishing industry and manufacturing. These three industries together accounted for 74% of emissions, but only 15% of total gross value added and 14% of employment in the Australian economy. Conversely, the commercial and services industries accounted for 70% of all employment and about 62% of gross value added, but only 5% of emissions. However, these and other industries, as well as households, can be seen as indirectly responsible for the emissions of the electricity industry.

In view of this, various techniques have been used to apportion emissions from electricity generation to the industries using electricity, since the electricity produced is ultimately consumed by industries and households. This has been done using the National Inventory by Economic Sector Gas Accounts produced by Department of Climate Change and Energy Efficiency (DCCEE). This work is described below under 'Greenhouse gas induced by final demand'.

Greenhouse Gas Induced by Final Demand

The measure of greenhouse gas directly emitted by Australian industries and households and the changes in emissions levels over time is a key element of the data used in developing and evaluating policy. This measure is often referred to as the production approach as it measures emissions that occur directly from Australian production and directly from Australian households (*e.g.* the combustion of fossil fuels in private vehicles).

It is also possible to look at emissions occurring through the final consumption of goods and services by Australian households and governments. For example, the cumulated emissions from the production of manufactured food products, including from agricultural production, manufacturing processes, transport and retailing, is attributed to the final consumer. This shifts the focus of the analysis to the demand-side, that is, to the emissions required to satisfy final demand, including emissions embodied in imports and exports. Ultimately, industries exist to satisfy final consumption in Australia and abroad. The international transfer of environmental burdens by a country can be understood by considering its environmental balance of trade.

The relationship between the production and consumption approaches to measurement of GHG emissions.

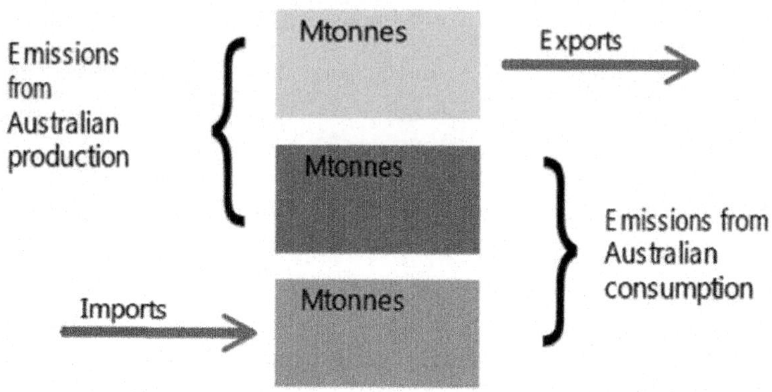

*Diagram based on Swedish Environmental Protection Agency report 5992, 2010. The climate impact of Swedish consumption.

Fig. : Production and consumption approaches to GHG measurement*.

The ABS has developed experimental estimates that identify and measure emissions according to the consumption approach using environmentally extended input-output analysis. This analysis shows how much greenhouse gas emissions are produced by Australian resident businesses and households; how much of these emissions are associated with goods and services leaving the country through exports; how much emissions are generated elsewhere through imports; and how much emissions are occurring both nationally and internationally in order to meet the demands of Australian consumption.

The direct and indirect emissions induced by final demand category for 2008-09. These experimental estimates show that of the 759 Mt of GHG emissions induced by the Australian economy, 531 Mt (or 70%) were induced to satisfy domestic final demand, while 228 Mt (or 30%) of GHG emissions were induced by exports. Approximately 90% of emissions induced under the mining category (and 41% of emissions induced under the manufacturing category) are induced by exports. In 2008-09, mining and manufacturing were the two most significant contributors to total GHG emissions induced by exports from Australia.

Of total emissions induced by categories of final use, 45% related to household final consumption expenditure (mainly through 'manufacturing-food, beverages and tobacco', 'electricity, gas, water and waste services' and 'commercial and other services').

Table : Experimental estimates of direct and indirect GHG emissions induced by categories of final demand (MT) - 2008-09.

Direct and indirect emissions	Household final consumption	Government final consumption	Investment and change in inventories	Total domestic use	Exports	Total use
Agriculture, forestry and fishing	13	0	5	19	37	56
Mining	2	0	5	7	68	75
Manufacturing						
Food, beverages and tobacco	43	0	1	45	27	72
Textile, Wood, paper and printing	6	0	1	7	1	8
Petroleum, coal and chemical products	15	2	2	19	12	31
Non-metallic mineral products	1	0	0	1	0	1
Metal products	1	0	3	4	43	47
Machinery and equipment	17	0	34	51	6	57
Total manufacturing	83	2	41	126	90	217
Electricity, gas, water and waste services	76	0	19	95	0	95
Construction	0	0	60	60	0	60
Transport						
Road	5	1	1	6	4	10
Other transport	19	3	0	23	15	38
Total transport	24	3	1	29	19	48
Commercial and services	101	36	13	150	13	162
Total direct and indirect emissions by Final use category	299	42	144	485	228	712
Direct Emissions by households	46			46		46
Total direct and indirect emissions Australia	345	42	144	531	228	759

While figure below provides a summary of the data results from this modelling exercise, the following figures each aim to draw out particular elements of the summary data.

The induced emissions by final demand according to the industry producing the final demand products. The most significant contributors are the manufacturing, and commercial and services industries.

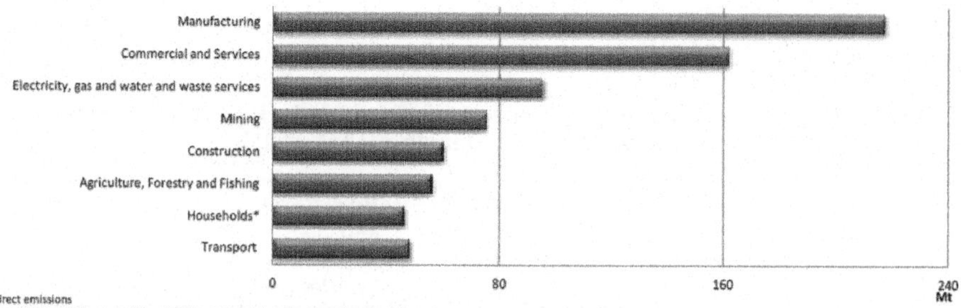

* Direct emissions

Source: ABS Input-Output modelling of GHG emissions induced by categories of final demand

Fig. : Experimental estimates of GHG emission induced by final demand, by industry, 2008-09.

Note that emissions induced by final demand relate only to those emissions associated with the final consumption expenditure of households and governments, and of gross capital formation and exports. It includes emissions embodied in intermediate inputs but excludes the emissions associated with that industry output subsequently consumed as an intermediate input by other industries. For example, the total emissions produced by the electricity generation industry are much larger than emissions induced by its final demand. This is because much of the output of the electricity generating industry is consumed as an intermediate input by other industries and therefore recorded as part of emissions induced by these other industries. Similarly, the transport industry records induced emissions that are less than the total emissions produced by this industry. Direct emissions by households relates to emissions arising from households' transport activities, combustion of gas for heating and petrol used for lawn mowers, etc.

A breakdown of GHG emissions induced by final demand for various manufacturing products. Typically, domestic demand is the main driver of induced GHG emissions across the various manufacturing industries - the notable exception is the manufacture of metal products which reports a large proportion related to exports. The observations are consistent with Australia's export profile.

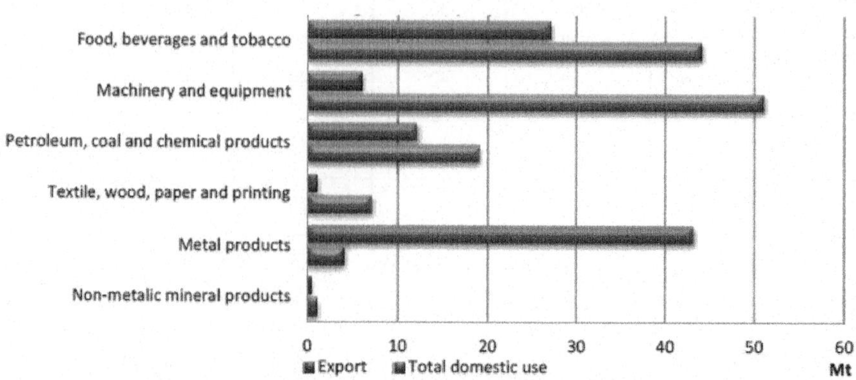

Source: ABS Input-Output modelling of GHG emissions induced by categories of final demand

Fig. : Experimental estimates of GHG Emissions embodied in Manufacturing Product (Mt), 2008-09

For induced GHG emissions for household final consumption the highest induced emissions for 2008-09 relate to goods and services produced by the commercial and services industry (101 Mt); electricity, gas, water and waste services (76 Mt); and food, beverage and tobacco product manufacturing (43 Mt).

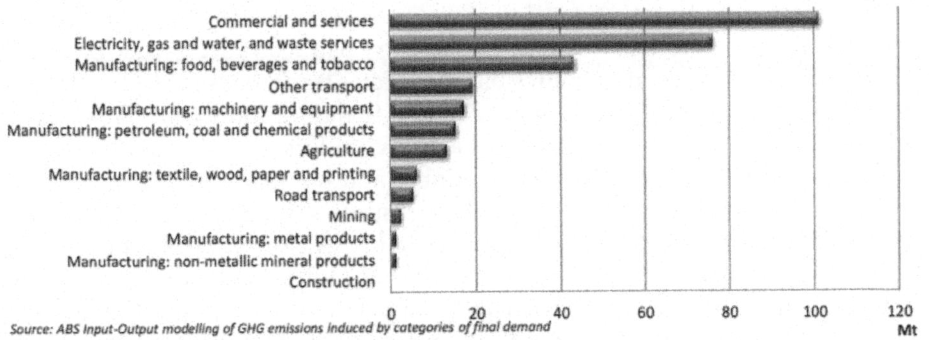

Fig. : Experimental estimates of GHG emissions induced by household final consumption (Mt), 2008-09.

Figure below shows that among Australia's exports in 2008-09, emissions were induced to the greatest extent by goods and services produced by the mining industry (68 Mt); metal products manufacturing (43Mt); agriculture, forestry and fishing (37Mt); and food, beverage and tobacco product manufacturing (27Mt).

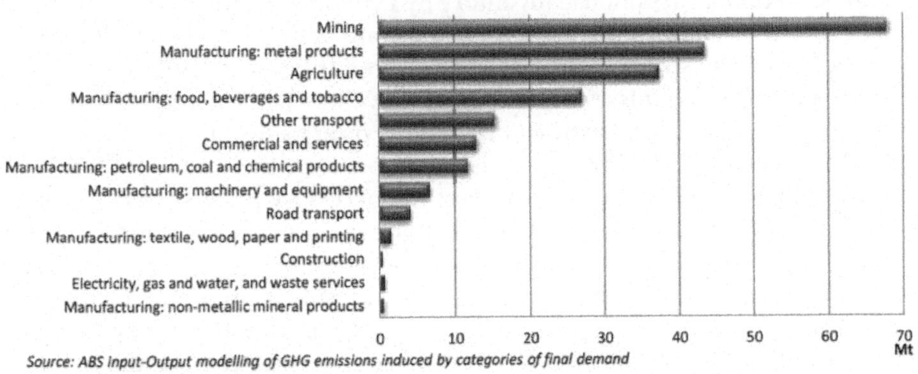

Fig. : Experimental estimates of GHG emissions induced by exports (Mt), 2008-09.

The following paragraphs provide further insight to the sources and methods used by the ABS in developing modelled estimates of greenhouse gases induced by categories of final demand.

This study uses the 2008-09 suite of I-O tables for the Australian economy in conjunction with data for GHG emissions by industry as supplied by DCCEE to model a consumption-based (*i.e.* final demand-side) view of Australia's GHG emissions. It uses the standard environmentally extended Leontief model to bridge the gap between the production and final demand sides of the economy. Importantly, it brings into focus the global GHG emissions implications of Australian consumption, regardless of whether that consumption is satisfied by domestic production or by imported products.

For the purposes of this study, the scope of emissions extends to all GHG emissions under the Kyoto framework comprising all IPCC energy sectors (including stationary energy and transport); industrial processes; solvent and other product use; agriculture; waste; and LULUCF. It uses the Australian Greenhouse Emissions Inventory as its primary data source but reallocates transport and electricity activity data onto an Australian and New Zealand Standard Industrial Classification (ANZSIC) industry basis, *i.e.* transport activity undertaken by a producing unit in the mining industry is allocated to mining and not to the transport industry. The DCCEE emissions data are available in respect of 40 sectors, *i.e.* at the 3-digit ANZSIC level for manufacturing and at 2-digit ANZSIC for the other industries, while the Australian I-O matrix uses a 111 Input-Output Industry Group (IOIG) classification. A key decision faced in undertaking this work was whether to model induced GHG emissions across the 40 sectors of the DCCEE presentation or across the 111 IOIG-based industries of the Australian I-O tables. The decision to use the 111 IOIG industries was adopted partly in order to avoid the aggregation error inherent in the alternative approach, which was to construct a smaller I-O table to match the 40 sectors of the DCCEE emissions data. A further reason was to ensure the modelled data were fully compatible with the SEEA and SNA frameworks.

The ABS performed a number of other adjustments to the emissions data from the National Greenhouse Inventory by Economic Sector. These adjustments are designed to support full integration and comparability with data of the SEEA and the SNA. The specific adjustments required to convert DCCEE data onto a SEEA basis relate to emissions induced by travellers while abroad; international bunkering (related to international transport, principally shipping and aircraft) and the reallocation of electricity emissions from an activity basis to an ANZSIC industry basis. The first two adjustments are necessary because the scope of the SEEA relates to the activities of all units that are resident of the economic territory, while Kyoto Protocol-based data relate to emissions taking place in a defined territory. The two bases differ because Australian residents may emit GHG abroad and similarly non-residents may emit within the Australian territory. Electricity GHG emissions data are reallocated from an activity basis to a standard ANZSIC industry classification basis. For example, the electricity self-generated by an enterprise in the manufacturing industry is included in the manufacturing industry, not the electricity generation industry.

The sum result of these various adjustments is a body of GHG emissions information that can be directly compared and integrated with estimates produced according to the SNA and the SEEA. This better informs decision-making across environmental and economic domains and substantially enhances the usefulness of the various data sets involved.

Nevertheless, the approach used here has two important assumptions that potentially affect the quality of the results. Firstly, in respect of GHG emissions, imported products are assumed to be produced using production functions that are identical to those used for locally produced products of the same type. Given the reliance on coal for electricity generation in Australia, the likely impact of this assumption is an overstatement of emissions embodied in imports. If this is the case, the data produced here will understate reported net exports of emissions. The assumption could be removed by incorporating global production functions and regional I-O models, though this would involve a considerable amount of additional data and would complicate the model significantly. The second important assumption is that all consumers of electricity pay the same price per unit for their electricity. Neither of these assumptions is expected to be entirely valid, but to date no rigorous analysis of possible biases in the results has been undertaken. In both cases, the assumptions can potentially be addressed through the use of more sophisticated models to take these factors into account, subject to the availability of data and resources, these developments may possibly be reflected in future editions of this work.

Energy Use as a Major Driving Factor of GHG Emissions

Energy derived from the burning of fossil fuels contributed 67% of Australia's GHG emissions in 2009-10 and fugitive emissions (*e.g.* gas escaping from coal mines and oil wells) from fuels contributed a further 7% (Kyoto Protocol basis including LULUCF). Given the dominance of fossil fuel combustion as a source of emissions, energy policy and research is directed at a re-engineering of industry production processes to be more energy efficient and to relymore on renewable sources of clean energy. In addition, there are policy issues around the future availability and prices of petroleum products.

The basis of much climate change mitigation policy in Australia is centred on economic instruments such as taxes. There are clear inter-relationships between policies that utilise economic instruments involving prices and taxes, and physical flows of various energy products and of related GHG emissions. In order to understand and to manage these inter-related phenomena, it is essential that our information on economic performance, energy supply and use, and flows of GHG emissions be directly comparable. The SEEA Central Framework directs the compilation of energy accounts and GHG emissions accounts in a way that allows direct comparison and integration of these sets of information with each other and with economicinformation of the SNA.

Information drawn from the SEEA-based Energy Account, Australia (ABS cat. no. 4604.0) can be used for monitoring the overall development of the Australian energy industry and in tracking the progress of policies to support clean technologies. Data produced within Energy Account, Australia are directly comparable with data from the broad suite of SNA and SEEA-based outputs. The production of GHG emissions accounts on a SEEA basis delivers a body of data that can be seamlessly integrated with information contained in Energy Account, Australia.

In practice, however, the data for energy and air emissions may come from different sources which are prepared using different concepts to meet different regulatory needs of governments. In this case, the SEEA has a role as a data integrating framework. Experience has shown that the resolution of inconsistencies is often a difficult and time consuming process, but it can be done - with positive benefits for the producers and users of data. The confrontation of data from different sources and the resolution of inconsistencies is an ongoing process within the ABS and other agencies.

Carbon Stock Accounts

The initial focus of the United Nations Framework Convention on Climate Change (UNFCCC) was to reduce fossil fuel emissions, this being the single biggest source of human induced GHG emissions. Under the guidance of the IPCC, a flows based global accounting system was established . Since the initial global climate change negotiations, land-based mitigation opportunities have received increasing attention by policy makers and researchers, for example the Australian Government's Carbon Farming Initiative.

Further thinking on the topic has recognised the need for a holistic view of carbon that extends the current flows-based framework to recognise the unique characteristics of different stocks of carbon.

BEST AVAILABLE TECHNIQUES AND BREFS

The BREFs are a series of reference documents covering, as far as is practicable, the industrial activities to the EU's IPPC Directive. They provide descriptions of a range of industrial processes and for example, their respective operating conditions and emission rates. Member States are required to take these documents into account when determining best available techniquesgenerally or in specific cases under the Directive.

Best Available Technology

Best available technology (or just BAT) is a term applied with regulations on limiting pollutant discharges with regard to the abatement strategy. Similar terms are *best available techniques, best practicable means* or *best practicable environmental option*. The term constitutes a moving target on practices, since developing societal values and advancing techniques may change what is currently regarded as "reasonably achievable", "best practicable" and "best available".

A literal understanding will connect it with a "spare no expense" doctrine which prescribes the acquisition of the best state of the art technology available, without regard for traditional cost-benefit analysis. In practical use, the cost aspect is also taken into account.

"But no person shall be subject to the foregoing penalties for any act done in the exercise of any right to which he is by law entitled, if he prove to the satisfaction of the court, before whom he is tried, that he has used the best practicable means, within a reasonable cost, to render harmless the liquid or solid matter so permitted to flow or to be put into waters."

Best practicable means was used for the first time in UK national primary legislation in section 5 of the Salmon Fishery Act 1861 and another early use was found in the Alkali Act Amendment Act 1874, but before that appeared in the Leeds Act of 1848.

The *BAT* concept was first time used in the 1992 OSPAR Convention for the protection of the marine environment of the North-East Atlantic for all types of industrial installations.

Some doctrine deem it already acquired the status of customary law.

In the United States, BAT or similar terminology is used in the Clean Air Act and Clean Water Act.

European Union Directives

Best available techniques not entailing excessive costs (BATNEEC), sometimes referred to as *best available technology*, was introduced in 1984 with Directive 84/360/EEC and applied to air pollution emissions from large industrial installations.

In 1996, Directive 84/360/EEC was superseded by the Integrated pollution prevention and control directive (IPPC), 96/61/EC, which applied the framework concept of *Best Available Techniques* (BAT) to the integrated control of pollution to the three media air, water and soil. The concept is also part of the directive's recast in 2008 (2008/1/EC) and its successor directive, the Industrial Emissions Directive 2010/75/EU published in 2010.

According to article 15(2) of the Industrial Emissions Directive, emission limit values and the equivalent parameters and technical measures in permits shall be based on the best available techniques, without prescribing the use of any technique or specific technology.

The directive includes a definition of best available techniques in article 3(10):

> "best available techniques" means the most effective and advanced stage in the development of activities and their methods of operation which indicates the practical suitability of particular techniques for providing the basis for emission limit values and other permit conditions designed to prevent and, where that is not practicable, to reduce emissions and the impact on the environment as a whole:
>
> - "techniques" includes both the technology used and the way in which the installation is designed, built, maintained, operated and decommissioned;
> - "available" means those developed on a scale which allows implementation in the relevant industrial sector, under economically and technically viable conditions, taking into consideration the costs and advantages, whether or not the techniques are used or produced inside the Member State in question, as long as they are reasonably accessible to the operator;
> - "best" means most effective in achieving a high general level of protection of the environment as a whole.

BAT for a given industrial sector are described in BAT reference documents (BREFs) as defined in article 3(11) of the Industrial Emissions Directive. BREFs are the result of an exchange of information between European Union Member States, the industries concerned, non-governmental organisations promoting environmental protection and theEuropean Commission pursuant to article 13 of the directive. This exchange of information is often called the Sevilla process because it is steered by the Institute for Prospective Technological Studies of the European Commissions' Joint Research Centre, which is based in Seville. The process is described in detail in Commission Implementing Decision 2012/119/EU. The most important chapter of the BREFs, the BAT conclusions, are published as implementing decisions of the European Commission in the Official Journal of the European Union. According to article 14(3) of the Industrial Emissions Directive, the BAT conclusions shall be the reference for setting permit conditions of large industrial installations.

United States Environmental Law

The Clean Air Act requires that certain facilities employ Best Available Control Technology to control emissions.

...an emission limitation based on the maximum degree of reduction of each pollutant subject to regulation under this Act emitted from or which results from any major emitting facility, which the permitting authority, on a case-by-case basis, taking into account energy, environmental, and economic impacts and other costs, determines is achievable for such facility through application of production processes and available methods, systems, and techniques, including fuel cleaning, clean fuels, or treatment or innovative fuel combustion techniques for control of each such pollutant.

The Clean Water Act (CWA) requires issuance of national industrial wastewater discharge regulations (called "effluent guidelines"), which are based on BAT and several related standards.

...effluent limitations for categories and classes of point sources,... which (i) shall require application of the best available technology economically achievable for such category or class, which will result in reasonable further progress toward the national goal of eliminating the discharge of all pollutants. ...Factors relating to the assessment of best available technology shall take into account the age of equipment and facilities involved, the process employed, the engineering aspects of the application of various types of control techniques, process changes, the cost of achieving such effluent reduction, non-water quality environmental impact (including energy requirements), and such other factors as the Administrator deems appropriate.

A related CWA provision for cooling water intake structures requires standards based on "best technology available."

...the location, design, construction, and capacity of cooling water intake structures reflect the best technology available for minimizing adverse environmental impact.

International conventions

The concept of BAT is also used in a number of international conventions such as the Minamata Convention on Mercury, the Stockholm Convention on Persistent Organic Pollutants, or the OSPAR Convention for the protection of the marine environment of the North-East Atlantic.

SUSTAINABILITY CHALLENGES FOR INDUSTRY

Today's society is faced with limited resources, expanding population, economic and social welfare disparities, increasing consumption and pollution, climate change, and degradation of human and ecosystem health. These challenges require innovations and policies to find and follow a sustainable path, allowing humanity "to meet current environmental and human health, economic, and societal needs without compromising the progress and success of future generations" (WCED and NRC reports).

The chemistry enterprise includes chemical and allied industries, their associated trade associations and professional societies, government laboratories, and the academic educational and research institutions that train the chemical workforce and advance scientific knowledge.

The chemistry enterprise has many roles in sustainability. It provides chemicals, materials, and technologies that improve the safe and efficient use of energy and natural resources and is responsible for delivering these in a way that protects human and environmental health. Chemistry—in labs, classrooms, and industry—is a central science for the development of sustainable technologies and innovations.

Chemical processes often require significant inputs of materials, water, and energy that are derived from limited resources. These processes also are potential sources of emissions and waste materials. However, the high-volume supply chains within the chemical industry provide opportunities to leverage and magnify improvements in material sourcing, process efficiency, waste reduction, co-product utilization, and the use and end-use of products.

In addition to reduced energy use in manufacture, chemical industry products also can improve sustainability in downstream uses. For example, the use of foam insulation in industrial, commercial, and home settings enables energy savings and associated drops in greenhouse gas emissions. A 2009 analysis found that every pound of CO_2 emitted by chemical industry operations was offset by more than two pounds of CO_2 emissions savings associated with the use the resulting products. The authors estimated that 2005 GHG emissions would have been 8-to-11 percent higher in a world without the chemical industry (ICCA).

A report by the National Research Council (NRC) in 2005 identified areas wherefocus would advance sustainability, including education; green and sustainable chemistry and engineering; life cycle assessment; toxicology; renewable fuels and chemical feedstocks; energy intensity of chemical processing; and separation, sequestration, and utilization of CO_2. While there has been progress in the ensuing years, those identified needs still remain relevant today. In addition, there is a need to address the professional system in which chemists and engineers work. These systems of incentives, regulations, markets, grant cycles, educational background, and corporate, academic, and government structures and climates has tremendous influence on the sustainable solutions that chemists and engineers can provide. More recent NRC reports have continued to examine these sustainability issues and largely reinforce those previous chemical industry-specific findings.

A 2008 ACS workshop recommended a framework that would address non-technical challenges to sustainability within the chemistry enterprise. NRC has noted that sustainability issues are complex and inter-linked, demanding a kind of holistic thinking and problem solving. NRC also laid out specific recommendations for federal agencies, noting that effectively managing sustainability can add value and efficiency across the government.

Chemists and engineers are embracing sustainability challenges through innovation and are using a life-cycle perspective to minimize or eliminate potential environmental and health implications of their technologies. In schools, curricula are including more instruction on green chemistry, green engineering and life cycle thinking; researchers are advancing the science and tools needed; and academic facilities are teaching by example in reducing the environmental footprint of their campus operations. However, the professional system in which scientists and engineers operate could be improved to further incentivize and support these efforts.

ACS believes that support for research to promote sustainability, green chemistry, and green engineering, combined with incentives for the adoption of sustainable technologies and new regulatory strategies that promote sustainable products and processes, will be instrumental in meeting the challenges of protecting human health and the environment, meeting our societal and energy needs, enhancing national and homeland security, and strengthening the economy.

Sustainability and Chemicals

According to the American Chemistry Council, "the business of chemistry [in the United States] is a $450 billion enterprise [about 26 percent of the global chemical production] and is a key element of the nation's economy. It is the nation's largest exporter, accounting for ten cents out of every dollar in U.S. exports. Chemistry companies invest more in research and development than any other business sector." As a result of trends in fossil fuel supplies, as well as compliance with chemical regulatory policies, business drivers for the chemical industry have evolved significantly over the past 50 years. There is an increasingly competitive landscape. Once a major net exporter, the U.S. chemical industry is now essentially a net importer (trade went negative in 2000–2001). These forces combined with transparency requirements, liability risks, and health indicators make sustainability goals, along with innovation, increasingly integral components of a company's ability to compete in the marketplace. Go to the web site of any global top 50 chemical companies—from the top three, Dow Chemical, BASF, and DuPont who each have 2004 sales in the $30-40 billion range, to number 45 on the list Lyondell Chemical with 2004 sales of about $6 billion—and there will be a statement of commitment to achieving sustainability goals. For example, the following statement appears on the Lyondell web site:

We aim to achieve excellence in every aspect of our economic, social, and environmental performance. We are committed to operating our world- wide business in a way that brings the greatest benefit to all of our stakeholders (employees, customers, investors, communities) and builds a sustainable future for generations to come,

One of the ways in which Lyondell says they work toward achieving these goals is by: "Investing in product and service innovations that use natural resources, as well as social and financial resources, in an efficient, effective, and economic manner over the long-term." Achieving such goals in the business world, including the chemical industry, is often referred to as the "triple bottom line." Many chemical companies are now even part of investment indexes such as FTSE4Good and the Dow Jones sustainability indexes that seek to raise business standards and investor awareness by tracking the financial performance of leading sustainability-driven companies.

However, the trend toward decreasing or at least flat research and development spending by the top 50 chemical companies and in industry as a whole makes it difficult to advance the science and technology needed to support such sustainability goals. Going forward, the chemical industry is faced with a major conundrum—the need to be sustainable, a desire to support science and technological innovation, and a lacking investment in fundamental research and development to make it all possible.

Despite these overall conditions, many companies are making the extra effort to advance sustainability goals. Pharmaceutical companies in particular have been quite successful. In 2002, prescription and over the counter drug companies invested more than $32 billion in discovering and developing new medicines, marking the thirty-second straight year the industry has increased its investment in R&D. As part of this effort, many companies are turning to green chemistry—"the design, development, and implementation of chemical processes and manufactured products to reduce or eliminate substances hazardous to human health and the environment"—and applying the twelve principles to redesign their active pharmaceutical ingredient (API) manufacturing processes. In this way, they have been able to dramatically reduce wastes generated. This success is highlight by the fact that five of the 52 winners of a U.S. EPA Presidential Green Chemistry Award are members of the pharmaceutical industry: Lilly (1999), Roche (2000), Pfizer (2002), Bristol Meyers Squibb (2004), and Merck (2005), which have all reported such kinds of improvements when green chemistry principles are applied.

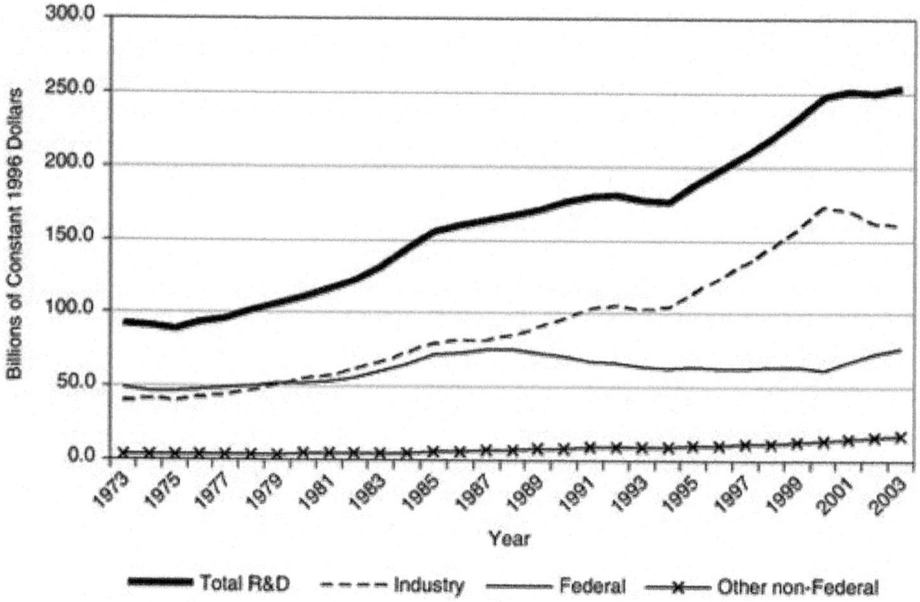

Fig. : U.S. research and development funding, by source of funds: 1973–2003.

Green chemistry thus offers a viable path for achieving sustainability goals across the chemical industry. That is, there is the potential to develop industrial technologies that could provide goods, products, and services in a way that does not reduce the supply chain of resources, harm the environment and human health, or limit the opportunities and choices for future generations.

More recently, a complimentary set of green engineering principles was developed. Together with green chemistry, the application of these principles ideally provide one or more of the following benefits:

- Lower costs of chemical processing
- Require less energy
- Produce both basic and fine chemicals in a way that is less hazardous to both humans and ecosystems.
- Be a source of chemical substitutes that improve quality of life.
- Provide a way of ensuring prosperity and social wealth.

At the same time, there are barriers that exist in developing and implementing these green chemistry and engineering principles. A study by RAND identified a set of barriers which fall into four main areas:

1. Lack of research, technology development, and new process engineering;
2. Industrial infrastructure problems and integration barriers;
3. Up-front investments required; and
4. Lack of coordinated actions by means of regulations, incentives, and government purchasing.

Complementary barriers have been identified elsewhere, such as inadequacy of education and training of chemists and chemical engineers as well as managers who direct them, the difficultly with measuring progress in green chemistry, and the lack of tools for effectively comparing green chemistry with conventional approaches.

Overcoming such barriers requires defining a "green" or "sustainable" agenda for the chemical industry. For example, the WTEC (World Technology Evaluation Center) Panel Report on Environmentally Benign Manufacturing identified the environmental research challenges facing the polymer processing industry, and the Vision2020 roadmap discusses environmental considerations for the CPI as a whole. In 2003, the UK Royal Society of Chemistry released the report *Benign and Sustainable Chemical Technologies,* which identified research opportunities to support sustainability in several areas, including: raw materials, agriculture and chemistry; new chemical routes (solvents and chemicals); catalysis, biocatalysis, and materials; new processes and process strategies; and environmental biotechnology.

MINING: BIG PROGRESS BUT NEW CHALLENGES AS SECTOR SEARCHES FOR SUSTAINABILITY

The mining and minerals industry has made major advances towards sustainability but the sector faces new challenges as governments in the global South reassert control over their natural resources yet lack the capacity to ensure mining can contribute to sustainable development.

Fig. : The Collahuasi copper mine in northern Chile is a zero-discharge operation, with much of its water being recycled.

These are among the findings of a 10-year review – published today by the International Institute for Environment and Development (IIED). It urges the sector to move from improving its standards in principle to implementing them in practice. It highlights new opportunities for governments and the mining industry to work together and engage with communities and other stakeholders to improve the sector's performance.

Researcher Abbi Buxton reviewed progress the mining sector has made over the past ten years, since its leading companies joined NGOs in calling for theMining, Minerals and Sustainable Development initiative to provide a landmark independent review of the sector's role in sustainability and set standards for it to meet. She will share its findings at IIED's Fair ideas conference in Rio de Janeiro on 16-17 June.

"The 2002 MMSD report was a game-changer," says Buxton. "For the first time, mining executives committed to act to maximise their sector's contribution to sustainable development, and they adopted the MMSD agenda as a robust and credible way to do this. Ten years on, however, the results are mixed and new challenges have emerged."

Buxton's report shows how the International Council on Mining and Metals - an umbrella organisation of leading companies such as Rio Tinto and Anglo American – has succeeded in implementing many of MMSD's recommendations for industry. But complementary measures proposed for governments, the small scale mining sector and communities have not matched this success.

"Some of the world's biggest mining companies have undergone major shifts in terms of their understanding of and commitment to sustainable development, and what they can do to contribute to it," says Buxton. "They have made major improvements in their environmental policies at a strategy level and have a gained a better idea of what good practice looks like. The challenge that remains is to implement these policies on the ground."

The report notes that despite the improvements larger companies have made, the capacity needed to make this happen is still lacking, particularly in mid-tier and smaller companies and in governments and communities. Bad practice persists despite good intentions across the industry.

"The past 10 years have been about setting standards, identifying best practices and demonstrating to stakeholders what sustainable development means," says Buxton. "The next 10 have to be about implementation. The ambition must also build from reducing environmental and social impacts to also sharing the benefits of mining to promote social and economic development. That means moving from 'do no harm' to a net positive impact."

The report describes how global rules for best practice in the sector have emerged in the decade since the MMSD report. But it shows that more often than not, there is a lack of implementation, independent verification, public reporting, or consequences for non-compliance.

One area in which there has been minimal progress over the past 10 years, says the new report, is in the capacity of governments in developing nations to ensure that mining contributes to sustainable development.

This poses fresh challenges in light of the new trend for such governments to reassert control over their mineral resources in response to the boom in commodity prices on the global market over the same period. Governments are doing this by, for instance, changing their tax laws and mining policies or acquiring equity stakes in companies that are active in the sector.

The report adds that mining companies need not see this trend a threat. It can be an opportunity for them to rethink. The social contract that has emerged in the past 10 years has seen companies increasing their responsibility while government and community capacity remains weak. There is now a opportunity to redefine roles and responsibilities and the social contract for achieving sustainable development in the sector.

"Mining companies have often played a role that governments should have played," says Steve Bass, head of IIED's Sustainable market Group. "As states move to reclaim their roles in community and economic development, there will be opportunities for improved partnerships between governments, mining companies and local communities. These new partnerships will be critical to ensure that mining contributes to local and national development in an environmentally sustainable way."

The report notes other big trends that have emerged since the MMSD report was published in 2002 and will affect the mining sector's journey to sustainability. These include:

- The rise in prominence of investors and operators from China, India and elsewhere – that did not play a major role in defining the sector's direction 10 years ago.
- The dramatic increase in community expectations, including demands for Free, Prior and Informed Consent (FPIC), which governments and companies must respond to like never before.
- The growing scientific evidence for climate change and its implications for the mining sector, which to-date has adopted a 'wait and see' approach.

POLICY DRIVERS FOR SUSTAINABLE CHEMISTRY

Chemicals

The presence of harmful chemicals in both indoor and outdoor environments poses risks for human health and the environment. Control of chemicals in a hospital setting is important to minimize risks to patients and employees, as well as to prevent harm to the surrounding community. For this *Sustainability Roadmap*, we consider chemicals in terms of indoor air quality management, environmental services and chemical waste management.

Indoor air quality management ensures a comfortable and healthy indoor environment for building occupants through the management of environmental tobacco smoke, ventilation, occupant comfort, and contaminants. As with energy and water management, many of these features and practices can be incorporated into the facility during building design and construction, but facilities of all ages and designs can make improvements through commissioning, retrofitting, staff training and minimization of products that contribute to indoor contaminants.

For environmental services departments, some cleaning chemicals can present a variety of human health and ecological concerns, contributing to poor indoor air quality and containing chemicals linked to cancer, reproductive disorders, respiratory ailments, eye and skin irritation, central nervous system impairment, and other human health effects and have been shown to be a primary cause of work-related asthma, particularly in nurses and cleaning staff. These conditions may be eliminated through the use of environmentally and health-sensitive cleaning products, equipment, and methods that are proven to be effective at infection control.

Finally, chemical waste management addresses the harmful impacts of chemicals on exterior surroundings by reducing or eliminating their emission and ensuring their proper disposal.

Importance

While chemicals play an important role in health care facilities as sterilants, disinfectants, solvents, chemotherapeutics and other pharmaceuticals, preservatives, paints, sealants, and so on, they also can pose a threat to human health and the environment. According to Health Care Without Harm (HCWH), an international nonprofit membership organization that campaigns for a health care sector that promotes the health of people and the environment, "industrial societies are experiencing an increase in diseases and conditions linked to daily environmental exposure to hundreds of hazardous chemicals." HCWH views current laws as inadequate to protect the public from these exposures and testing of chemicals with unknown health effects as slow and incomplete. These factors contribute to the likelihood of potentially dangerous chemicals being widely dispersed throughout the environment and compel health care facilities, as institutions invested in public health, to respond with proactive measures that lead to the development of safer practices and products.

Compliance and Policy Considerations

To manage chemicals safely, hospitals are required to enact policies for handling chemicals, training and educating staff, and providing strict oversight. Policies should dictate standard best operating procedures and may include a lab safety manual, OSHA's hazard communication policy, a hazardous materials management plan, spill policies, EPP policies, IAQ management plan, and others specific to having a particular chemical in the facility (*e.g.*, EtO, glutaraldehyde, mercury, solvents, etc.). Once the chemical is deemed a "waste", it is regulated by several U.S. Environmental Protection Agency (EPA) regulations that dictate the disposal of all hazardous chemicals.

Organizations can anticipate future chemicals regulation by putting in place policies of their own, and watching what regulations are slowly being introduced around the country that will directly or indirectly affect chemicals management in health care facilities. For example, in California, Proposition 65: *The Safe Drinking Water and Toxic Enforcement Act*requires, at least annually, publication of a list of chemicals known to the state to cause cancer or reproductive toxicity. Several states ban the use of fluorescent bulb crushers. And, while not yet regulated, several chemicals are of particular concern, including mercury, Biphenyl A, flame retardants, and di (2-ethylhexyl) phthalates.

Managing chemicals according to strict regulatory compliance is the first step in a facility's internal chemicals management policy. However, managing *beyond compliance* should also be the goal. Chemicals purchasing and management policy can help minimize the risks of having chemicals on-site. The objectives of a chemicals management policy are to

1. Seek the least hazardous (or least toxic) chemical to get the job done. Understanding chemical components of products and/or the hazards of using a particular chemical is best begun during the product evaluation process.
2. Use the smallest amount of the chemical necessary to get the job done.
3. Use the chemical safely by properly training staff (following all regulatory guidelines of OSHA's hazard communication policy and all U.S. Environmental Protection Agency (EPA) regulations related to use and disposal of hazardous chemicals).
4. As a first step, require vendors to disclose hazardous chemical components of products and to provide an MSDS (material safety data sheet) for each item.
5. Employ, to the fullest extent necessary and reasonable, personal protective equipment and engineering controls.

Chemical Policy

A broad chemicals policy addresses a health care organization's commitment to and support of changes in national (and potentially international) chemicals policies. In the United States, many credible groups are working together to support policies that protect public health. To this end, the Centers for Disease Control and Prevention (CDC), the Agency for Toxic Substances and Disease Registry (ATSDR), the EPA, and others have created the National Conversation on Public Health and Chemical Exposures. The National Conversation is a collaborative initiative intended to strengthen the nation's approach to protecting the public from harmful chemical exposures.

Lab Chemical Management

The bottom line is there are many compliance issues to consider when buying, using, and disposing of hazardous materials and waste. If, after undertaking the process of looking for less hazardous or non-hazardous alternative products or services, it is found that use of a chemical is necessary, the EPA suggests using this sustainable chemistry hierarchy to select products and processes, with the goal of purchasing products and using processes that represent the highest level possible of these green actions:

1. Practice green chemistry (source reduction/prevention of chemical hazards).
    - Design chemical products to be less hazardous to human health and the environment.*
    - Use feedstocks and reagents that are less hazardous to human health and the environment.*
    - Design syntheses and other processes to use less energy and fewer materials (high-atom economy, low-E-factor).
    - Use feedstocks derived from annually renewable resources or from abundant waste.
    - Design chemical products for increased, more facile reuse or recycling.
2. Reuse or recycle chemicals.
3. Treat chemicals to render them less hazardous.
4. Dispose of chemicals properly.
5. *Chemicals that are less hazardous to human health and the environment are:
    - Less toxic to organisms and ecosystems.
    - Not persistent or bio-accumulative in organisms or the environment.
    - Inherently safer with respect to handling and use.

Chapter 4

Building Corporate Social Responsibility in Developing a Sustainability Management System Framework

CORPORATE SOCIAL RESPONSIBILITY

Technopolis Invests in Corporate Social Responsibility

Responsibility is a strategic choice and part of the company's ethical values. Responsibility helps to develop the added value, profitability and risk management of operations and to predict stakeholder expectations. Technopolis sees responsibility as something more than responding to increasing external requirements; it is an opportunity to enhance its competitive advantage, support profitable growth, and increase added value to stakeholders in the long term.

Responsibility is reflected on Technopolis' campuses in the healthy, safe, and eco-efficient space and services, motivated personnel, and a sense of community that supports success. These contribute to high occupancy rates and property values. With its responsibility activities Technopolis aims to improve environmental-friendliness, employment, employee wellbeing, and the openness of its operations. Technopolis steers its responsibility by developing Corporate Social Responsibility reporting according to GRI (Global Reporting Initiative) principles.

Corporate Social Responsibility at Technopolis involves investments in financial, social and environmental responsibility. Our social responsibility is comprised of the development of the well-being and competence of our personnel, respect for human rights and equality, and responsibility in our relations with stakeholders. We support the growth, development and networking of companies. With our Talent Match -service we also help international professionals seek employment in the surrounding community. Read more about Talent Match.

The national environmental impact of the building stock is significant. Technopolis aims to reduce its ecological footprint and has set demanding environmental targets for 2011-2016 Read more about our sustainability strategy. We make green investments annually to our existing buildings. Furthermore, possibilities for energy efficient solutions and the use of renewable energy sources are studied in new construction projects already in the planning phase. Technopolis also pursues environmental labels and building ratings Read more aboutTechnopolis' LEED and Green Office projects.

Improving a building's environmental performance has positive impacts. It maintains the building's condition, enables increasing of building's value and managing the development of operational and maintenance fees. In addition to the employee health services offered on our campuses, our healthy and safe premises have positive impacts on health, well-being and comfort, and thus our customers' productivity. Professional building maintenance and management and functional technical systems ensure good indoor air quality and uninterrupted operation for our customers in our premises.

Corporate Social Responsibility Reporting

Technopolis publishes its Corporate Social Responsibility report annually, and the information presented therein corresponds with the financial period, January 1 - December 31. Technopolis' Corporate Social Responsibility report for 2014 complies with the GRI G4 guideline's "Core" level and GRI 3.1 guideline application level B. The responsibility report describes Technopolis' financial, ecological and social responsibility in accordance with the principles of the GRI (Global Reporting Initiative) G4 and G3.1.frameworks, the most recent Construction and Real Estate Sector Supplement (CRESS) version, and EPRA's Sustainability Best Practice recommendations for responsibility reporting. In addition to its Corporate Social Responsibility report Technopolis participates annually in the international real estate sector GRESB sustainability benchmark.

GRI

Sustainability requires openness. The Corporate Social Responsibility report is an assessment and communications tool that allows an organization to inform all its stakeholders how its actions support sustainability. Reporting focuses on three dimensions: financial, social and ecological sustainability. The Global Reporting Initiative network (GRI) has created corporate social responsibility reporting guidelines for this purpose, that can be used by all organizations regardless of size, industry or geographical location. Reports composed in accordance with the GRI reporting framework can be used for a range of purposes, including:

1. Comparing and evaluating an organization's corporate social responsibility with regard to legislation, standards, regulations and voluntary initiatives,
2. Showing how an organization's corporate social responsibility goals influence its operations and vice versa, and

3. Comparing an organization's operations with regard to its own and other organizations' operations.

EPRA

Technopolis also implements the European Public Real Estate Association's (EPRA) best practice ~recommendations on Sustainability Reporting. EPRA helps listed companies in the real estate sector to prepare for the likely introduction of more stringent mandatory sustainability reporting regulations. EPRA's Sustainability Performance Measures and accompanying principles for Sustainability BPR are based on the GRI Construction and Real Estate Sector Supplement (CRESS) and have been developed by the EPRA Sustainability Reporting Committee in consultation with the wider EPRA membership.

Technopolis has achieved Bronze award accreditation in EPRA Sustainability BPR Awards. The awards aim to ensure greater reporting consistency and transparency which allow better sustainability comparisons.

GRESB

GRESB is an independent real estate industry responsibility benchmark survey measuring eight different areas on two dimensions. They include identification of risks and opportunities, stakeholder cooperation, monitoring and environmental management systems, operational responsibility indicators, responsibility policy, responsibility management, building ratings, new building projects, and significant renovation projects. GRESB improves responsibility transparency and stakeholder cooperation and provides investors with a tool for benchmarking and monitoring the responsibility of operations.

Technopolis took part in the GRESB (Global Real Estate Sustainability Benchmark) sustainability comparison for the first time in 2014, earning the highest total points out of all Nordic office property companies and receiving the highest Green Star score. In the Nordic benchmark group, Technopolis was the best in terms of, inter alia, responsibility policy, identification of risks and opportunities, monitoring of operations, environmental management systems, and responsibility management.

The GRESB Quadrant Model describes the participant's success in sustainability comparison through two dimensions. Placement in the quadrant illustrates how far the organization has progressed in integrating sustainability into its real estate portfolio.

SUSTAINABILITY AND CORPORATE SOCIAL RESPONSIBILITY

CSR is not a static concept—it is a moving, evolving target, said Norine Kennedy of the U.S. Council on International Business. According to Kennedy, there is no solid definition of CSR; however, it is not a replacement for the governmental role and responsibility in meeting challenges of sustainable development.

Sustainable development within business promotion is expanding rapidly in several directions. Some interpret corporate responsibility to mean what companies should do above the call of law; others think it should be legally mandated at the national or international level; others, again, take the position that it is already here and we are already doing it, said Kennedy.

The scope of corporate responsibility varies country by country, region by region, interest group by interest group. At a minimum, it includes environmental issues but it also takes on social, ethical, governance, health, and other issues. Potentially, it is a very broad concept to cover, and it is a challenge for the business community.

Millennium Development Goals

As a follow-up from the world summit on sustainable development in Johannesburg in 2000, the United Nations developed Millennium Development Goals (MDGs) with the implications for corporate responsibility, environmental, and health issues. One hundred ninety-one UN member states endorsed the Millennium Declaration. There are 18 MDGs grouped around eight goals, most of them having 15–20 objectives.

The main notion of MDGs is that it is not just governments, but also other interest groups in society that are expected now to carry out the commitments. It is clear in the international arena that companies are increasingly expected and, in some cases, required to take on roles and responsibilities that are traditionally those of governments. Today's world and its markets are globalized, and the international impacts are unmistakable, said Kennedy. What happens internationally matters to companies in the United States. There's not a one-size-fits-all solution for corporate responsibility, which makes it quite a challenge as we are looking for an internationally-agreed-upon approach.

There are several references to health-related issues in MDGs, such as reduction of the mortality rate of children under 5 years by two-thirds, reduction of maternal mortality by three-quarters, and attempting to decrease the incidence of human immunodeficiency virus (HIV)/acquired immunodeficiency syndrome (AIDS), malaria, and other major diseases. Other health-related issues targeted in MDGs are safe drinking water and concern for slum dwellers' health. Finally, under the heading of a global partnership for development, there are two points:

(1) access to medicines in cooperation with pharmaceutical companies and the private sector, and

(2) make available benefits of new technologies, especially information and communications technology.

Another CSR incentive called the World Summit on Sustainable Development focuses on implementation and execution that is synchronous with the finance and trade negotiations of Monterey and Doha. According to the WTO, the November 2001 declaration of the Fourth Ministerial Conference in Doha, Qatar, provides the mandate for negotiations on a range of subjects, and other work including issues concerning the implementation of the present agreements. At the Summit on Financing for Development in Monterey, Mexico, delegates from participating nations pledged new resources for development and to adopt the policies needed to ensure that these resources are well used. In Monterey, President Bush underscored the link between good governance, good policies, and human well-being when he put forward his concept of the millennium challenge account, noted Kennedy.

This new type of assistance will go only to developing nations that are governed wisely and fairly, are strongly committed to investing in health and education, and that follow sound economic policies that encourage entrepreneurs and spur growth. Because of the World Summit on Sustainable Development, the UN becomes even more open to nongovernmental organizations (NGOs) and business. Therefore businesses now have the opportunity to be moreengaged in UN discussions and bring forward business experts and practitioners who are involved in partnerships, noted Kennedy. It is also a challenge to businesses how to measure and report business performance in corporate responsibility and other areas where they have been active as members of the business community.

The World Summit on Sustainable Development work program refers to corporate responsibility in the following four places:

- Sustainable patterns of consumption and production that enhance corporate environmental and social responsibility and accountability through actions such as voluntary initiatives, standards, reporting, dialogue, financial institutions engagement, cleaner production initiatives
- Sustainable development in a globalizing world that actively promotes full development and effective implementation of intergovernmental agreements, initiatives, partnerships, regulations, and continuous improvement in corporate practices in all countries
- Health and sustainable development, a linkage between health and environmental protection, reduction of environmental health threats, access to health care services, safer technologies for drinking water and waste management, reduction of occupational injuries and illnesses, a link between public health promotion and reduction and elimination of HIV/ AIDS, tuberculosis, and malaria, phasing-out of lead in gasoline and paint
- Strengthening of institutional frameworks that promote corporate responsibility and accountability and exchanging of best practices

The Role of International Organizations in Promoting Corporate Social Responsibility

International organizations play a major role in promoting better governance, and better economic processes in general, said Kathryn Gordon of the Organization for Economic Cooperation and Development (OECD). The OECD has a very distinctive way, a consensus-based way, of promoting better governments—governance among its member countries. A consensus development at the OECD is based on soft law instruments, meaning nonbinding statements of values and principles. To make these soft law instruments meaningful, OECD engages in consensus-based peer reviews about how these values and principles are implemented in different national policy contexts.

The instruments provide guidance for both government and corporate responsibilities in the investment area. On the government responsibility side, the instruments express the core investment values of transparency, nondiscrimination between foreign and domestic investors, and investment protection. On the corporate responsibility side, the OECD Guidelines for Multinational Enterprises provide guidance for international business. It is a comprehensive code of conduct that covers such areas as environmental management, human rights, anticorruption, and supply chain management. The OECD guidelines implementation procedures involve a distinctive and unique combination of voluntary and binding elements. Observance by business of the guidelines is voluntary, but the OECD governments assign a binding commitment to promote the principles of the guidelines among multinational enterprises operating in, or from, their territories.

At a minimum, corporate social responsibility includes environmental issues, but it also takes on social, ethical, governance, health, and other issues. *– Norine Kennedy*

Challenges Ahead

Today's businesses face multiple challenges in terms of corporate responsibility. Businesses have to keep up with the new initiatives on a wide range of fronts such as voluntary, regulatory, stakeholders, partnerships, and others. Credible and meaningful indicators of how companies are contributing to the quality of life and how they are implementing corporate responsibility can be challenging, said Kennedy. These challenges are occurring at the same time the concept of corporate responsibility is evolving in several different directions. She noted that one single approach or definition is not going to meet the various needs; given the number of contexts in which businesses work, it is appropriate and healthy to have different approaches. Businesses need to watch very carefully the codes of conduct and guidelines of the MDGs and the World Summit for Sustainable Development, to come forward to talk about what has to be done and how it is going to be delivered, and to continue to argue for the enabling frameworks for corporate responsibility at the international level, noted Kennedy.

Some discussion participants noted that CSR functions effectively when there is committed leadership in corporations. They noted that corporate leaders need to transform into leaders who will move corporate responsibility efforts forward. Business schools, where those leaders are educated, are in a position to influence the transformation. Additionally, society, shareholders, and employees need to become more vocal about what they are expecting from business leadership, thus becoming a determining factor about what corporations are focusing on and what their objectives are, noted some participants.

Corporate Social Responsibility in the Context of Regulation

According to Orts, CSR is an orientation to business enterprise that claims a company has more than just an economic duty to shareholders and owners of the company; it is also a social entity that entails moral obligations and imperatives that go beyond legal requirements and compliance.

Many people in society at large, and especially the business community, do not believe that CSR is a good idea. For example, Orts noted that Milton Friedman is famous for saying that we should not have CSR because the constraints should be given by the government, so a company should maximize profits as much as it can, and the law should provide it with constraint. The proponents of this view of CSR dispute that there is a tension between economic arguments about the need for businesses, especially public corporations, to focus on the bottom line, namely, shareholder value. On the other hand, there are ethical arguments, and every particular company needs to identify its ethical obligations that are either going to be constraining or a part of its definition as a business. Orts noted that the American Law Institute in its principles of corporate governance suggested that the primary objective of a company is to make profits for shareholders, but it still has to follow the law even if it is not cost effective.

According to Orts, Stanford University professor David Baron clarified the ethical argument for social responsibility by distinguishing between what he called CSR and corporate social performance. True CSR involves an allocation of a firm's wealth toward some view of the public good motivated by normative, that is, ethical principles. Strategic CSR (or mere corporate social *performance*) involves actions that appear to be motivated by higher social purposes and are, in fact, motivated by profits, noted Orts. Extreme cases amount to simple deception or "greenwashing". Therefore, when we talk about CSR in the context of environment and health, we are talking about true CSR, said Orts. Additionally, Orts quoted Paul Hawken's book *The Ecology of Commerce* in which the author says that business is a primary participant in destroying the world, and if businesses continue as they are going, there is not going to be a wildlife preserve, wilderness, or indigenous culture left.

Many companies today realize that they are responsible for the future of the world, and they no longer accept the maxim that the business of business is business only, noted Orts. Their new premise is: Corporations are the dominant institutions on the planet today; therefore they have to help address social environmental issues that affect humankind. This premise is increasingly becoming the ethically driven view of many large multinational companies. Globalization of the world implicates the idea of global corporate citizenship in globally constituted civil society and brings in the need for clarity about how to enforce different regulations; to whom should the multinational company owe their loyalties—whether they should be an American company or a global company—and what would be a global company's responsibilities in a broader civic sense. According to Orts, corporations are citizens of a global society and therefore owe a duty to participate in that general society.

Corporations are the dominant institutions on the planet today. Therefore, they have to help address social environmental issues that affect humankind.

– *Eric Orts*

Many governments in the world have basic capacity for addressing the complex societal challenges. Global problems such as global climate change, ozone layer depletion, biodiversity loss, depletion of fisheries and forests, hazardous waste, transportation disposal, migrating microbes, and invasive species, as well as local air and water pollution are global issues, and have a health component. Often it is not possible for individual governments to address them.

Legal Reform Strategies to Enhance Corporate Social Responsibility

We live in a very complex society in which the government is not going to be able to answer all the questions about what the standard rules are. Therefore, companies have to take a bigger role in this process and establish creative legal strategies, so called reflexive law, that go beyond the command-and-control approach, asserted Orts. Informational regulation in the form of mandatory disclosure of information similar to the toxic release inventory could also be used to enhance CSR. Environmental contracts are among other strategies that may help improve CSR. The idea of environmental contracts is that companies can have partnerships and work independently with NGOs or with other governments on specific issues. Laws could be passed to help promote this strategy. If a company made a contract about a specific issue, it might create more progressive and creative solutions to these problems than if it relied solely on the Environmental Protection Agency or the U.S. Congress, noted Orts.

Corporations may chose to be socially responsible and get involved in addressing certain social or health issues with precision and competence. However, these choices have to be made without losing sight of the fact that the primary interest for a company is an economic one, concluded Orts.

CORPORATE SOCIAL RESPONSIBILITY: ROLES OF GOVERNMENT, THE PRIVATE SECTOR, AND CIVIL SOCIETY

The Cause of Corporate Social Responsibility as a Phenomenon

The modern phenomenon of CSR is closely linked to that of globalization, stated Kernaghan Webb of Carleton University in Ottawa. Globalization increased movement of people, goods, ideas, and corporate activity across borders. In a globalized marketplace, the underlying premise is that organizations should behave with equal respect to people and the environment wherever they are. Advances in telecommunications (*e.g.*, the introduction of the Internet), NGO activity, and media scrutiny mean that an organization's activities can be critically tracked and followed more easily than ever before, regardless of their location. Because governments do not have the ability to fully address environmental and social problems on their own (particularly in developing countries), the idea that corporations should take on some of this responsibility has gained currency. In effect, CSR is largely a response to state incapacity, stated Webb.

The result is growing expectation that firms should be economically, environmentally, and socially responsible wherever they operate, even if government regulations are inadequate or poorly enforced. These expectations apply to small, medium-size, and large firms, and all sectors: pharmaceutical, mining, refineries, chemicals, and so on.

Efforts by many organizations are underway to develop flexible, practical, standardized CSR approaches for a global economy. Intergovernmental-level initiatives include the Global Compact, the International Labour Organization declarations, OECD guidelines, the World Bank, and others. Individual governments such as the United Kingdom are taking lead roles, as well. Other initiatives include investment, standards, industry, and those that are NGO-driven or faith-based. Although the initiatives indicate considerable engagement on this issue of CSR, there is considerable variability from one to the other in terms of the actual content, scope, comprehensiveness, interoperability, and take-up, said Webb

As a result, even efforts made in good faith may suffer in the confusing abundance of initiatives. The lack of standardization can discourage business from good behavior and it can also discourage consumers, investors, and governments from rewarding good behavior. Competing initiatives can slow down the momentum of the CSR movement, cautioned Webb.

Corporate Social Responsibility Initiatives and Law

Even though laws and international conventions have limitations, they will remain the foundation for environmental and social protections in our society, said Webb. In part, as a response to these limitations, the private sector and NGOs have developed CSR-oriented voluntary codes and standards as supplements. For optimal effectiveness, governments need to stimulate and structure these various initiatives.

There are strengths and limitations to laws. The strengths are that command-and-control regulatory approaches articulate societal positions on important issues and are the products of democratically elected legislatures in democratic countries. The laws are enforced by specialized government agencies and backed up by the courts. Command-and-control approaches have made considerable progress in improving the lives of people around the world. However, they have expensive, protracted development and enforcement processes, as well as jurisdictional constraints on subject matter, approach, and scope, noted Webb. Developing countries are particularly vulnerable to inconsistent and inadequate implementation and enforcement, in large part because of the inadequate budgets in place to fund such activities. Another downside to conventional command and control regulatory approaches is a tendency toward very inflexible and formal approaches that can lead to adversarial and legalistic behavior—a "going by the book" attitude toward compliance. This tendency can impede the development of optimal solutions to particular public policy problems, said Webb.

Further, with technologies moving so quickly, the law system is frequently put in a situation where it is one or two steps behind, no matter how hard governments try to stay on top of issues, said Webb. If limitations at the domestic level are challenging, they are considerably more problematic at the international level. Even though international laws have contributed to a lot of progress and provide the international regime of human rights, environmental protection, worker protection, and commercial activity, issues such as national sovereignty and the reluctance of states to agree to participate in, ratify, or implement international laws, slows down greatly the effectiveness of international laws to address social and environmental problems. Also, the divide between developing and developed countries is a particularly intransigent challenge for the international community because there is very little enforcement capacity at the international level and within developing countries, said Webb.

There are many possible ways to address the aforesaid issues and challenges. The International Organization for Standardization (ISO)—a nongovernmental body, although governments and private sector and others participate in it—generated the ISO 14000 series of standards, in particular, 14001, an environmental management standard. The ISO approach is intended to supplement legal regimes. It does not work as well when there is no effective legal regime because a management system works optimally in conjunction with a set of legislative or regulatory obligations, said Webb.

Environmental organizations have taken a lead role in developing a number of international voluntary certification regimes that apply to the CSR area. For example, the Forest Stewardship Council initiative started by the World Wide Fund for Nature was a response to the fact that the international intergovernmental community could not come up with an international forest protection convention. This type of response is a paradigm change for NGOs because they moved from being rule takers, participating in someone else's processes, to taking the initiative themselves and realizing the power they have to do so. NGOs realized that they can have retailers as allies who can frequently act as surrogates for consumers, and they can require, on behalf of their customers, that their suppliers agree to the standard or will not get business from them. Businesses are seeking "social licenses," meaning individual companies are developing environmental, worker, and community-oriented codes for business reasons. At the same time, communities are entering into "good neighbor" agreements with companies. For example, in the United States, coal companies are the subject of good neighbor agreements where community members are allowed to go into the plants to do inspections and to check the company's records. Just like laws, the new developments have limitations. For example, in the United States the Responsible Care Program for chemical manufacturing sector seems like an effective industry-initiated voluntary initiative, yet 90 percent—not 100 percent—of the chemical industry and chemical producers in United States are involved in the program. In other words, because it is a voluntary program, there is no legal mechanism to require 100 percent industry participation. Primarily, the challenge with voluntary instruments lies in implementation and enforcement associated with conflicts of interest and transparency and accountability issues.

The new approaches should be applied as supplements to the laws, not as replacements for them, said Webb. In some cases, they may act as precursors, and sometimes industry asks that these voluntary initiatives become law.

The corporate social responsibility is not a new idea. It extends back to the late 1600s when Quakers were the first users of corporate social responsibility, said Webb. Even though they were interested in trust and ethical behavior, they found that there were strong business dividends for engaging in that behavior.

According to Webb, CSR can be described as attempts by businesses to balance and integrate their economic, social, and environmental responsibilities in a way that minimizes societal harm and optimizes societal benefit while providing wealth to business owners and shareholders. It includes, but is not limited to, philanthropy; it assumes compliance with law and is generally seen as a voluntary, non-mandated set of activities. The pressure for CSR is becoming stronger because businesses are realizing that failure to consider the interests of their workers, surrounding communities, civil society organizations, and customers can have negative consequences to their reputation, and it may have a negative legal and commercial effect. Businesses are realizing that working with stakeholders can have benefits, not only enhancing their legal license to operate in a community, but their social license to operate as well (in the sense of community acceptance).

Increasingly, in addition to governments using their legal pressures, customers, lenders, insurers, investors, and shareholders are providing additional pressure for businesses to do the right thing. But it is very important to remember, cautioned Webb, that if businesses do not make money, they are not going to be able to engage in social responsibility; therefore, wealth creation does remain the main objective for business.

The Role of Governments in Corporate Social Responsibility

Today's societal problems necessitate concerted efforts of government, the private sector, and civil society. CSR represents the private-sector contribution to the efforts, but it is not a panacea; there are limitations on governments, intergovernmental legal instruments, to be able to address problems voluntarily. Governments can structure and encourage CSR as a supplement to conventional approaches, and organizations like the ISO can play an important role as a bridge between laws, intergovernmental instruments, community expectations for substantive obligations, and reporting standards. ISO is just one piece of the puzzle, said Webb.

Government encouragement of CSR stems from the understanding that CSR activities can assist governments in meeting societal needs. A country or an industry sector can be negatively or positively affected by individual firms' behavior. CSR can be a competitive advantage for a country. For example, the United Kingdom approach to CSR represents the most sophisticated model. It has realized the importance of CSR and has taken great effort to institute pension disclosure laws, support ethical trading initiatives, and encourage development of many other related initiatives, including those in the standards area. Among developing countries, Brazil is the CSR leader, using standards to encourage good business behavior.

The Role of the ISO Social Responsibility Standard

Webb suggested that ISO could play an important role in lessening industry confusion and increasing acceptance by developing a social responsibility standard. This would be a third-generation standard building on the management system standards that are already in place, and it would be a guidance document, not a specification document; that is, it would not be the subject of certification, but it still could build on existing approaches. Considering that there are more than 600,000 facilities around the world that have been certified to ISO 9001 or ISO 14001, ISO is well placed to develop a standard concerning operationalization of social responsibility ideas, said Webb. ISO seems to represent the most accepted international rule infrastructure. One hundred forty-five countries around the world—the majority of them are developing countries—participate in it. However, an ISO social responsibility standard would be only part of the solution, one piece of a puzzle, cautioned Webb. It would be one more additional element compatible with the existing initiatives and laws, concluded Webb.

THE CHALLENGE OF CORPORATE RESPONSIBILITY

To augment the Essential Knowledge Project, this essay discusses the perplexities and challenges of corporate social responsibility (CSR). The essay features the "what," "why," and "how" of CSR. As they help formulate CSR standards and give voice to organizations, public relations practitioners can use experts' carefully considered thoughts as well as research findings to determine the best plan of action. Internally and externally, practitioners can help build a foundation for image/reputation management, brand equity, relationship management, issues management, and crisis management.

"Are you a good corporate citizen?" That rhetorical question was asked by Ben W. Heineman Jr., senior vice president for law and public affairs at General Electric Corporation, writing in the Wall Street Journal (6/28/2005). His question helps set the tone for the topic of corporate social responsibility as a vital part of the Essential Knowledge Project. The column explained how GE sets a challenging goal—high performance with high integrity—for successful companies and, indeed, for 21st-century capitalism". High performance and high integrity are good for the bottom line. Citizenship requires a "rigorous, unwavering compliance with the law." It blends strict adherence to capital performance with integrity to never allow that commitment to corrode those principles. Quality judgments support, rather than defeat, a commitment to financial standards and to the globe where the organization works, including commitment to "reduce greenhouse gases and increase energy conservation".

What is CSR?

When discussing CSR, some prefer corporate responsibility (CR) to avoid the assumption that it is limited to "social" concerns (strategic philanthropy and community relations). Others use "social" responsibility to avoid the stigma that this topic only relates to businesses driven to place profit over social principle. Those who prefer "social" without the modifier "corporate" recognize that non-profits and governmental organizations are and should be held to responsibility standards. By whatever name, interests of organizations cannot long be at odds with mutual interest and common good.

Consideration of corporate social responsibility is as old as organizations themselves. The central question always has been this: Does each organization, as it strives to achieve its mission and vision, add value to the society which franchises its existence? Long before consideration focused on business, especially large corporations, questions about CSR addressed the rationale and acceptability of government and church.

History: A Long-term Perspective

Despite this timeless concern, the topic came to receive serious discussion and management policy development during the 1950's in nations with large corporations—or those working to overthrow governmental and capitalist colonialism. In developed nations, social movement activism and principles of social democracy offered not only a rationale but also a voice for the concern that mass production/mass consumption societies had spawned large corporations that were slowly dominating the standards of business performance in ways that distorted not only their rationale in society but also the wholesomeness of the private sector. Efficient and rational business practices could in fact harm the society that franchised the organizations to operate.

In the 1960s, interest in CSR became feverish as activists at all points of the ideological and geographical compass called for higher standards of business and government performance. Non-profits asked to share business profits. Academics argued over the differences between the modern and post-modern organization.

As Basu and Palazzo (2008) observed: "The last three decades have witnessed a lively debate over the role of corporations in society". This scrutiny revealed how CSR standards are defined by the ideology of each society. Public relations practitioners contribute to this ideology as they discuss business performance internally and externally. Critics have reasoned that CSR can become insulated, self-serving, and self-affirming—often to the detriment of the society where it was supposed to be a social, political, technical, and financial benefit. Public relations can help organizations craft their standards and implement them through actions and statements.

One of the primary discussants in that 30 year-long dialogue, observed 17 years ago: "The idea of corporate social responsibility has failed to help create the good society. Long seen by academics and managers alike as the missing link in capitalism, the concept of corporate social responsibility has not delivered on its promise".

Debate Around Friedman's Arguments

In one way or another, all discussions of CSR recall the famous, or infamous, claim of economist that CSR is bunk. He sparked decades of controversy by arguing that the only responsibility of publicly held companies is to increase profits—the efficiency paradigm of organizational excellence. Companies should pay only as much wage/salary as necessary to operate efficiently and pay taxes reluctantly. Some today laud his sentiments, and indeed many empirical tests have not found a positive relationship between CSR activities and major corporate financial performance indicators such as profit. Agle *et al.* (1999) stressed the methodological challenge: "Corporate social performance is notoriously difficult to quantify".

Others argue that Friedman's view of the role of companies too narrowly addresses the key issue. Stovali, Neill, and Perkins (2004) fought the traditional interpretation of the Invisible Hand of Adam Smith, which serves to legitimize the maximization of shareholder wealth, and as a result, shareholder-dominant corporate governance. A broader view of the Invisible Hand considers a "sympathy principle," or the ability and propensity of human beings to consider the interests of others. This view may be more aligned with the dominant concept of CSR and promote a broader, multiple stakeholder approach to corporate governance. If sympathy suggests only philanthropy, we add that CSR demands empathy, an outside-in way of thinking and planning to help management reflect on its role in and impact on the society where it operates. Critics believe that Friedman failed to understand the positive advantages to be gained from CSR: Reduce business costs and bolster profits.

Public Relations Practitioners' Views

Public relations practitioners added their voices to this controversy. Senior public relations practitioner, John L. Paluszek (1995) reasoned that advocates of CSR can be committed to profits and the community where they are generated,

Business is increasingly in society not only in its traditional role of improving the standard of living—by generating jobs, offering products and services and paying taxes—but also *via* an overlay of sensitivity that supports employees, empowers customers and investors, and relates to the needs of local, national and international communities.

The daunting question raised by this debate is this: Does the price of enjoying the franchise to operate in a society where profits are generated include increasing operating standards so that the organization adds value to that society beyond merely making a profit? Clark (2000) quoted public relations giant Arthur Page: "All business in a democratic country begins with the public's permission and exists by public approval". In fact, he might say the same of business in "undemocratic countries." If it lacks public support, it operates on a questionable franchise.

The legendary John W. Hill (1958) (co-founder of Hill & Knowlton) advised us to understand sound management as the initial step toward appreciating the roles and challenges facing public relations professionals.

It is not the work of public relations—let it always be emphasized—to outsmart the American public in helping management build profits. It is the job of public relations to help management find ways of identifying its own interests with the public interest—ways so clear that the profit earned by the company may be viewed as contributing to the progress of everybody in the American economy.

Another senior practitioner, Raymond Ewing (Allstate Insurance) (1987) reasoned, "A corporation exists for the optimization of the satisfactions of its stakeholders".

This history of CSR discussions leads to many conclusions, including the following ones:

- Knowing and achieving legal and ethical standards are not inherently opposed to one another, but they are not one and the same;
- Knowing and achieving engineering and operating standards that advance quality of work life, product quality and the community well-being where organizations operate is fundamental to CSR;
- Understanding and appreciating that the franchise to operate is granted by and required to serve the positive interests of society;
- Adopting standards that blend the public and private interest is basic to achieving aligned and mutual interest.

Thus, higher standards of employee or consumer safety help manage risks and reduce the likelihood of crisis related to employment and consumption.

Source of the Controversies

The Role of Business in Society

During the 1980s, discussion of CSR progressed logically from responsiveness to responsibility and rectitude: The ability "to act with rectitude, to refer their policies and plans to a culture of ethics that embraces the most fundamental moral principles of humankind". The heart of CSR is a firm interdependence between business and society.

Although his statement is fraught with the irony surrounding the tobacco industry, George Weissman (1984), former Chairman and Chief Executive Officer of Philip Morris, Inc., set corporate responsibility into its sociopolitical context: "Like property, the corporation is a creation of the state. It gets its charter from the government. To survive, the corporation subjects itself to regulation by government, and to serving the needs of government and the commonwealth." This point "is fundamental to understanding the concept of corporate responsibility in its current context; to understanding that we are not dealing with a fad; and to understanding that we are dealing with the fundamental existence and survival of the corporation". Businesses are expected to be involved in community affairs, formerly the exclusive responsibility of government, reasoned Chrisman and Carroll (1984) in their efforts to reconcile profits, business goals, and society.

CSR is basic to each organization's character, good, bad, or in between. Its character depends on how well it can align its interests with those of consumers, audiences, and publics. For this reason, CSR is not merely a communication challenge; it is a management planning and performance challenge.

Public relations can help an organization know the standards needed to achieve CSR through issue monitoring. Such issue monitoring facilitates an outside-in approach to considering what makes an organization good – the standards and expectations by which it is judged. An outside-in approach makes pragmatic and ethical sense because activists and other critics play a significant role in the raising of CSR standards and in evaluating private sector performance.

Platitudes are likely to foster as well as frustrate efforts to discover which standards of corporate responsibility truly make a difference in the quality of relationships with key stakeholders. The challenge is to know how the organization's interests align with those of its stakeholders. One goal of such discussions is to achieve the common good and to enhance community.

Meeting stakeholder expectations is a vital part of achieving the legitimacy needed to operate. Sethi (1977) coined the legitimacy gap between organizational performance and the expectations of myriad stakeholders. At least in part, questions about legitimacy express "concern for a corporation's image". Is it a crisis (therefore the rationale for crisis management and communication) when an organization's image is damaged because its legitimacy is questioned because it fails to meet established CSR principles?

Heath (1997) featured the legitimacy gap as the rationale for his discussion of the four pillars of strategic issues management (SIM):

- Strategic business planning which includes foresight judgment, budgeting, and choices among options that lead to organizational success within a sociopolitical environment.
- Issue monitoring which can help organizations understand the alignment of interests, points of misalignment, as well as power resources and challenges by various stakeholders.
- Corporate responsibility, the standards by which an organization is judged.

- Issues communication, the dialogue to better understand facts, develop appropriate evaluations, create vital identifications, advance reputations, and foster public policy positions that achieve mutually beneficial interests. Such discourse defines terms, establishes principles and standards, and thereby creates the socially constructed meaning that guides CSR.

SIM sees CSR/CR as the socially approved expenditure of material and human resources to advance the strategic business plans in ways that meet stakeholder expectations. When firms' operations offend stakeholder expectations, legitimacy gaps motivate stakeholders to correct those operations. Such logic applies not only to the private sector, but also to non-profit organizations and government agencies.

Without oversimplifying the topic, four general themes seem to dominate the discussion of business CSR/CR:

- Profits efficiently earned may not constitute a sufficient CSR standard.
- Companies should engage in strategic philanthropy whereby good acts improve operating conditions and quality of life—and serve society.
- Companies should partner with non-profits and government agencies to solve social, economic, and psychological problems in society.
- Companies should meet or exceed stakeholder expectations of performance standards needed to satisfy the moral rectitude that business contributes to the long-term business climate by collaborative decision making and operating in the public interest.

In such matters, CSR might be defined as corporate strategic philanthropy, even narrowed to "effective public relations" where images are massaged by wordsmiths who tout the excellence of the organization and build accommodating relationships with stakeholders. It can be viewed as cause-related marketing. At best, it can be conceived of as knowing, achieving, and communicating about higher standards of performance in the public or community interest.

Profits and CSR: Partners or Opponents? (Are We More Able to Generate and Use Revenue Wisely by Meeting Stakeholder CSR Expectations?)

Does the cost of achieving CSR decrease, increase, or have no effect on "profit" -the cost-benefit residual? What communication activities add impact to CSR performance? In addition to making general claims that corporate responsibility is good for business and observing the trend that companies are integrating environmental values and competitive strategizing, many empirical studies offer data that demonstrate the relationship between profits and CSR.

Most studies have identified a positive relationship (although not always linear) between CSR activities and organization performance as measured by various indicators such as shareholder returns, profit, or marketing impact. For example, the data from Ogden and Watson (1999) suggested that despite the cost of improving the quality of customer service performance, "shareholder returns respond in a significantly positive manner to such improvements".

Beliveau, Cottrill, and O'Neill (1994) found that the CSR-profit relationship is not linear and varies by industry as well as by different company performance indicators. For instance, some industries are rewarded for being innovative more than others. In addition, stock market measures lead in CSR performance, while accounting measures lag CSR. Including sociological factors (such as social impact of an industry) can help explain CSR impact on performance measures. Finally, perceptions of management reputation can have a social and economic advantage.

Sen and Bhattacharya (2001) discovered that CSR performance can increase marketing clout if it relates to product quality and/or consumers' personal preference views on key social issues. This study suggests that aligned interest is crucial to and accountable for positive impact of CSR, but also cautions that such facts don't count if key stakeholders don't know about the organization's accomplishments.

Awareness counts. Sen, Bhattacharya, and Korschun (2006) used survey data to conclude that knowledge of companies' CSR (primarily defined as strategic philanthropy) is fairly low. When consumers are more aware of what companies are doing philanthropically, that knowledge increases desire to purchase from the company, and makes it more attractive as a place to work and in which to invest. The motives (positive or negative) attributed to companies for this action affect the perception of and motivation to respond favorably to the company.

Rochlin, Witter, Monagahn, and Murray (2005) discovered that "by building a business strategy that aligns social, environmental, and economic performance with long-term business value, corporate responsibility becomes part of core business and is tied to long-term value creation for both business and society".

Other studies have shown how CSR decisions relate to people's attitudes and behaviors toward organizations. CR can sell products but is best when CR decisions lead to increased sales and reputation management, rather than when sales drive CR. Based on ranking data, the following list was compiled, ranging from most important to least important factors in making decisions relevant to a specific company: business practices, community support, employee treatment, quality, environment, service, price, convenience, and stability. Analysis revealed that:

- 34% of the respondents would avoid buying a product or service from a company they perceived as unethical.
- 16% seek information to consider a company's business practices and ethics as part of their decision making.
- 50% intended not to purchase a product or service from a company they considered not to be socially responsible.

Ellen, Webb, and Mohr (2006) found consumer opinions on CSR performance are complex rather than simple—either serving economic or social ends. Customers are more positive when they see CSR as being values driven and strategic. They think badly of companies whose efforts are attributed to egotism or as merely accommodating.

Using social identity theory, Cornwell and Coote (2005) found that if supporters of a non-profit organization know of a mutually beneficial relationship between it and a company that knowledge and the identification with the non-profit positively predict consumer purchase intentions. Barone, Miyazaki, and Taylor (2000)found a positive connection between cause-oriented marketing and consumer relations.

CR also has shown value by reducing costs and gaining support. Carroll (1991) argued that CR pays for itself. Other assumptions relevant to the bottom line are these:

- CR reduces the propensity and rationale activists have to call for excessive and punitive legislation/regulation and the cost of such mistakes.
- CR protects organizations, at least for a while, during a crisis and can reduce various costs, such as litigation and related punitive damages.
- CR increases the likelihood, on the part of non-profits and governmental agencies, that they will get funding they need because they are accomplishing a mission which stakeholders support.

These factors translate into axioms that Carroll (1991) argued serve the good of the company and community: Do good for the community, be ethical, obey the law in spirit as well as principle, and make a profit (or generate revenue regardless of the type of organization) that is put to proper purpose.

Carroll (1991) used these principles to formulate the pyramid of corporate social responsibility:

- the foundation is economic responsibility (to produce an acceptable return on investment)
- legal responsibilities (to act within the legal framework),
- ethical responsibility (to do no harm to its stakeholders and operating environment),
- philanthropic responsibility (or more proactive, strategic behaviors that can benefit the firm and society, or both):

"Be a good corporate citizen; contribute resources to the community; improve quality of life". To accomplish CSR, effective management requires the appropriate balance between an orientation toward owner/shareholders, and employee stakeholders, customer stakeholders, and community stakeholders. To make society "good" becomes a reality only when this high expectation "becomes the aspiration and preoccupation of management".

DEFINING CSR: AN EVOLVING CONCEPT AND STANDARDS

A Variety of Definitions

As is often the case on thought provoking topics, the CSR literature has produced a variety of definitions, each of which features the advantages to be accrued from CSR.

- Kotler and Lee (2005) defined it as "a commitment to improve community well-being through discretionary business practices and contribution of corporate resources". These authors found an increased commitment to give more, to report on the giving, to set high social goals for organizational success, and to use such details to build corporate reputation and brand equity.
- Basu and Palazzo (2008) reasoned that CSR is "the process by which managers within an organization think about and discuss relationships with stakeholders as well as their roles in relation to the common good, along with their behavioral disposition with respect to the fulfillment and achievement of these roles and relationships". CSR is "the continuing commitment by business to behaving ethically and contributing to economic development while improving the quality of life of the work force and their family as well as the community and society at large".
- Mahon and McGowan (1991) adopted common-good principles of CSR: "it is clear that most authors mean corporate social responsibility to include behavior and actions beyond merely profit making that serve to improve the conditions of society and individuals within that society".
- Ihlen (2005) suggested that Bourdieu's (1986) concept of social capital offers a constructive approach to the power elements within a relationship between an organization and those on whom its success or failure depends. Whether instrumental, symbolic, or purely relational, the quality of each relationship rests on whether one entity wants another to continue operation in the current fashion.
- In his entry on Corporate Social Responsibility in the Encyclopedia of Public Relations, Rawlins (2005) proposed that CSR means "doing well by doing good." That means, for instance, that organizations with high CSR become the employers of choice (an excellent place to work), a neighbor of choice (community where it operates is pleased to have it operating there), and vendor of choice (avoiding bad product design and safety issues—giving full value). The real challenge is for organizations to be accountable beyond financial obligations.

Main Themes

However platitudinous various CSR definitions are, at least four realities are ever present in management discussions and strategic planning regarding CSR:

- Every organization operates in a multiple stakeholder arena where each stakeholder is likely to hold different expectations of how it should operate.
- No absolute standards of corporate responsibility exist; they are defined (socially constructed) by each generation.
- Executives are outraged by accusations that they prefer unethical business practices. For this reason, how discussion transpires within an organiza-

tion accounts substantially for its positive or negative impact on helping management to be reflective: take an outside-in approach to their planning, operations, and evaluations.

- Calls for operating in the public interest, or community interest, often requires profoundly complicated analysis that defines this interest as an interlocking set of multidimensional determinants of mutual interest.

Such considerations demand that CSR is more than strategic philanthropy or community relations, such as efforts to sponsor little league teams and engage in goodwill fundraising. It is more than the sort of odious public relations where accommodation and being nice is seen to be more effective than engaging in policy development and implementation that achieves a true community of interest, now ever more global and focused on sustainability.

Expectations of how organizations should perform change over time. "Reshaping the corporate citizenship debate poses challenges to both the advocates and critics of corporate social responsibility. But it is where the debate should occur". Advocacy over standards is likely to occur within as well as outside of each organization. The broadest set of ethical concerns arises from the organization's economic, social, and political interests.

Why is CSR Important?

Answers to this question, partially addressed, depend on how CSR is defined and what obligations that definition hold for an organization's mission and vision, its standards of corporate performance, as fundamental to its planning and operations—and the conceptualization of the societal obligation of the organization to satisfy its franchise. As Basu and Palazzo (2008) concluded, "Such a process view of CSR locates the phenomenon as an intrinsic part of an organization's character (*i.e.*, the way it goes about making sense of its world), with the potential to discriminate it from other organizations that might adopt different types of sense making processes".

One of the strongest CSR roles of public relations is to participate in the social construction of the meaning that defines and evaluates CSR standards by type of organization and during a given era. In contrast to Milton Friedman's narrow and conservative concern, a more reflective answer to the question of CSR's importance can be seen in the metaphor that once offered the challenge: What is good for General Motors is good for America. The contrasting point, and the theme of post-modern discussions of corporate responsibility, is this: What is good for society is good for General Motors.

Framed this way, CSR is important for an organization's success for two primary reasons:

(1) To enhance its reputation as being morally bound to rectitude, a rational discretionary choice bringing in economic benefits, a means for boosting brand equity and sales, and

(2) To advance the organization's credibility and character in public policy battles and during the early stages of a crisis.

Mutual, Aligned Interests and Moral Argument

CSR is the foundation for achieving mutually aligned interests and winning the moral argument about the social relevance of the organization. The modern approach to organizational management was to communicate in ways that shaped markets to the advantage of businesses – an inside-out approach to relationships. A post modern approach suggests that an outside-in approach is more capable of creating and sustaining relationships by achieving truly mutual and aligned interests.

This attention to mutual and aligned interests forms a moral argument for CSR. This argument, similar to the business citizenship perspective, also known as normative stakeholder management, states that organizations consider themselves to be duty bound and deeply embedded in the strength and wholesomeness of community. A large part of corporations' success comes from the values, expectations, and principles of the wider society within which they operate and which franchises them to operate. In a sense, a social contract exists and conforms corporations to society's objectives. Within the corporate social performance (CSP) framework, Carroll (1979) stated that corporate social responsiveness is measured by the degree to which management responds to the social sphere by enacting each of the firm's social responsibilities.

Systematic consideration of CSR can be used to reduce friction and increase harmony with stakeholders and increase the organization's strategic business advantage. To do so requires issue monitoring and critical thinking which are second nature to effective strategic issues management as public relations:

- Ascertain the standards of corporate responsibility held by key stakeholders.
- Compare those standards to those preferred and used by the organization.
- Determine whether differences exist and, if so, whether they strain the relationship.
- Ascertain whether differences in facts account for the disparity in expectations.
- Decide whether value differences constitute the disparity between the organization and its key stakeholders.
- Budget for change options, whether communication strategies, public policy efforts, or redefined strategic business strategies to respond to stakeholder expectations.
- Alter performance or operating standards to lessen the legitimacy gap.
- Take a communication or public policy stance based on correct facts or preferred values when the community interest would be better served.

- Eliminate misunderstanding and disagreement by supplying facts or redefining standards vital to the community interest.
- Incorporate preferred standards of corporate responsibility in strategic business planning, and communicate with key external stakeholders.
- Integrate standards into individual, unit, and corporate performance review, including efforts to achieve total quality management.
- Use improved standards of corporate responsibility to achieve competitive advantage.
- Integrate these standards into product, service, and organizational reputation messages.
- Achieving mutually beneficial interests is not easy in a multiple-stakeholder- environment.

Not all stakeholders see the world in the same way. Interests often conflict. Priorities differ. High standards of corporate responsibility foster aligned relationships and can avoid costly conflict. Good performance is a bottom-line issue.

Cost Reduction and Marketing Advantage (Rational and Economic Arguments)

CSR is important because it is the foundation for reducing cost and gaining marketing advantage.

- Cost reduction (outcome): CSR can reduce cost. Conflict can raise costs, through litigation, legislation, and regulation. These can be dysfunctional means for making decisions of public policy when outrage drives the discussion of stakeholder expectation.
- Marketing advantage (outcome): By achieving appropriate standards of CSR, an organization has a more favorable image and is preferred by customers and other commercial stakeholders. Thus, CSR can increase revenue. CSR is more important when customers can vote their performance expectation with purchase dollars. It has advantage when market conditions prevail so some can achieve market advantage through higher standards. David, Kline, and Dai (2005) offered this challenge: "consumers' knowledge of CSR practices of an organization is a function of corporate communication activities". This knowledge can lead to positive perceptions of the company, as well as purchase intention and behavior. Perception of an organization's willingness to engage in discretionary CSR practices and moral/ ethics predict belief that the organization has strong corporate social values. Using the theory of reasoned action, Werder (2008) reasoned that knowledge of CSR initiatives can foster salient beliefs about the company, but do not predict attitudes toward it or behavior intention. Luo and Bhattacharya (2006) reasoned that CSR initiatives with stated financial, marketing, and communication objectives can enhance company visibility, enhance customer satisfaction, and generate financial

gain. Impact is greater when CSR standards are perceived to be more altruistic than self-interested. As Porter and Kramer (2002) claimed, "The acid test of good corporate philanthropy is whether the desired social change is so beneficial to the company that the organization would pursue the change even if no one ever knew about it".

These two outcomes build a rational or economic argument for CSR. CSR offers a rational argument for managers who seek to maximize their performance by minimizing operational and financial constraints, especially in today's globalizing world, where activist organizations feel empowered to enact change. This view of CSR builds on what is best about the efficiency and rational market argument supporting the private sector. In addition, CSR is an argument of economic self-interest for business because it allows companies to reflect the needs and concerns of their various stakeholder groups and thus gain its societal legitimacy and maximize its financial viability.

Indeed, Clarkson's (1995) ten years of research project on corporate social performance found a clear distinction between stakeholder issues and social issues. Similarly, Freeman (1984) suggested the stakeholder framework argues the firm is not responsible to society at large, but to specific stakeholders. Different from the moral argument, this view is similar to strategic philanthropy, or instrumental stakeholder management. CSR should focus on specific activities and target publics that are most desirable in terms of loyalty, trust, and goodwill, consumer purchase decisions, and the goals of the corporation so that ultimately the shareholder value is increased in ways that affirm the franchise to operate.

Globalization/Free Flow of Information and Changing Trends (Reflectiveness and Stakeholder Expectations)

CSR is the foundation for understanding and meeting the challenges of global stakeholders. In the last half of the 20th Century, activist publics formed to change every aspect of business and government. What tended to start as local concern and networking increased to global proportion. With the Internet, virtually every aspect of business and government around the globe can receive instant and detailed attention 24/7.

For this reason, the stakeholders of any organization play an increasingly important role in its standards of CSR. This stakeholder reciprocity/common good rationale for CSR was introduced by Freeman (1984) for strategic management and effective public affairs. This approach reasoned that key stakeholder expectations were crucial to building harmony with targeted organizations.

Basu and Palazzo (2008) highlighted the stakeholder driven approach to CSR is fundamental to a content approach whereby stakeholder standards are important for an organization to know and implement. Planning and performance combine to achieve those standards. The motivation for achievement rests on the principle that reputation counts; it can bring revenue and reduce costs.

This stakeholder rationale motivates authors such as Fombrun (1996) to merge public policy battles and crisis response. Stakeholder adjustments are important. The foundation of successful CSR rests on outside to in thinking: effective CSR goes beyond the question "what's in it for us" to focus on the higher moral position, "what's in it for them" or even for "us"?

Related to the stakeholder argument, social capital is thought to be a "stockpile of goodwill" that can aid an organization in addressing and recovering from crisis. It can also increase the likelihood that it will be a desirable relational partner in many ways such as a preferred provider of products or services.

Discussing social capital and its specific role in collective action regarding climate change, Adger (2003)reasoned that such topics "inform the nature of adaptive capacity and normative prescriptions of policies of adaptation. Specifically, social capital is increasingly understood within economics to have public and private elements, both of which are based on trust, reputation, and reciprocal action". Such dialogue can be framed as matters of the public good and ability of organizations to make appropriate adaptations to change. As a resource challenge is framed in terms of economics, CSR (as related to social capital) must come to grips with this reality: "...adaptation processes involve the interdependence of agents through their relationships with each other, with the institutions in which they reside, and with the resource based on which they depend". By this analysis, "social capital is an important determinant of human well-being, along with the traditional factors of production and natural capital". As a force in organizational and societal decision making, social capital can be seen as both bonding through shared meaning and governmental intervention and networking through shared interests and expectations. Such reasoning agrees with Bowie's (1991) challenge that a stakeholder CSR features "reciprocal duties".

In similar fashion, Clark (2000) concluded that CSR and public relations "have much in common". Her communication-management approach (CMA) works "to join the most compelling arguments made by CSR and public relations, that is, to use the knowledge of identifying stakeholder groups and a corporation's responsibility to them with the ability to strengthen these relationships through effective communication".

To achieve the fully functioning quality of each relationship, the client organization must have the management procedures and cultural commitment that lead to admiration rather than condemnation. No efforts to foster effective relationships can preclude the need for the organization to be good, as a first step to being a good communicator.

How to Implement CSR

How organizations implement CSR depends on how they define it, whether as a moral obligation and a rational approach to stakeholder satisfaction. It serves best when it is part of organizations' culture, planning, and management. It has implications for budgeting, return on investment, and measures of effectiveness. As mentioned above, public relations practitioners not only participate in the dialogue to define CSR standards but they also play a crucial role in helping markets, audiences, and publics to be aware of the standards client organizations are willing and able to implement.

Plan of Action: Three-factor Model of CSR Implementation and Application to Public Relations

CSR requires a comprehensive approach that, according to Basu and Palazzo, features the classic troika of human nature:

- Cognitive: Matters of identity and legitimacy that define what and how firms think
- Linguistic: Matters of justification, positioning, and transparency that define what firms say
- Conative: Matters of posture, consistency, and commitment that define how firms behave.

Practitioners can participate in the cognitive, linguistic, and conative aspects of their organizations to foster the alignment of mutually beneficial interests in society.

Cognitive: Public relations, through issues monitoring, can play a vital role in helping the organization to know and think about changing CSR standards and the means for achieving them. The reality, however, is that beyond scanning for changing stakeholder expectations and helping to create a matrix of multi-disciplinary players to learn about and analyze such changes, public relations is not expert on many of the matters that are at the core of CSR standards and performance management. Others in management must commit to a strong CSR program. Accountants must recognize, appreciate, and implement higher financial management standards, as must general counsel, engineers, process experts, human resources specialists, nutritionists, environmental impact specialists, to mention only a few of the key disciplines.

Effective public relations and CSR requires every discipline in an organization to understand how an organization can improve, how that improvement enhances stakeholder relationships, and how it can be communicated. Such planning often requires practitioners to convince management that stakeholders are calling for higher engineering standards and processes to achieve employee or product safety, or even more daunting—sustainability. Practitioners may not know what is required in terms of engineering standards or accounting practices. But, practitioners can ascertain that strains occur when key stakeholders expectations are not being met—or when they are being met but the stakeholders do not know that fact.

Linguistic: Exploring the interconnection of communication and management (public relations and CSR),Clark (2000) observed that these two disciplines "have similar objectives: both disciplines are seeking to enhance the quality of the relationship of an organization among key stakeholder groups. Both disciplines recognize that to do so makes good business sense". Making this conclusion, she also noted that "questions as to the chosen message and how it affects the reputation or perception of an organization as responsible remain". Relevant to reputation and issue position are the terms that define a good organization and the socially responsible position on key issues. In such matters, meaning matters.

From this linguistic perspective, public relations can play a leadership role in understanding the terminology or linguistic changes in the communities where each organization operates—outside-in thinking. As the famous language theorist Kenneth Burke (1973) observed, we are interested in co-created meaning. Human experience can never free itself from the terminology operating at a given moment that filters views of physical and social realities (including standards of corporate responsibility). As Burke reasoned, "We must use terministic screens, since we can't say anything without the use of terms; whatever terms we use, they necessarily constitute a corresponding kind of screen; and any such screen necessarily directs the attention to one field rather than another". Looking as he did so well for the chinks in the armor of communication, he warned: "If language is the fundamental instrument of human cooperation, and if there is an 'organic flaw' in the nature of language, we may well expect to find this organic flaw revealing itself through the texture of society".

No useful discussion of CSR can ignore the terminological challenges, and the role of public relations in such efforts. "In any process of institutionalization, meaningfulness is never 'given' but has to be struggled for, has to be secured, even against the resistance of others". Public relations helps to define key terms that become part of the general dialogue which influences how the ideology of and evaluation of CSR performance is conceived in each society in any era. Therefore, practitioners should be especially trained and positioned to understand, appreciate, and respect the development of idioms that are current in private and public sector thinking and decision making.

Conative: Discussing the conative dimension of CSR, Basu and Palazzo (2008) reasoned that it focuses on matters of posture, consistency, and commitment that define how firms tend and prefer to behave. Organizations enact the standards of CSR in all that they say and do. The operational enactment of each organization's standards of CSR occurs by "putting their money where their mind and mouth are." This enactment opens the door for another key role of public relations: To help the organization communicate internally and externally its commitment to various standards and goals and the actions that demonstrate that commitment. Best practices companies believe transparency includes stating their CSR goals and then reporting how well they meet those goals. Such statements can help reduce any legitimacy gap by demonstrating how organizations meet or exceed the expectations of others.

In this way, effective public relations can foster mutually beneficial relationships, which Heath and Coombs (2006) reasoned exhibit the following characteristics:

- Openness and transparency: Letting others see whether the organization has sound CSR principles and whether it meets them
- Trustworthiness: Demonstrating that the organization uses CSR principles to be seen as reliable, non-exploitative, and dependable.
- Cooperation: Enacting collaborative decision making regarding what standards should be met and the measures needed to achieve them
- Alignment: Showing that the organization is responsible, responsive, and able to achieve rectitude through shared interests, rewards, and goals with its stakeholders
- Compatibility of views/opinions: Co-creating (socially constructing) through dialogue the standards and implementation of CSR
- Commitment: Planning and operating in ways that achieve a balance between the interests of the organization and those of the persons whom the organization affects and whose support the organization needs for its success

Although CSR and public relations are not identical they must be interdependent to be effective: Being the good organization as prerequisite for being a good communicator.

In truth, few articles on CSR focus on how it can be communicated. Clark (2000) sought to correct this flaw by showing the parallels between CSR and communication management approaches to public relations which features problem identification, conflict resolution, and relationship quality.

We may add scope to Clark's argument and show public relations to have a dynamic role in the processes by which organizations advance their CSR performance, in pursuit of strengthening the relationships between them and their stakeholders. In this way, public relations can help organizations to make society more fully functioning by solving collective problems rather than merely managing relationships in ways that can accommodate various entities to one another. Having a good relationship does not mean that organizational standards of CSR will be raised. In fact, an antagonistic relationship is more predictive of the kind of change that has led to higher standards of CSR. They have been forged through heated debates (issue communication focusing on fact, value, and policy) by various groups engaged in power resource management, at various times with the good of society genuinely not under consideration but merely the good of a sector of the society, perhaps to the disadvantage of some other sector.

As cases, Enron and other businesses in crisis demonstrate that often the organization with what seems to enjoy the best relationships with stakeholders may not actually—and substantively—be the ones adhering to high CSR standards. The paradox is image driving substance rather than substance driving image.

The quality of such relationships needs to be based on solid principles and the willingness and ability of each organization to meet or exceed stakeholder expectations over the long term. Sometimes, and cases are ample to demonstrate this point, organizations foster positive relationships to seem to be good when those relationships actually mask a crisis in the making. Once the crisis manifests itself, the relationships are not only demonstrated to be based on false assumptions and performance standards, but the damage of falsely built relationships makes crisis response and recovery difficult, and even impossible. A sound foundation of CSR commitment can help an organization recover from crisis. In sharp contrast, if the crisis results from inadequate or fraudulent CSR commitment, it is very likely to be even more damaging.

Black and Hartel (2004) conceptualized a model which they tested empirically to better understand the connections between their concepts and the way that public affairs and CSR can make organizations more effective.

- Public relations can support value attuned behavior and dialogue.
- CSR can define and support ethical business behavior and accountability.
- More socially aware organizations exhibit a higher commitment to and ability to achieve genuine dialogue and engagement with stakeholders,
- Organizations with higher CSR exhibit a caring atmosphere (commitment) in the workplace and higher standards of business ethics and
- "Foster employee beliefs about the value of accountability".

In such matters, organizations under attack may smooth relationships instead of making true change in ways to reduce the honest hostility of critics. Advocacy as a cognitive variable can, and probably should, be seen as a positive communication option since it addresses from many perspectives the key issues upon which standards of corporate responsibility are forged and framed.

Implementable Strategies and Tactics

However hard an organization strives to meet high CSR standards, that effort can be squandered if targeted publics do not know of this accomplishment as evidence of the organization's commitment to CSR. Reputation, crisis response, relationship development and other benefits of public relations depend on how well it can communicate about an organizations' CSR performance.

Fundamentally, public relations practitioners can and must bring discussions of CSR into management decision making to refine and implement mission/vision statements, budgeting, performance standards, and a philosophy or culture of working for a fully functioning society that defines and supports the organization. It must engage in outside-in thinking, issues monitoring, by which it aids other disciplines as well as brings its own expertise to play on knowing the standards expected by various stakeholders. It must engage with other voices to discover, consider, evaluate, and implement facts and opinions that individually and collectively advance the common good. Finally, it must not only help to create and define the culture of CSR excellence internally, but also report the organization's achievement and commitments externally. Finally,

Specific action steps identified by Werther and Chandler (2006) and discussed by other authors offer short-term strategies including top down commitment, the creation of a CSR framework, CSR position statement, CSR ombudsman, CSR audit and report, and awareness creation. After these basic conditions are created, organizations may engage in more long-term strategies such as stakeholder involvement, corporate governance, and managing key messages. Not all of these strategies should or can be managed by public relations practitioners—but they can add value to internal discussions and planning and are essential to external communication.

Short-term Strategies:

- Top down commitment: The integration of CSR into an organization's operational culture should start with top management, preferably the CEO (or the other top organizational leader such as a vice president of corporate responsibility), who should collaboratively establish the necessary components of an effective CSR policy and institutionalize it within the firm—and use the standards for the assessment of individual, unit, and organizational effectiveness. To this end, the leadership must take a stakeholder perspective rather than input/output approach to strategic management.
- CSR framework: Structural support for CSR is needed. CSR should be at the same executive level as other key corporate governance issues. To gain visibility and sponsorship, ideally a CSR officer position should be taken by a company executive and must be backed by the CEO.
- CSR position statement: This statement, with the CEO's active support, should involve the organization's key stakeholders and their perspectives and contain a conflict resolution process that seeks mutually beneficial solutions. It also provides practical guidance by reinforcing the importance of CSR through rewards and sanctions and by specifying how CSR is to be implemented on a day-to-day basis. From a business perspective, Hess, Rogovsky and Dunfee (2004) suggested that the CSR process needs to connect to the company's core values and core competencies, the response to moral pressures, and set clear objectives and means of measurement.
- CSR audit: CSR audit with published results will further awareness among internal and external stakeholders. Using survey methodologies in consultation with stakeholders has been considered an effective method in this procedure. One of the earlier efforts in public relations to facilitate the CSR auditing process was reported by J. Grunig (1979) who used situational theory to explain communication behavior and attitudes of publics arising around issues of CSR. Essentially that study helped managers to select certain social issues for attention and auditing as well as to devise a communication program to inform publics about responsible corporate acts.

- CSR annual report: This report should incorporate triple bottom line thinking: financial, environmental, and social performance. For an "ideal" CSR report, Middlemiss (2004) suggested the following:
- Takes account of the demand (stakeholders) and not of the supply (company).
- Focuses on management systems and doesn't list too many detailed indicators.
- Doesn't consist of more than 50 pages and is complemented by the internet and special publications.
- Reveals its key messages within, at most, 30-minutes reading time.
- Is written in a business like fashion and sparsely illustrated.
- Financial report (report to shareholders or donors/followers): Financial stakeholders appreciate firms' ability to generate profits in ways that demonstrate preferred standards of responsibility. Some mutual funds are designed and managed to feature CSR standards to justify why organizations deserve investment dollars. Such claims are strengthened when they demonstrate how improved ethical standards increase profits, especially by avoiding increased amounts of regulatory constraint, limiting legal liability, increasing market share, or operating efficiently. Sometimes, companies realize that by improving their engineering standards, for instance, they can use more of the feedstock materials they purchase while reducing environmentally damaging wastes and emissions, thereby avoiding unnecessary regulatory constraints. Companies can tout the reduction in raw materials used, improved processes to lessen environmental impact, lowered accident rates, and other practices that mark financial and ethical improvements. They can avoid crises associated with discoveries such as child labor practices, sweatshop conditions, and economic colonialism.
- Online Reports: Printed reports have traditionally been circulated to tout an organization's CSR standards and accomplishments. Many printed documents never get circulated or only reach those who are not particularly interested in such reading. Today, many organizations place such reports, or at least summaries of such reports with the full report available by link, on their home pages. In this venue, organizations tout their standards of CSR with text and even pictures (including video). Rather than languishing on shelves of public relations practitioners, online versions are available 24/7. The interested reader can even follow links to find additional comments by organizations that positively comment on the focal organization's standards and achievements. Links can be used to bridge between a business and its non-profit partners in CSR. It is wise as well to continually report on progress and be willing to combat against unsupported or biased claims to the opposite. This online venue can allow for CSR accomplishments to be attached to related messages such as marketing and organizational history.

- Employee communication: Another means for communicating corporate responsibility lies in the array of vehicles, including executive statements and the intranet, used to reach employees. Organizational culture is vital to the way employees conduct themselves and perform their work. This culture is crucial to the ability of an organization to achieve CSR.
- Advertising and other promotional options: These tools can integrate CSR claims into product and service advertising, as well as arguments by government agencies for funding and by nonprofits for fund-raising. Such claims demonstrate how an organization's actions truly benefit its stakeholders.
- CSR ombudsman: This position serves as a key component of the continuous internal reinforcement necessary for an effective CSR policy.
- Awareness creation: Rewards and measures are key in creating an organizational culture that is sensitive to CSR. Many articles report that key stakeholders are less aware of CSR performance than the sponsoring organization should prefer.

Medium to Long-term Strategies:

- Stakeholder involvement: Here public relations practitioners can, for instance, change the focus from financial reporting and investor relations to include social issues and relationship building with all key publics. Smith (2003) indicated that such stakeholder engagement should be at the core of designing any CSR strategy. Lack of awareness about a firm's obligations to its stakeholders can produce a legitimacy gap. Therefore, stakeholders' engagement is more likely to lead to informed management thinking and decision making. Given the foreseeable difficulties in this process (such as a lack of skill set and openness of the management), diverse perspectives of stakeholders and management, and the unwillingness of certain stakeholder groups to engage in such a dialogue, some firms hire talent from the non-profit sector.
- Corporate governance: Transparency and accountability are the key words. Shareholder activism is increasing and drives reform in the area of corporate law.
- Manage the message: Strategic CSR requires that stakeholder expectations are met in reality and that excessive self-promotion should be avoided otherwise CSR efforts fail to demonstrate the other interest or mutual interest orientation needed.
- Measure social performance: Smith (2003) suggested developing appropriate metrics, essentially benchmarks for measuring social and environmental performance and goal setting.
- Communication strategy guidelines: Framing strategies for communicating about corporate responsibility, Sims (1992) offered the following sage bits of advice:

(a) Be realistic and do not promise what the organization or industry cannot deliver,

(b) encourage organization or industry-wide input into the standards and the best means for accomplishing them (as well as the stumbling blocks),

(c) allow diversity,

(d) allow whistle-blowing,

(e) provide ethics training,

(f) recognize the ambiguity that is inherent in ethical standards and their implementation, and

(g) integrate ethical decision making into employee and operating unit appraisal.

DEVELOPMENT OF A CSR MANAGEMENT SYSTEM FRAMEWORK

CSR Management Framework

Nidec's CSR management system has regulated our business conduct in Japan and Asian/Southeastern Asian countries, where many of our core businesses manage manufacturing operations. The system's principles have been supported by a health, safety and environmental management framework, based on the acquisition and maintenance of international certifications including OHSAS 18001 and ISO 14001. It also focuses on improvements with regard to human rights, labor practices and ethical issues in line with our self-imposed guidelines that incorporate the essence of the Electronic Industry Code of Conduct (EICC Code).

Still and all, the changing face of globalization will continue to present new challenges for businesses across the world. Particularly given the fact that the breadth and depth of corporate social responsibilities required of multinational companies is under a fresh spotlight, we have built a shared understanding that steering closer toward internationally accepted norms for accountable business conduct would be the most viable and responsible option.

In line with this thinking, we have moved on to revamping our CSR management system in collaboration with a major international certification body, taking a new set of measures to ingrain a unified code of conduct into every level of the Nidec Group companies.

The revised CSR management system provides a framework fully compliant with the EICC Code established by the Electronic Industry Citizenship Coalition, whose cross-industry initiatives are starting to gain recognition among manufacturers beyond the realm of the electronics industry. The new management framework incorporates a third-party assessment scheme and monitors whether the company engages in business with sufficient consideration of human rights, labor, ethics, health, safety, and the environment. It also evaluates and ensures the soundness of the management system itself.

The EICC Code, integrating the main elements of OHSAS 18001 and ISO 14001, meets our need for introducing a management system to sites that have yet to obtain these certifications. Constantly updated to reflect the key elements of global challenges, the EICC Code will guide our sustainable CSR management in a manner consistent with the trend of the times. The EICC Code is oriented toward a global standardization of work environments unimpeded by legal and cultural differences between countries and regions, with particular emphasis on human rights, labor practices, and ethics. By aligning our own operations with the provisions of the evolving EICC Code, we seek to raise the quality of our CSR management to a level comparable to international standards in a practical sense.

The transition to the aforementioned new system will be executed based on a three-year plan in accordance with a long-term, gradual action policy. It will exclude sites that have already obtained OHSAS 18001 and ISO 14001 certification and are in virtual compliance with the EICC Code. In our first year (FY2014) under the new CSR management system, a PDCA program, including audit, risk assessment, and training conducted by an outside international certification authority, will be carried out for production sites in Asia/Southeast Asia employing an approximate total of 1,000 employees. With the dissemination of a proper evaluation and improvement cycle based on this program, we intend to apply it to the entire Nidec Group's management system to satisfy high-level CSR requirements.

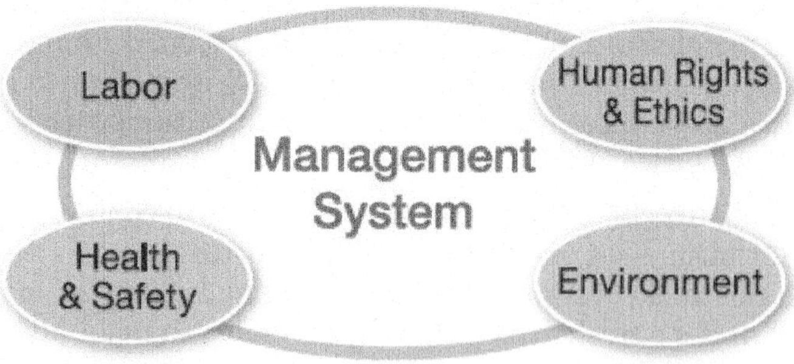

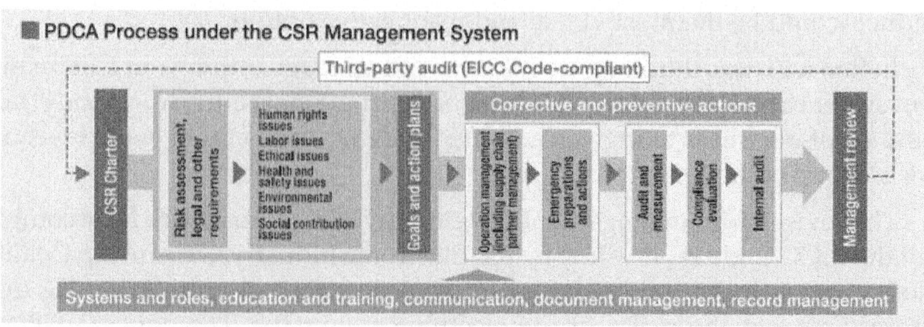

*1. OHSAS 18001: An international standard that defines the requirements for occupational health and safety management systems.

*2. ISO 14001: An international standard that defines the requirements of environmental management.

*3. EICC (Electronic Industry Citizenship Coalition) Code of Conduct: A code of conduct drawn up by the electronic industry for business activities covering the whole supply chain.

CSR Management

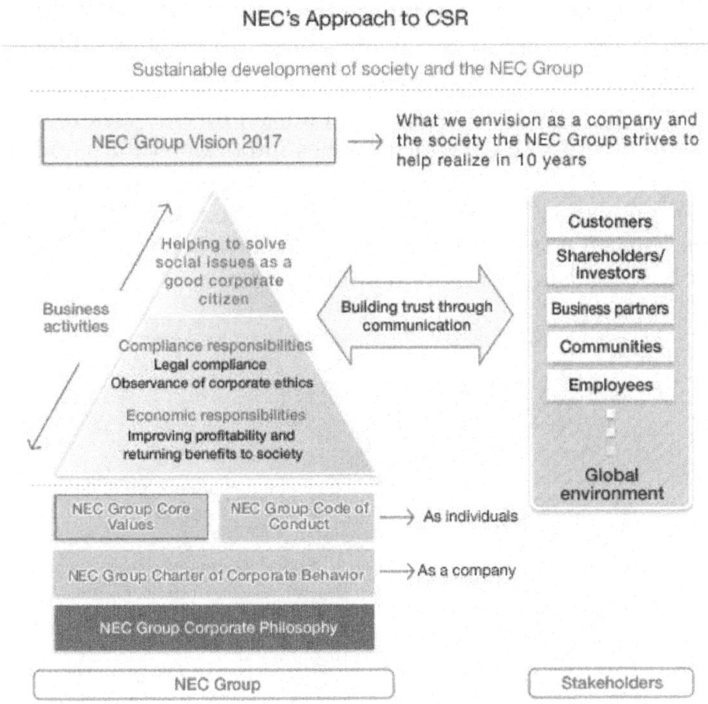

The basic elements of CSR are to fulfill compliance-related responsibilities (ensuring observance of corporate ethics, including legal compliance) and economic responsibility (improving profitability and returning profits to society). Fulfillment of these responsibilities is an essential condition for the continued existence of an enterprise. Recognizing this, NEC believes that the most important social responsibility expected of companies is to support customers and contribute to solving various issues faced by society through business activities. Disclosure of CSR activities and related accomplishments to stakeholders to fulfill the company' s accountability to stakeholders, as well as improving business operations and earn the trust of stakeholders through communication, are also important social responsibilities.

By steadily implementing measures to fulfill these responsibilities on a daily basis, our belief is that we can increase the corporate value of the NEC Group and enable the "sustainable development of the NEC Group and society."

At NEC, we regard management based on this concept as "CSR-driven management."

CSR Management Promotion Platform and The NEC Way

In April 2008, the NEC Group formulated the NEC Group Vision 2017, which represents what we envision as a company, and the society the NEC Group will strive to help realize in 10 years, based on the Corporate Philosophy. As we head toward the year 2017, the 40th anniversary of the declaration of NEC' s original "C&C" (the integration of Computers and Communications) vision, we aim "to be a leading global company leveraging the power of innovation to realize an information society friendly to humans and the earth."

Separately, to realize our Corporate Philosophy and Vision, we formulated the NEC Group Core Values, which defines important values and principles of conduct that every NEC Group employee should embrace. The four values of "Passion for Innovation," "Self-help," "Collaboration" and "Better Products, Better Services," which is our founding principle, have been cultivated and handed down over NEC' s history of more than 100 years.

NEC Corporate Philosophy, Vision and Core Values

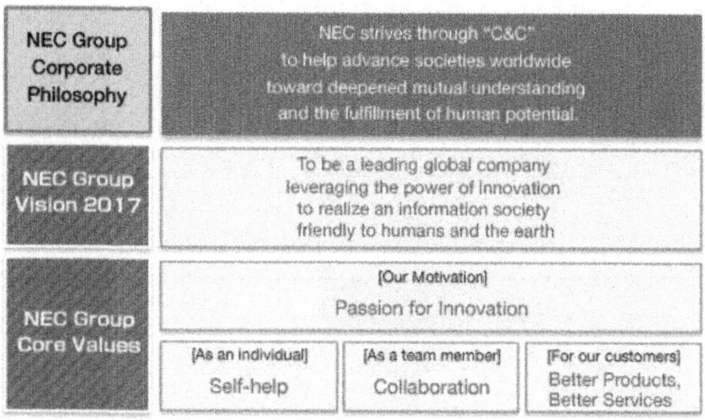

Additionally, we have organized NEC Group business activities into a cohesive framework called The NEC Way. This includes the NEC Group Charter of Corporate Behavior, NEC Group Code of Conduct as well as the NEC Group Corporate Philosophy and NEC Group Core Values. As a platform for promoting CSR management, The NEC Way will be shared and put in practice throughout the NEC Group.

The NEC Way: Framework for NEC Group Management Activities

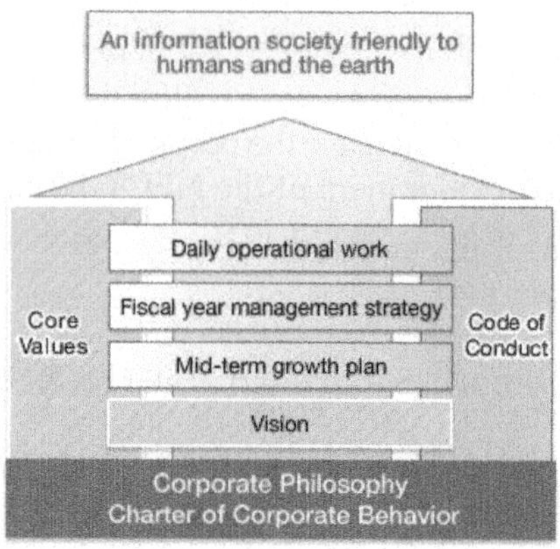

Three Basic Policies for Promoting CSR-driven Management

Based on NEC' s approach to CSR, we established three basic policies in April 2004 for promoting CSR-driven management (partially revised in September 2007).

1. Strengthen risk management and compliance: NEC is strengthening risk management and compliance to ensure that NEC products or services, employee actions or other factors do not result in incidents or wrongdoing that cause problems for customers or other stakeholders.
2. Contribute to solving social issues through business activities: As a good corporate citizen, NEC is taking active steps to help solve social issues through innovation in its business activities.
   - Build Reliable Information and Communications Infrastructure (Next-Generation Networks (NGNs), etc.)
   - Provide Multifaceted Security
   - Include Everyone in the Digital Society
   - Address Climate Change and Environmental Preservation
3. Promote communication with stakeholders: NEC actively discloses everyday activities to stakeholders in publications such as the NEC CSR Report to fulfill its accountability to them by explaining such activities. By incorporating stakeholder feedback into business activities and using it to improve CSR-related initiatives, NEC aims to form relationships built on trust with stakeholders.

FRAMEWORK

CSR Promotion Committee

Chaired by the executive with responsibility for CSR, this committee has as its principal members the president, senior executive vice president, the heads of each Business Unit (BU) and the executive responsible for NEC Corporation corporate staff divisions. Meeting as a rule twice a year, it is a forum for discussions on important matters relating to the promotion of CSR-driven management.

Risk Control and Compliance Committee

This committee has as its principal members the executives responsible for NEC Corporation corporate staff divisions. Meeting as a rule every month, it is a forum for studying and discussing key matters relating to the promotion of risk management and compliance across the NEC Group.

Important matters discussed by both of these committees are reported to the Executive Committee and to the Board of Directors.

CSR Promotion Unit

This unit oversees and coordinates the promotion of CSR across the NEC Group. Its constituent divisions are the CS Promotion Division, the Environmental

Management Division, the Social Contributions Office and the CSR Promotion Office. Along with the Risk Control and Compliance Division and the Financial Reporting Internal Controls Division, this unit constitutes the "CSR function" for NEC Corporation.

*Note: The above divisions in the CSR function promote CSR-driven management within NEC both through their own efforts and through close cooperation with various other divisions, departments and offices responsible for business planning, corporate communications, investor relations, human resources, business support, legal affairs, IT strategy, materials procurement, and other areas.

CSR Promoters

Managers responsible for CSR promotion are in place at all Business Units and NEC Group companies. Within each organization, CSR Promoters play a central role in developing and managing CSR-related systems and mechanisms.

Organizational Framework for Promoting CSR-driven Management

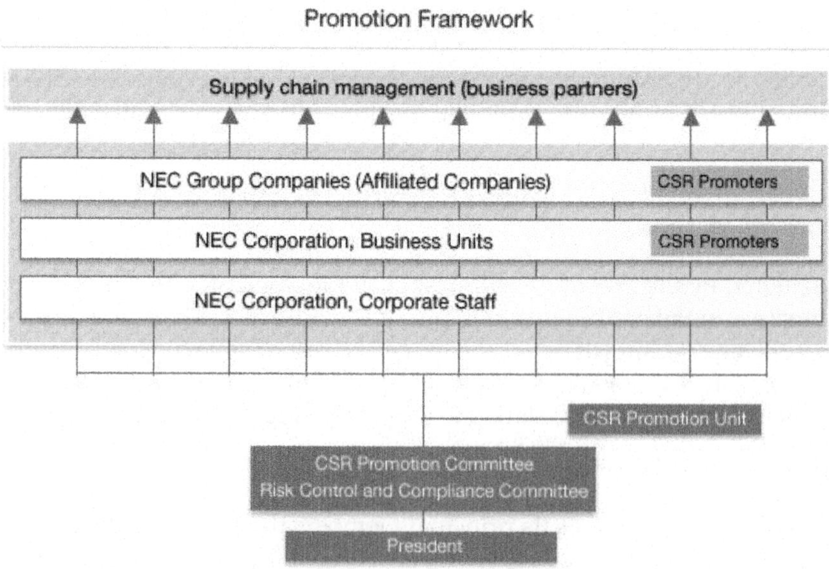

MANAGEMENT KNOWLEDGE AND COMMITMENT (SOFT FACTOR)

Barriers for an Efficient Management of Knowledge

After giving a working definition based on his organisation's understanding of Knowledge Management (KM), the author outlines his 'prospected' method to process KM within his organisation. Based on first-hand experience, in this case study main barriers to implementing the 'prospected' KM process are presented, including some ideas and ways to deal with selected barriers. As the author's experience derives from an intercultural context, this chapter addresses issues to consider when working in such an environment. As findings, main cultural challenges are identified as: different influences of hierarchies; language barriers even within a single organisation, differences in individuals' skills, priority of ad-hoc activities and different accessibility and usage of technology.

Aim and Assumptions

This chapter aims to deliver an overview of barriers and selected attempts to elude these barriers to establishing and maintaining knowledge management (KM). This paper is a descriptive and inductive case or field study-approach intended to provide building blocks for more effective KM implementation.

Main research questions are: "What hinders knowledge management especially in an intercultural context in southern Africa?" and "What are experiences in trying to overcome these hindering barriers?"

As a German in southern Africa, the author implicitly uncovers 'cultural differences' but also uncovers barriers concerning technology, content, processes and routines, some of them interlinked to 'cultural differences'.

Anywhere, each employee and employer has a unique identity, formed by education, family, experiences and other external factors. Thus, the term 'cultural differences' can be used for differences between individuals - no matter how similar the persons under consideration seem to be - as well as differences between groups of people.

Comparing a whole country's culture with another - especially considering southern Africa's ethnical heterogeneity - would only result in invalid generalizations. Additionally, as an organisation consists of different departments, sub-organisations and sub-systems, they are never homogenous entities to begin with. Examples given in this chapter neither stand for a whole organisation nor a country's singular 'culture' - taking into account the nuances of both. Some of the experiences and examples quoted here will be very specific, and/or derive from subjective experiences and thus must never be reduced to overarching generalisations. Instead, it seems appropriate to speak of a higher or lower probability of barriers to emerge; or of different characteristics of barriers.

Organisation's Definition and Aim of KM

Different models and schemes describe knowledge management within an organisation. Within the presented organisation, an understanding of knowledge management was slowly created, agreed upon and broken down into the following components:

- **People**: as providing the information and looking for and receiving pieces of content

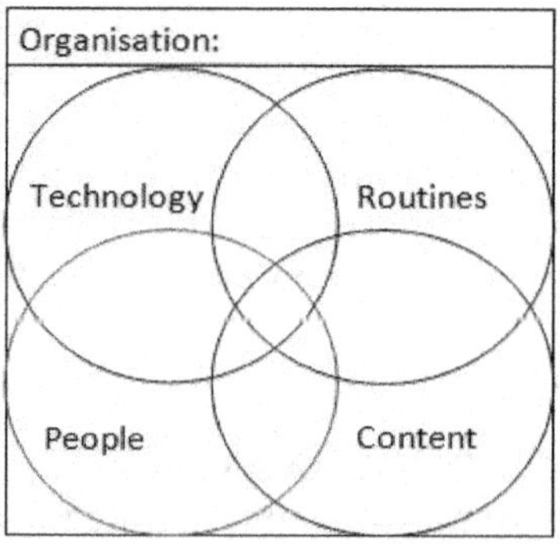

Fig. : Scheme of KM that was agreed upon in the organisation.

- **Content**: the 'real' pieces that carry information that can generate knowledge
- **Routines** and procedures: secure the ways to provide, collect, forward and access existing and new information
- **Technology**: tools to create, exchange, store and make available these pieces that carry information
- In the **Organisation** people, content, technology and routines are co-existent.

The more formalized **task of knowledge management** - at least decided so in the organisation at hand - is **to lead and execute activities** to support and enhance single components and sub-components. It is also to support and enhance the connectedness of people, content, technology and routines within the organisation and to take (at least co-) ownership of the process. Looking at Diagram 1, KM should make interfaces permeable and identify and support intersections.

An organisation's definition of knowledge management is not always understood the same throughout the organisation. This agreed upon definition of KM is therefore important - but is only the first step towards a transparent and knowledge-sharing organisation. This definition will need clarifications and adjustments again and again as new needs become visible.

Processing Knowledge Management: From Inventory to Intervention

As till then nobody in the organization has been explicitly responsible for knowledge management it seemed appropriate to firstly get an overview on what is available in the organisation. Thus the prospected first activities of the advisor were described as leading a way through the following:

This primary step, according to Luckhardt (no year), "… comprises of classification, selection, acquisition, indexing and storage of knowledge resources" , describing the status quo in the organisation. Next, this inventory needs to be analyzed. It would be best to compare this status quo with a prospected situation in order to identify bottlenecks and fields for intervention and amelioration.

However, this ideal situation would require a viable description of a prospected situation - which seldom exists, at least not in detail.

1. Collection / Overview of Knowledge Inventory: The knowledge inventory should list and connect all necessary information about the above mentioned: people, routines and procedures, content and technology. Thus, the knowledge inventory is a meta-information centre. Collecting and summarizing this knowledge inventory already is a critical first step where barriers will be encountered.
2. Expert's Analysis: The expert knowledge manager can be used to identify the first signs/avenues for enhance-ments - just from analysing what is given in knowledge inventory. Considering that the suc-cessful implementation of KM can only be achieved when all players are properly involved in the process, the expert's external analysis is only an initial step in defining KM activities.
3. Participatory Analysis: Having the personnel aboard and giving them the space to reflect on their own situation, their own input and their own needs provides very valuable hints. In most cases, participation will strengthen the process and the chances for a change. Participatory processes also help external advisors to understand how the organisation 'functions' from within.

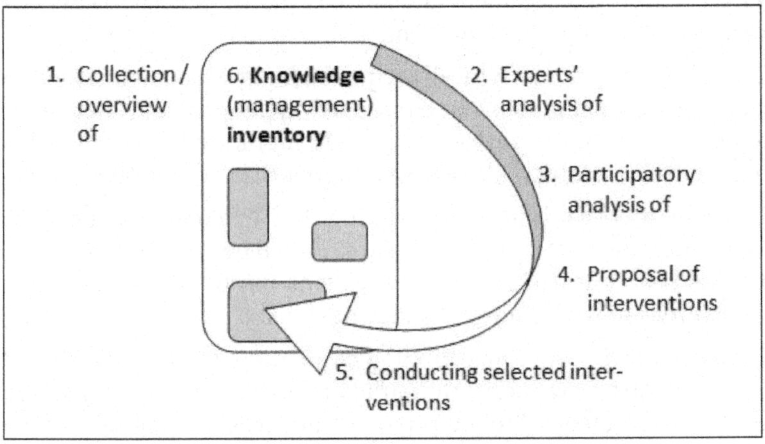

Fig. : Scheme of a Formalized Knowledge Management.

4. Proposal of Interventions: As a next step, personnel involved should work on creating ways to improve knowledge management in the future. Summarizing ideas that have been developed in a participatory manner, and proposing alternatives to resolve bottlenecks and realize enhancements, is one of the main tasks for a knowledge manager. These alternatives may consist of various ap-proaches, like implementing new routines, collecting new information, using new technology, etc.

5. Conducting Selected Interventions: After – in best case: participatory – prioritisation and decision on activities on how to enhance the management of knowledge, these activities should be implemented – thus creating a change in the inventory.
6. Knowledge Management System: The formalized process of updating technologies, routines, organisational structures and personal skills would then be called 'Knowledge Management System'.

Reality Check: Barriers for Knowledge Management

The kind of 'ideal' process – which may appear rather mechanistic and perhaps even naïve. In reality, working with people is never like a control loop that entails simply scrutinizing problem areas and then re-adjusting these for change.

As mentioned, this chapter should primarily be seen as a southern African case study, the examples mentioned below have a higher likelihood to be relevant in southern Africa, but could also help to avoid surprising revelations elsewhere.

Barriers in Technology

High-end and elegant software solutions could make life easier in many regards. Software for data interchange, archiving, information sharing, communication, work flow management and so on could be quick and easy solutions to restructuring knowledge management.

Technological solutions typically require a budget, however. And this can easily become a giant constraint. Even if free software is available, there is often a lack of hardware; lack of bandwidth and lack of IT literacy when it comes to handling the software, thus making costs rise even higher than comparable 'pay'-software.

Moreover, an organisation can be caught in a technological trap, caused by a long gone deci-sion for special software. Reversing this old decision once the software has been implemented could become impossible – due to financial or reputational reasons or because of a lack of skills.

In some countries, special technologies are more ubiquitous than others. *e.g.*, most people in Africa are more familiar to and used to work with mobile phones compared to computers: "Mobile phone is the African computer. If we want to train health workers in Africa we can't ignore the mobile phone". According to Greenwood, Louise (2009) and own experience even money transfers work *via* mobile phone, Chigona (2007) describe extensively used mobile chat applications.

Barriers in Content

To collect content for the knowledge inventory can be hard work – no matter where in the world you are. Transforming implicit knowledge into explicit information is an activity in which special skills and often creativity are needed. Some communication and information proc-esses are very difficult to describe. Few possible content *e.g.* a description on how to gain special and restricted information, could even be illegal or against organisational rules. Other examples could include unauthorized informal meetings or exchange *via* software that is not authorised within the organisation (*e.g.* Instant Messengers).

Another barrier is linked to individual skills: delivered content may simply not be understood. Maps only make sense if people know how to read them - which is not always the case in southern Africa. To work with digital or analogue audio files as well as video files, people must have the experience or know-how on using a suitable media player.

Barriers in Routines and Procedures

Some processes and procedures that are only claimed to exist, *e.g.* regular departmental meetings. In addition, work plans or strategy and progress papers may only represent ideal situations that have no link to reality. Some routines may not be recognized as routines by employees - such as everyday joint coffee breaks among staff. Some processes may work in certain cases but not in others or never again - which makes them unreliable. For example: you need to ask a particular person for access to the staff library but if this person does not respond to your request you are stuck without the info you require.

Some work is carried out without planning - which may lead to inventing the wheel again and again. This could be due to a culture of 'last minute' or 'hands on' crisis management. Thus strategic and planned work frequently has a low priority, while variable ad-hoc processes become 'routine', and 'quick and dirty' becomes being 'business as usual'.

Barriers in Organisation

A knowledge-sharing culture in an organisation that is badly role-modelled by those highest in the organisation's hierarchy can hinder knowledge management. High ranking staff may consider themselves to be more important than others - manifesting in not sharing informa-tion. Hierarchically-structured organisations appear to be the standard in southern Africa. Here, staff is deeply influenced by line managers' behaviour. Thus, if high ranking members of the organisation are unreliable or don't follow up on activities or do not care, middle and low ranking staff will not either.

Also, "structures are multi-layered, polyvalent, and often contradictory ... (and) ... maybe invisible even to those who inhabit them." For instance, the head of organisation could explicitly be telling everybody to use and support knowledge management - but may not apply this instruction to his- or herself. This would strengthen a culture of 'saying but not taking it serious' or 'not practising what you preach' - which hinders all processes, including knowledge management. This culture of 'unreliability' within an organisation makes it difficult to cooperate and succeed.

The cooperation ideal - when it comes to planning and decision making and wanting person-nel's ideas to be included in the decision making - can lead to a deadlock. This can be due to the fact that staff members do not understand what the decision is all about and are afraid of admitting this (selective) ignorance. As a response personnel could then choose the strategy of delay. Another reason for refusing to take action on anything or to be decisive can be the uncertainty of the line manager's wish.

Personnel fluctuation seems to play a huge role in NGOs especially in southern Africa. In 20 months from about 100 colleagues 34 quit.

Organisational survival - especially in the face of budget and funding restrictions –is often the main focus, thus strategic issues like 'knowledge management' do not have high priority.

From a system theory point of view, any organisation aiming for autopoiesis, recreates itself again and again, and even wants to avoid changes - including changes that concern the sharing of knowledge. The persistency of an organisation also hinders hiring innovative - and thus more likely KM-friendly - personnel.

Interesting would even be to consider information as a currency. Personnel 'pay' with bits and pieces of information - but they will not give it for free.

Barriers in Personnel

The human factor is a key factor. Many of the above mentioned issues are connected to individual behaviour.

A first challenge is that personnel might not have any idea or understanding of what 'knowledge management' is all about; there is no or at least no matching definition of KM in the personnel's mind.

Providing and sharing information can be hindered by a lack of motivation: employees do not receive or do not understand the surplus that comes along with cooperating on knowledge management. All they see is that they have to give information to others - which, from the individual point of view, does not make sense at all, because keeping information secret and unshared can help to secure a job.

Some display of information is meaningless for personnel, *e.g.* even highly qualified staff do not know how to read a map. Messages may not be properly understood –as prime example, in South Africa there are 11 official languages, English being a second language for the majority of the population. This easily leads to chaotic communication and incorrect transmission of information.

Personnel - even those of high rank - might have difficulties dealing with knowledge-sharing technology - or technology in general. And some of those personnel will not admit to this lack of skills.

When it comes to working on an online questionnaire, for example, some staff do not differentiate between phrasing aspects, how different questions are formulated, and techno-logical aspects, how the submission of the answers will work electronically. Thus, people responsible for technologies feel the pressure of working on enhancement of content and phrasing.

Non-cooperative attitudes of line management and colleagues lead to stagnation, resignation and avoidance of the active search for information.

The priority of focusing on-time problems makes it difficult to focus on strategic activities like knowledge management. Personnel often do not seem to have the time for knowledge management procedures during day-to-day work.

As well as each organisation, also each individual – no matter whether operative or strategic staff – has different interests and hidden agendas which could be in opposition to a transparent knowledge management system. People are afraid to lose their job; people are afraid of giving 'secret' information; people do not trust each other. Personnel or co-workers easily suspect a hidden agenda on the part of the knowledge manager – even if this does not exist. This suspicion could lead to reservation and non-cooperation.

The knowledge manager him- or herself may even have a hidden agenda – like carrier planning – that could hinder the proper enhancement in knowledge sharing. He or she might prefer to have a visible output of his / her activities, instead of 'only' influencing the organisation's knowledge sharing culture, which cannot be captured in statistics. Ferguson states: "In 'development discourse', the fact that there are no statistics available is no excuse for not presenting statistics, and even made up numbers are better than non at all."

When proposing procedures for knowledge management, different co-workers could support different parts of the procedure, those being parts they can personally benefit from. Different, but also necessary parts of processes and routines are jeopardized by the same person. *E.g.* cooperating when it comes to creating an online archive for research articles but not providing the own collection of research articles.

From a more political point of view, aiming to gain power could be a main motivator for members of the organisation. Thus, the sharing or not-sharing of information is a sub-function in the quest for power. Sharing information can only be motivated by win-situations for the provider of information.

HANDLING BARRIERS IN INTERCULTURAL CONTEXTS

As a manager or advisor in KM in intercultural contexts – but also in 'normal' contexts – different strategies to deal with barriers will come up. Find here selected examples plus a proposal considering the hierarchical focus.

Punctual Attempts and Failures and Successes

Identifying barriers isn't that easy as it seems, sometimes co-workers would deny that there is a barrier – this must be an indicator to watch out. Reacting to barriers in most cases will not necessarily solve the problem but lead to the next barrier. Thus being prepared to reflect and to learn will always be necessary, for there will never be an easy pre-describable circumvention for barriers. Here selected experiences from southern Africa.

Approaching Barriers in Technology

When it comes to technology, KM is to address usability aspects ('Keep it simple') and the need of a proper implementation plan – including learning sessions.

The introduction of a new technology is often expected to solve all problems. When there is no budget for this technology people could reckon that all problems are due to this financial restriction. Proper installation and configuration of software and the proper usage often are not seen as a (cost) factor when it comes to deciding on which software to acquire. There is hardly an understanding of what an implementation plan serves for. Cooperating in creating an implementation plan thus ends in being left alone and trying to deal with the acquired software and doing the configuration all alone – after struggling to be allowed to.

The financial barrier I tried to face by proposing to start with free software – which often can be less complex when it comes to usage. However, proposing cheap or free software is no easy task, especially when there is something like a 'must-have' and already established standard. *e.g.* there is no chance to change to Open Office.

To reinforce transparency and internal communication an intranet was wished for. As there was a decision on the usage of an eMail server application the search for the intranet software was shortened: the eMail application incorporated a sites-functionality. Thus no-body was against the advisor's proposal to use this available application with no additional costs. The usage of the very same application for archiving and file sharing could not (yet) be established organization-wide – maybe due to the missing proper pre-implementation plan or maybe due to a technology trap: different other (free) hosting/archiving/file-collaboration sites were already used in the organisation. The usage of the calendar within the same appli-cation is slowly increasing – because a main player of the organisation is using it and he forces personnel to look into his calendar when it comes to making appointments with him.

Proper usage of existing software is very heterogeneously across personnel, average staff hardly knows how to apply quite 'usual' features of standard software like MS Word and MS Excel. Thus teaching lessons for *e.g.* how to create a diagram, how to do a mail merge, how to format a word document seemed a reasonable strategy. Only some persons were willing to take part in such one-by-one teaching lessons. But the ones that took part – at least three of them – now get along by themselves.

Approaching Barriers in Content

Exemplarily look at the issue of 'directions'. As European I was surprised to find no exact map with all main sites that are supported by the organisation, nor a map to the venues where 'youth training lessons' take place. Trying to get the detailed addresses did not work – be-cause hardly anybody is into this 'address business' or proper addresses simply do not exist. People do ask for directions, where to turn to the left, where to turn to the right.

Additionally, after having created pretty exact maps, those were of no use because often people are not used to read maps. At least we now can generate directions from maps and thus service those who are not used to read maps.

The usage and making public of audio files or video files were to be enhanced, as audio files are produced at least every second week for the national radio stations. It proofed to take more than one year for me to get a hold of some of these audio files, and another half year to establish a routine to have those uploaded by the producer himself to be used as podcast *via* rss feed.

Approaching Barriers in Routines and Procedures

One task of Knowledge Management is to support, simplify and formalize routines and procedures. When it comes to firstly collecting an inventory or at least get an overview of existing routines and procedures it happened that some routines were nicely described and claimed to be carried out in regular basis. But the descriptions only existed on paper, in reality the procedures merely end up to be carried out ad-hoc and without caring about a described process. Consequently a basic idea was to try to establish departmental or/and personal 'to-do-lists' as a fundamental first step to have working plans. In one department I supported this list was never touched by anybody but me.

The attempt to generate procedures for special circumstances, *e.g.* What to do when a new staff member joins, was more successful, policies and requirement lists and sign-off needs were formalized on paper and these documents were made public in the intranet.

At moment there is an idea to formalize departmental meetings, to have an extendable standard schedule and a standard format to make minutes of these meetings. However, the frequency of departmental meetings taking place largely differs from department to department.

Approaching Barriers in Organisation

In southern Africa – as well as possibly elsewhere in the world – hierarchy is a mayor structuring category in organisations. Line managers, directors, executive officers do mainly influence the carrying out of work of all personnel. Their role model is needed to make steps forward to exchange information and knowledge, to create transparency.

Even after several agreements to do so, until today the minutes of the directors' meetings are not in the intranet. The only way to receive strategic information about the organisation often is informal chats – and thus they are a main KM tool. Enhancements on the formal site come from the announcements section of the intranet which is now used more and more often.

Establishing anonymous quick polls as incentives for personnel to access the intranet proved to be quite successful. At the moment we have 85% of overall personnel across the country visiting the intranet at least once a month. Between 40 and 70 poll answers are registered weekly. The establishment of an anonymous suggestion box on the intranet also proofed to make sense, personnel now can shout out their discontent. In 13 months there were 78 entries.

To face the 'ad-hoc culture' within the organisation KM tries to get the phrase 'Implementation plan' into the mind of the decision makers. KM also tries to role model reliability, create an aura of trust and support and thus tries to multiply the willingness to share info and enhance transparency.

KM should help to deal with personal fluctuation. A formalized exit interview form was established to get at least the reasons for quitting and use those answers for mitigating further quitting. A proper leave and handover process only takes place if the line-manager handles this. When asked and incorporated, KM gives advice on archiving documents of the person that quits.

Approaching Barriers in Personnel

Lack of skills to use technology can be faced by training sessions and the establishment of technology which is usable. Lack of understanding what KM (and an advisor) is all about can be faced by discussion rounds and presentations. Lack of willingness to share proper information can be faced by creating incentives. Lack of strategic thinking can be faced by repeated confrontation with the need to incorporate strategic thinking.

All these approaches can be performed, but results will always be uncertain. Even the incentive to learn and enhance one's own skills is not automatically working. But it can make sense if the KM persons keep on supporting wherever possible and thus make visible that additional skills can enhance one's own job situation.

Different hierarchical levels could make use of different enhancements in skills. It seems that management skills and personnel management could make perfect sense in some cases. KM here approaches where possible with advice on formalizing and structuring routines like *e.g.* departmental meetings and usage of departmental calendars and mailing lists.

The ad-hoc focus can be addressed by repeatedly in advance emphasizing upcoming important events like annual reports or implementation of additional software. KM here should – when incorporated into this process – assist in preparing these upcoming events.

Hidden agendas seem to be one main barrier for proper participation in KM. Especially in intercultural context the strategy of directly addressing these hidden agendas – when recognized as such – must cautiously be used, especially persons in a higher hierarchy are not used to be addressed openly. One alternative is to 'unhide' one's own agenda and thus start creating an aura of trust.

Hierarchy Focus Approaches

Here, two alternative approaches for KM from a more hierarchical focused perspective are presented - keeping in mind that hierarchy plays an important role in southern African countries. Maybe the third alternative as a kind of synthesis can do better.

Top Down - Leadership Gives Strong Mandate

Pressure from top management to cooperate with KM could be an adequate measure to enforce intra-organisational cooperation. However, this pressure needs to lead to ownership of (selected) personnel in KM processes. In reality, personnel will always find ways to foil commands. Economically speaking, for some personnel the expected added value from strictly obeying the orders is lower that the expected value from not strictly obeying the order.

Opposing targets could be: securing one's own power; securing one's own work space; minimizing the own amount of work; and maximising influence.

Top-down approaches could even increase defensive demeanours of employees that are suspicious of things and processes they are not familiar with. Top-down approaches in particular need authentic role modelling from top management to ensure the credibility of knowledge management.

Bottom Up - Building Trust

On the other hand, the person responsible for knowledge management could build trust, relationships and identify a network - even if it is informally - of interested persons. Those persons then are not forced to cooperate but are interested to cooperate - such self-motivation and ownership makes cooperation much easier, no matter where.

The co-workers personal value added could be the 'special know how' they receive as a cooperating employee. As 'first movers' those persons would receive special support by the advising knowledge manager, and would acquire additional skills in dealing with technology and procedures. And those persons - together with the responsible person for knowledge management - can be seen as role models and motivation for all staff.

However, the gap between verbally expressed agreement and real action can also be bigger than expected in a bottom-up scenario. And it may take time to identify those who are interested in cooperating.

Viable Mix

Knowledge management and the persons responsible for it need some kind of mandate from the top levels of the organisation, at least to start processes and cooperation. A clarification of assignment between top management and advisor KM will make sense. As the human factor and motivation are key, there additionally is a need to build trust within the operational ranks. Within the process of enhancing knowledge management, an important factor is acknowledgement for the work of all participants.

'Processing knowledge management: from Inventory to Intervention', a very first step (Step Zero) needs to be added, even before the collection and over-view of knowledge inventory: Building trust; receiving a kind of mandate; explanation to personnel.

In reality, all successive steps of the process need to be repetitive and from time to time new 'pieces of information' or processes have to be added to the inventory.

Top down and bottom up need one another. In some situations it is crucial to have the CEO's mandate. In other situations, the commitment of the employees is necessary. The process of establishing effective knowledge management needs reliable participation from top level as well as the operational level. If the establishment of knowledge management is to go beyond punctual enhancements, it needs participative and sustainable cooperation that includes all levels.

Dealing with decision makers who are not familiar to KM requires patience and diplomacy. A viable route might be that a decision is prepared by the expert. Personnel must still be incorporated, however - after making clear the pros or chosen alternatives by the expert. If avail-able, including the Line Manager in the cooperation of personnel will also help.

In the Bertelsmann paper it is stated that interacting in an intercultural environment should always be effective and appropriate. From my point of view, under some circumstances it might also be effective to sometimes act inappropriate:

If the knowledge manager or advisor has an 'external' position, he or she can make use of this externality. There is not always a need to adapt all habits and routines that are already established within an organisation. As an external, he or she can act more freely, and can even have the freedom to run activities that would normally be considered as faux pas- from operational or from top level management. This can help to raise awareness on the issue of knowledge management and might come closer to achieve the set goals.

Globalized Knowledge and Outlook

Knowledge is ubiquitous - this is what we are told in these days of internationalization and globalization. The author remains dubious about this sweeping generalization that leaves little room for contextual factors.

Bandwidth in African countries may improve; internet access prices are decreasing; and mobile phones as receivers could increasingly be available - but as long as income, especially in rural areas, remains as it is, no person and no small organisation will be financially able to access this global knowledge.

The digital divide is not only infrastructural, it is also a financial, skill and educational divide. IT literacy is not in high gear all over the globe, especially not in rural Africa.

Assuming global knowledge is available and accessible, it will often only be of minor use in local organisations. Knowledge about knowledge sharing technologies, about transparency, and about proper knowledge management may make perfect sense in some parts of the world it may be impractical or inapplicable other parts of the world. For instance, elegant and perfect software solutions are a non-practical idea in a rural African organisation; video-conferencing does not work when nobody knows how to establish a connection, whereas SMS/text conferences that work in southern African could be viewed as ridiculous in European contexts. In the global knowledge inventory - the internet might be defined as - solutions for rural organisations' problems are (still) rare.

In many cases, relevant knowledge is not explicit. Organisations often do not have the chance to make this information explicit either, *e.g.* due to a lack of individual language skills, a lack of verbal or drawing skills, or a lack of skills to describe in proper structure. And making information explicit can also be hindered by ethnical, moral, cultural or legal constraints or labour law provisions. In specific circumstances, personnel chose to keep silent in order not to compromise themselves or others or disrupt the status quo.

Even if English as a lingua franca is spoken all over the globe, there will always be additional language barriers in particular contexts.

Outcomes for Knowledge Management

When cooperating can be a challenge in international and intercultural contexts, advising on knowledge management can be a challenge as well. How to support managing knowledge if the knowledge manager could be excluded – by different barriers – from this knowledge?

Thus, not each and every KM activity may result in visible or direct outputs. The output may be implicit and not measurable, *e.g.* a rise of the awareness on knowledge management. The knowledge manager may not even realize how many changes he or she has triggered.

The performed activities of KM in the organisation under consideration sometimes ended up to be isolate. However steps forwards were performed. For the involved co-workers the acquainted knowledge will be sustainable, for the organisation the sustainability of the per-formed KM activities can be achieved if the activities will keep on being executed regularly, even if staff is changing, like *e.g.* updates in the intranet.

Patience, personal relations and networking within the organisation seems to be the first step for successful advising on KM – and not just in intercultural contexts.

Additionally a knowledge manager in an intercultural context especially needs to keep in mind:

- The language barriers hinder a common understanding and a proper communication, even within the organisation,

- o Creating an inventory can be a good starting point but the KM process should be flexible enough to quickly focus on - sometimes even ad-hoc - identified bottlenecks,
- o Formalizing processes won't always work, informal processes could play an even bigger part in southern Africa compared to Europe - often these informal processes work fine.

Never ending study area can be the importance of informal processes. Studying the importance of enhancement of skills for operational and for higher level personnel could make sense - to be able to strengthen further development cooperation activities. Comparing different case studies helps to falsify wrongly made assumptions from singular observations.

And always to be kept in mind: people, organisations and societies have survived for centuries without western knowledge management. Changes cannot - and should not - be established within a short time frame. Cultural identities won't change simply because people from outside tell what they think is better. Even external advisors won't avoid to learn from working in an intercultural environment, thus - next to experiencing possible conflicts - also emending "... the prospects of business and private environments that are labelled by cultural diversity."

Knowledge Management: The Essence of the Competitive Edge

Currently, no universally accepted definition of knowledge management exists, but there are some basic concepts to be explored, and considered. Simply put, knowledge management undertakes to identify what is in essence a human asset buried in the minds and hard drives of individuals working in an organization. Knowledge management also requires a system that will allow the creation of new knowledge, a dissemination system that will reach every employee, with the ability to package knowledge as value-added in products, services and systems. However, knowledge management goes far beyond the storage and manipulation of data, or even of information. There is a need for a more structured approach to transferring knowledge to decision-makers before it is needed and enabling access to information in a timely fashion.

To remain competitive in the future, organizations will need to abandon their ideas of information hoarding and embrace knowledge sharing. Competitive success will be based less on how strategically physical and financial resources are allocated, and more on how strategically intellectual capital is managed — from capturing, coding and disseminating information, to acquiring new competencies through training and development, and to re-engineering business processes. In view of these trends, and recognizing that knowledge has great potential value and because there is a corresponding failure to fully exploit it, some corporations have embarked on comprehensive knowledge management programs.

Above all, companies, governments, and large corporations often do not know what they know. Knowledge management attempts to prevent problems such as lost knowledge or lost income by establishing human and technological networks capable of harnessing a company's collective expertise and experience. The move from an industrially-based economy to a knowledge or information-based one demands a top-notch knowledge management system to secure a competitive edge and a capacity for learning. An organization's ability to quickly tap into wisdom gives it a competitive edge in the marketplace. As a result, knowledge is displacing capital, natural resources, and labor as the basic economic resource. Governments know this all too well.

If an organization wants to begin managing knowledge as a resource, what should it should it take into consideration? The organization needs to design and install procedures and processes to create, protect, and use known knowledge, while articulating the purpose and nature of managing knowledge as a resource and embodying it in other initiatives and programs. In addition, anyone considering a foray into the knowledge management arena needs to plan and build environments that will allow employees to feel safe to discover and release their own tacit knowledge. Knowledge management systems administer the skills and competencies that lie within an organization, and allow them to blossom by freeing people to be the best that they can be.

Through allowing deep personal reflection about one's own sense of purpose and deepest aspirations, and through cultivating individual and collective responsibility for the future, goes a long way to create an organization that has the creation of knowledge as an ongoing goal. When people bring their 'grand will' instead of their 'unfree will' to authentic inquiry into the questions of where are we going and what are we here for it can breathe life into an organization. It is the key to the future of strategy in an independent world.

Two Kinds of Knowledge

There are two kinds of knowledge: tacit, which is hard to articulate, versus explicit knowledge, which can be expressed in words and numbers and can be easily communicated and shared in hard form, as scientific formulas, codified procedures, or universal principles. Tacit or unarticulated knowledge is more personal, experiential, context specific, and hard to formalize. This kind of knowledge is difficult to communicate or share with others, and is generally inside the heads of individuals and teams. Since knowledge may be an organization's only sustainable competitive advantage, it is very important to capture tacit knowledge. How can we transfer it?

By putting mechanisms of socialization, mentorships, apprenticeships, and opportunities for face-to-face communication in place. Intranets and email can help it flow through an organization, but tacit knowledge often moves laterally through informal channels of communication, through communities of practice. For example, those groups that hang around the coffee machine are exchanging knowledge, just as are the smokers who huddle near the entrance to the building at break time. The information that is passed in this way is very important because it is useful for helping people to get their work done more effectively, in part, because nobody is willing to question or think about the process very much. Communities of practice must have their place in a comprehensive knowledge management effort.

Knowledge is intangible, dynamic, and difficult to measure, but without it no organization can survive; flows of knowledge are an organization's capacity to learn. They are all you really have.

The Knowledge Economy

Currently, governments around the world, multinational corporations, and a multitude of companies are interested, even concerned about the concept of knowledge management. The shift from a local and national economy to a transnational one is changing the way all organizations are doing business. Organizations and governments can no longer rely purely upon national approaches to maintain their profitable growth. More and more, companies and industries of all types must globalize in order to maximize their profits.

Indeed, even individual Canadian provinces have an interest in understanding the flow of knowledge within their confines, trying to become more aware of the structures that exist within their hierarchies and in the folds of their bureaucracies. There has been a shift from bureaucracies to networks because traditional hierarchical designs that served the industrial era are no longer flexible enough to harness an organization's full intellectual capability. For example, Prince Edward Island has agreed to become the first world test site for a new Knowledge Assessment Methodology (KAM) developed by the U.S. National Research Council in Washington, and coordinated by the Institute of Island Studies. What the tool does is assess Prince Edward Island's ability to compete in a knowledge economy. The new source of wealth is knowledge, and not labor, land, or financial capital. A knowledge economy requires real change and has a built-in necessity for less information and more knowledge. Thomas Stewart, Board of Editors of *Fortune* magazine has the following to say:

1991 was the crossover year when capital spending by U.S. companies was greater on telecommunications, copying and computer equipment than on industrial, construction, mining, and farming equipment. We tend to think of the mid-1990s as the transition point, but we actually made this shift, from a macroeconomic perspective, around 1991.

The knowledge economy differs from the traditional economy in several key respects: the economics is not of scarcity, but rather of abundance, and unlike most resources that deplete when used, information and knowledge can be shared, and actually grow through application. In addition, the effect of location is diminished. By using appropriate technology and methods, virtual marketplaces and virtual organizations can be created offering benefits such as speed and agility, and an a-round-the-clock global operation.

Do You Really Need KM?

Only if your department wants to stop constantly reengineering and downsizing: talented people are assets to be developed for a global 21st century. Partly as a reaction to downsizing, some organizations are now trying to use technology to capture the knowledge residing in the minds of their employees so it can be easily shared across the enterprise. No longer should companies have to worry that employees will walk out the door with valuable knowledge that it no longer has access to. Although many individuals may come and go, their learning is embedded for future use. Knowledge when locked into systems or processes has higher inherent value than when it can 'walk out of the door' in people's heads. Leveraging organizational knowledge is emerging as the solution to an increasingly fragmented and globally dispersed workplace. For many of these companies, knowledge management has become the next silver bullet. In effect, knowledge management is the successor to the reengineering and downsizing efforts that marked the last decade.

If you are interested in the knowledge grid, you probably need a knowledge management system. Knowledge cultures are those in which formal attention is paid to what some academics have called the "knowledge grid." It has four categories: "what we know we know; what we know we don't know; what we don't know we know; and what we don't know we don't know." If you understand that reuse of knowledge saves work, reduces communication costs, and allows a company to take on more projects, then knowledge management is for your organization.

What is needed is a similar mindset about the collection of intellectual assets: it belongs to everyone employed in a particular place, and it has to be shared. This kind of policy saves money right across the board and allows something just as valuable as money to be saved: TIME. Saving time means more work can be generated because you do not have to reinvent the wheel every time a new project is taken on. Competitive success will be based on how strategically intellectual capital is managed. Increasingly, senior executives are recognizing that knowledge and learning represent the preeminent source of sustainable advantage in a fast-moving, highly competitive world. They know it is no longer enough to leave critical knowledge sitting passively in the minds of individual employees in a period of radical change.

Capturing the knowledge residing in the minds of employees so that it can be easily shared across the enterprise. Workforce mobility, falling educational standards, and the rapid rate of business change mean that individuals can no longer be relied upon to provide consistent, comprehensive insight. Instead, the knowledge trapped within the employee base must be leveraged to the organizational level, where it can be accessed, synthesized, augmented, and deployed for the benefit of all. Organizations and individuals must learn rapidly and uniformly across different functions and levels of the organization. The best reason for an organization to develop a knowledge management system is to gain a competitive advantage in the marketplace: by turning intellectual assets into value through innovation. The real differentiator for those leading companies is knowing how to use innovation to create value and ongoing growth.

Does a KM System Need a Chief Knowledge Officer?

Only if your organization is serious about implementing a knowledge management program. Economic realities and competitive edge factors play a large role. If your organization or department is serious about implementing a knowledge management program, then appointing a CKO is one way of initiating, and coordinating such a strategy. In this post-industrial knowledge economy, no one can forget that knowledge is the most important resource; forgetting could be fatal. Knowledge, its creation, uses, and innovation is the only real competitive edge when governments are moving from industrial-based economies to knowledge-intensive ones. The best scenario would be that, like total quality management, knowledge management would become embedded in organizations, with knowledge a source of value. Of course, in the long-term the goal will see all members of the organization owning and driving the knowledge management system, and over time, seeing the need for CKOs decline.

However, today's CKOs are discovering that knowledge management comprises a large agenda and that making substantial progress takes time. In the beginning, a leader or coordinator is imperative for initiating, keeping up the momentum, deciding on software and systems (with the IT staff), distilling, codifying, and helping others learn to share their unique or tacit knowledge so that an organization can raise its knowledge capabilities. Therefore, appointing a CKO is likely the best place to start when embarking on a knowledge management program, and even if an appointee does not have the title of Chief Knowledge Officer. As long as your CKO has the requisite profile, performs activities that propel a knowledge management project forward, has access to the proper resources, and possesses success factors that work.

What's the Strategy?

There are two very different knowledge management strategies. In some companies, the strategy centers on computer hardware, and knowledge is carefully codified and stored in databases, so that it can be easily accessed by anyone in the organization. This is called the codification strategy. In other organizations, knowledge is closely tied to the person who developed it and is shared mainly through direct person-to-person contacts, as in gatekeepers passing information to mentees.

Emphasizing the wrong strategy or trying to pursue both at the same time can quickly undermine a business. Companies, such as Anderson Consulting, that follow a codification strategy rely on employees reusing information it has already created. By contrast, the personalization strategy relies on harnessing the tacit knowledge of experts in the organization. Companies that use knowledge effectively pursue one strategy predominantly and use the second strategy to support the first. Think of this as the 80-20 split: 80% of their knowledge sharing follows one strategy, 20% the other. It is tempting to think that the two knowledge management models can coexist in different business units within one corporation. Indeed, they can coexist, but only in corporations where business units operate like stand-alone companies. In a company like General Motors, where the car divisions have little to do with the credit and finance divisions, different models can in fact work in each business unit. Companies with tightly integrated business units, however, should either focus on only one of the strategies or spin off units that do not fit the mold. The issue is sometimes complicated by two additional concerns: the existence of multiple business units and the commoditization of knowledge over time.

Knowledge Management Types

1. Competency management is one of the fastest growing areas of knowledge management. It focuses on displaying information in sophisticated ways to understand labor trends and compensation. Telecommunications giant L.M. Ericsson in Sweden has already embraced the idea of tracking skills and competencies throughout its organization. The company has used knowledge management to post consistent performance gains throughout the 1990s.
2. Another type of KM that is attracting attention is knowledge sharing. A growing number of firms use Intranets and online forums to spread knowledge. Glance into the inner workings of Buckman Laboratories and you can get an idea of how knowledge can flow in every direction and how each person can become a willing participant in the program.
3. Competitive knowledge management is yet another area of KM that blends competency management and knowledge sharing. Since Arthur Andersen's Atlanta-based business consulting division established an Intranet two years ago, consultants at the firm post knowledge workplans, methodologies, research, proposals and resumes, so that others in the organization can tap into high-level expertise on an as-needed basis. It's called: not reinventing the wheel logic.

Knowledge Management Enablers

Organizational requirements for leveraging intellectual capital require that attention be paid to what have been recognized as Knowledge Enablers: structures and attributes that must be in place for a successful knowledge management program. Knowledge management programs need enabling factors if there is to be any real success. The following knowledge management enablers should be considered:

Learning culture leadership is absolutely critical to liberate the creativity of teams, and yet not have chaos. In a sustainable knowledge managed learning organization, leadership is often expressed as the self-confidence to navigate the unknown waters of the future.

Access to knowledge and making sure it is available provides value. The challenges are to facilitate access to the right content at the right time and place, be content rich and navigation lean, manage regulations and copyrights, and provide flexibility and ease of access.

Technological infrastructure that can link people enterprise-wide, support collaboration, and allow people in the organization to access all global resources. Information technology is most effective when it converts the tacit knowledge of an individual into explicit knowledge that is then accessible by all. Companies must think through their technological systems to ensure they have the capabilities needed for the 21st century because technology such as Intranets and advanced collaborative software have made knowledge management possible. Technology provides the tools to help the employees and the organization to develop and apply its management capital. Arthur Andersen's Michael Stone explains it this way:

> *Technology allows people to collect, find, filter and distribute information far more rapidly than ever before. It is now possible to move large volumes of information quickly, and institutionalize what has always been an informal and haphazard process.*

A learning culture embeds a learning attitude and places value on competency development. Historically, more training expenditures have gone to developing cognitive skills than to developing motivated creativity. Those days are gone now. Currently, the goal is to see intelligent people using innovation to create knowledge out of information. To do so, there needs to be a corresponding culture that values organizational behaviors and supports an environment of trust and collaboration. Changing the culture is imperative. To create a climate in which employees volunteer their creativity and expertise, managers need to look beyond the traditional tools at their disposal. They need to find ways to build trust and develop fair process. That means getting the gatekeepers to facilitate the flow of information rather than hoard it, and to offer rewards and incentives. Buckman Laboratories' Koskiniemi reports:

> *Successful knowledge sharing is ninety percent cultural, five percent tools and five percent magic. All the technology and tools in the world won't make you a knowledge-based organization if you do not establish a culture that believes in sharing. To encourage knowledge sharing, some companies are creating central knowledge repositories.*

Workplace culture is paramount because although management may pay lip service to the value of cooperation and its ability to facilitate organizational knowledge, sometimes that is not what they are thinking. There is an underlying belief that performance is really driven by "corporate stars" and that internal competition must be in place to attract these superstars since only under-performers can be found in cooperative, team-oriented workplaces.

Rewards and incentives signal what behaviors and outcomes are most valued by management. It should not be surprising that knowledge accumulation and sharing are not valued. Management sends strong signals through its compensation policies: different roles are perceived of value according to their allocated compensation. So be careful about sending mixed signals, and keep in mind that culture is more than just compensation, and it is responsive to influences other than paychecks. Management sends signals about what is important through its recruiting priorities, promotions, and, possibly more than anything, through its own behavior. These deeply embedded cultural assumptions are significant.

Knowledge management programs need all enabling factors, and not just one or two, if there is to be any real success. Knowledge tools, controls, and new organizational structures will go far to create a new culture in which informed decision making is valued, but explicit efforts to cultivate that culture are still needed. Knowledge and skills can never substitute for the motivation only an effective organizational culture can provide. And that is why it is important to have the knowledge process facilitated by a steady development of a knowledge culture, based on incentives, and strong management leadership that values, shares, and uses knowledge.

Expect Organizational Changes

Not surprisingly, knowledge management blurs the line between departments and operating divisions. Managers traditionally have kept a tight leash on their own department's data, and they have frequently interpreted it in narrow, rigid ways. Knowledge management collapses these boundaries, and allows people to use information across applications. An interesting representation of what is going on in business comes in the form of a quote from Paul Saffo at the Institute for the Future:

> *It is hardly news that the corporation as we know it is headed for the scrap heap of business history. Internal corporate structures are already mutating beyond recognition. Corporate boundaries are dissolving into commercial irrelevance as businesses explore entirely new modes of association and interaction.*

Dissolution of boundaries is an important goal because a knowledge management system cannot work through hierarchies. This means that while knowledge management does not have to be driven from the top, it does very much have to originate from the top. In order to justify financial investments in technology, CKOs must understand organization structure, motivation of people, and cross-boundary processes. Since it is a bottom-up and peer-to-peer tool that seeds an organization with intellectual capital, it is foreign ground for most organizations.

And do not isolate knowledge management. Some CEOs have put knowledge management at the top of their agendas. Others have not given it the same attention as they have given cost cutting, restructuring, or international expansion. In companies where that is the case, knowledge management takes place - if at all - in functional departments such as HR or IT. But companies that isolate knowledge management risk losing its benefits, which are highest when it is coordinated with HR, IT, and competitive strategy. While corporate directives to share information and create cross-functional teams can improve knowledge management efforts, real innovation and self-directed learning happen only at the individual and team level. Individual and team learning processes must become the true driver of organizational learning.

Why KM? What's the Big Deal?

There are so many reasons for instituting a knowledge management-intensive learning organization. Here are just a few:

1. There is an increase in employee satisfaction due to greater personal development and empowerment.
2. It helps you keep your employees longer and thereby, reduces the loss of intellectual capital from people leaving the company.
3. Saves money by not reinventing the wheel for each new project.
4. Reduces costs by decreasing and achieving economies of scale in obtaining information from external providers.
5. Increases productivity by making knowledge available more quickly and easily.
6. Provides workers with a more democratic place to work by allowing everyone access to knowledge.
7. You can learn faster with KM.
8. Learning faster means staying competitive.
9. KM software and technological infrastructures allow for global access to an organization's knowledge, at a keystroke.
10. An organization does not have to reinvent the wheel each time a new project comes along, and that saves time.

Knowledge Management and Future Planning

Scenario thinking can help us to see the blind spots, and help to create the future we want. One of the major problems with governments, corporations, companies, organizations, and private citizens is that they have no concept of the future and never think about its ramifications. Future studies must be figured into an organization's overall knowledge management system because to sustain a commitment over the course of months and years, people need to have awareness of the whole and understand the direction in which an organization is going. The challenge of organizational strategy and purpose is to revitalize and rethink the organization's business focus, and figure out where it is heading. To expect ongoing knowledge creation, it must have some relevance to the future you are creating. Therefore, a future element must be ever-present. Knowledge Management must somehow be connected to future studies for at least one significant reason and that is because to have a knowledge management system, it presupposes the ongoing creation of new knowledge.

Where are we going? What are we here for? People need awareness of the whole, and are truly interested in what direction the organization is taking. Gone are the days when employees could be expected to stay in one place of work forever. As well, levels of education have surely risen and good employees no longer have to stay where they find the organizational and leadership structure lacking. Leveraging organizational knowledge is emerging as the solution to an increasingly fragmented and globally-dispersed workplace. Organizations that do provide a goal to reach in the future can provide great incentive for a KM initiative. Effective leveraging lies within an organization's capacity for rethinking and recreating. One thing for certain is that when questions, big questions (questions that matter to the future of an organization) are asked and explored, it can be a very powerful force because it is essential for co-evolving the futures we want, rather than the futures we get.

Sustainability of a Knowledge Management Endeavor

To achieve sustainability, there must be a focus on learning, and learning how to harness the learning capabilities that lead to innovation. There are three fundamental processes that sustain profound changes such as the introduction of a comprehensive knowledge management system, and they are to improve business results; to enhance personal results; and to develop networks of committed people who like where they work and what they are doing.

For a knowledge management effort to work, it must be understood that no progress is sustainable unless innovators learn to understand why the system is pushing back, and how their own attitudes and perceptions contribute to the "pushback." Until they see this, there will be no development of systematic strategies for sustaining profound change.

The enhancement of personal results is often the first source of reinforcing energy for sustaining deep change. And, it is inherently satisfying to work in a network of committed people. Given the choice, very few people would not want to become part of a network where there is excitement, perseverance, innovation, experimentation, and commitment. As Dr. W. Edwards Deming used to say, "People seek joy in work." In this day of bottom line focus, when employees often assume that personal needs are subservient to the organization's needs, it is truly liberating to discover that the two can be aligned rather than in opposition. For those who have gone from a run-of-the-mill organization to one, which literally supports learning and knowledge for everyone, feel transformed. They learn to understand just how much they are worth, and what they can give to the workplace. They become intellectually-stimulated, energized, and emotionally-engaged.

For significant change to lead to sustainability, hierarchical control must be put aside. The emergence and development of informal networks must be supported so that people can share their tacit knowledge and help one another. Communities of practice can be geared so that people continually learn to help each other. The diffusion of innovative practices needs informal networks through which new ideas can spread in and across organizations. Such informal networks are almost always superior to hierarchical channels for spreading new innovations because these informal networks already exist; experimenting with new ideas requires help and time to think in a safe context; and because there is a strong element of credibility with people whom we know. Without active internal communities of practice, an organization cannot expect profound and sustained change.

Managers need to surrender control in order for learning capabilities to be enhanced. This will lead to greater business results by eliminating wasteful practices, and bring about the development of new business practices that lead to better results. And mental models need to be examined. The limits that block organizational learning initiatives have a great deal to do with the mental models embedded in the culture of an organization. For example: hoarders, hierarchies, and stars are the order of the day! For organizations to attain sustainability for their KM efforts, they must be prepared to develop new mental models for thinking about their workplaces and new ways to reach across internal and external boundaries.

Mental Models must be managed because they do prevent new and powerful insights and organizational practices from becoming implemented. The process begins with self- reflection, unearthing deeply held belief structures and generalizations, and understanding how they dramatically influence the way we operate in our own lives. Until there is realization and a focus on openness, real change can never be implemented. This is hard work, particularly because the pull of sustainable development as an aspiration may bring diverse groups of people together in collaborations that require breaking down old boundaries and hierarchies. However, once employees are forced to let go of the existing conventional wisdom of the organization, they can be innovative and creative, designing new approaches and learning to learn faster.

STRATEGIC PLANNING – THE CHOICE OF SUSTAINABLE STRATEGIC PILLARS

Two Cultural Pillars in Dow Chemical's Sustainability Strategy

Dow Chemical has been on a clear sustainability journey since the early 90s. However, in 2006 they set sustainability goals for year 2015that became game-changers for the entire company. While all of the goals are impressive, two in particular stand out. Dow committed to the daring goal of achieving at least three breakthroughs that will significantly help solve world challenges in areas such as energy and climate change, water, food, housing and health. Secondly, they vowed to contribute to community success and to achieve individual community acceptance ratings at 100% of the sites where they have a significant presence. All of the seven goals are bold. To ensure success in achieving them, Dow must make certain that their organizational culture will enable their efforts. Two interrelated cultural pillars emerge as essential: Collaboration and Trust.

Collaboration

Increasingly the global landscape is a web of relationships and associations. Corporations live in this integrated space along with many and varied stakeholders. The boundaries between who is 'inside' and who is 'outside' of any given organization have blurred. Dow understands this model of an open, collaborative community characterized by people and organizations coming together to create shared value. It is all about collaboration, according to Mark Weick, Director, Sustainability Programs and Enterprise Risk Management for Dow. "For years we have prided ourselves on being a safety and science-based culture. Now we are more focused on the human side of chemistry. Science is important, but we won't succeed unless people believe in the science." They understand that that the most important element isn't on the periodic table; rather, it is the human element.

Stakeholder engagement is not new for Dow. For example, Dow's Sustainability External Advisory Council (SEAC) has brought diverse outside-in perspectives on environment, health and safety, and sustainability issues for the company since its formation in 1992. The SEAC has had a major influence on Dow's approach to climate change, and has played a critical role in developing the 2015 Sustainability Goals. The Council is currently composed of thought leaders from around the world and is chaired by a senior Dow executive.

When the SEAC challenged them to 'think bigger' and put themselves in the shoes of a seventh-generation child, a porpoise, and the poorest woman on the planet in crafting their goals, Dow stepped up to the challenge. They asked themselves what they would have to do to earn the trust of the 'person on the street'. "We realized that we aren't just a science-based company. We are a company that looks for solutions. And the world needs big solutions," contends Weick.

Recently Dow started a 5-year collaboration with The Nature Conservancy (TNC) to help them incorporate nature into their business goals, decisions and strategies. In announcing the partnership, CEO, Andrew Liveris stated, "This collaboration is designed to help us initiate new approaches to critical world challenges while demonstrating that environmental conservation is not just good for nature – it is good for business." This collaboration will deepen the significance of relationship-building in the interest of sustainability and sound business practices in the Dow corporate culture.

The timing of this partnership also coincided with the launch of the International Year of Chemistry (IYC), designated by the United Nations as a worldwide celebration of the achievements and future of chemistry to take place around the globe throughout 2011. Fostering and celebrating international and cross-industry collaboration to solve complex global issues is one of the basic tenets of IYC, and this is a perfect example of collaboration in action.

Trust

Of course, any long-term collaborative relationship is built on trust. Companies that support transparent interactions with stakeholders can achieve the kind of trust which builds cooperation, loyalty and mutual respect. Trust is rooted in beliefs that the parties in relationship are competent, open, concerned and reliable. Recently Accenture released a white paper entitled Trust: Managing the Scarcest Commodity of All. The report claims that "Nothing is more fundamental to a healthy business than the belief by consumers, the public, investors, regulators, suppliers and stakeholders that they are doing business with a trustworthy organization.

And, Dow 'gets it'. Dow emphasizes trust-based collaborations as critical to their 'social license to operate' – a phrase that refers to stakeholders' acceptance and on-going approval of a company's operations in their communities. Weick stated that many companies still don't get it. "If they have not yet been confronted with resistance from stakeholders, they may still underestimate the significance of relationships and trust to their ability to succeed as a company." To reach their goal of achieving individual community acceptance ratings for 100% of their significant sites, Dow will continue to drive towards a triple bottom line, balancing the company, society and the environment.

The culture that Dow is cultivating is one in which 'sustainable' becomes an adjective, not an initiative. For example, they refer to sustainable operations, sustainable chemistry, and sustainable science. They are taking actions to ensure that protecting the planet- one of their stated values- is built into all of their processes.

Embracing the hunt for 'the big solutions' as well as committing to the creation of stronger, safer and sustainable communities are significant steps in Dow's decades-long sustainability journey. To succeed, Dow leaders realize that they must attend tenaciously to the organizational culture that they wish to create. As with any transformational change, the direction is clear but the destination is murky and the path to reaching it is not fully charted. "Some people feel paralyzed by this lack of clarity," Mark Weick said. "It is like the gazelle in the zoo. They stand immobile in front of a three-foot high wall that separates them from their food even though they could easily step over it. If they can't see where their feet will land, they won't take the step." Leaders must enable employees to step over their own proverbial walls by sending the right signals.

"Culture is fragile," according to Weick. "Change requires constant vigilance and nurturing. Leaders can send the wrong signals more easily than the right ones. And it takes time for people to wrap their minds around how to attack these world-changing goals." Nevertheless, it seems that the hard work is paying off. Their fourth quarter 2010 Sustainability Goals Update indicates that they are on track for reaching their 2015 targets.

Is it worth the effort? Mark Weick returns to his opening comment in response to this question. "We are a company that finds solutions and the world needs big solutions!"

Chapter 5

SUSTAINABILITY ASSESSMENT METHODS AND TOOLS

One of the key challenges of sustainable development is that it demands new and innovative choices and ways of thinking. Innovations in technology are challenging organizations to make new choices in their operations, products, and activities that impact the earth and people as well as economics. There is, however, no standard method for measuring the triple bottom line success of technological innovations at the design phase and the principles to achieve sustainability by themselves are insufficient to create the right framework for design towards sustainability. It would be useful to have a screening tool to assess how a new product or process under development would perform in terms of sustainability, or compared with a commercial process. Although there are various international efforts to measure sustainability, only a few of them have an integral approach taking into account environmental, economic, and social aspects. In most cases, the focus is on one of the three aspects. For example, Life Cycle Assessment (LCA) is used to evaluate the environmental performance of products, but it concentrates on environmental impacts only. As well, environmental impact assessment (EIA), a procedural tool for the design phase, only evaluates the environmental implications of decisions.

In order to fully evaluate the sustainability of new process routes, there is a need for a comprehensive evaluation of the environmental, economic, and social impacts of these new routes at an early process design stage. The chapter suggests using multicriteria assessment for sustainability assessment and demonstrates its use in assessing a novel carbon dioxide-based reaction route to dimethyl carbonate (DMC).

SUSTAINABILITY ASSESSMENT METHODOLOGIES

There are a number of sustainability assessment methodologies evaluating the performance of industrial facilities. The World Business Council for Sustainable

Development, the Global Reporting Initiative, and development of standards are key drivers for adopting sustainability management in industries.

The most extensive work in terms of sustainability assessment has been done by the Global Reporting Initiative (GRI). GRI is a nongovernmental organization that aims at driving sustainability and has developed an environmental, social, and governance (ESG) reporting framework to be used worldwide. GRI version 4 on Sustainability Reporting Guidelines defines the principles and indicators that organizations can use to measure and report their economic, environmental, and social performance. Many companies use these indicators while publishing their annual or environmental reports. GRI is committed to continuously improve and increase the use of the guidelines which are available to the public.

The American Institute of Chemical Engineers (AIChE) has defined the AIChE Sustainability Index (SI) to measure the sustainability performance of representative companies in chemical industry. The AIChE SI uses publicly available data on the companies' strategic commitment, sustainability innovation, environmental performance, safety performance, product stewardship, social responsibility, and value chain management to measure their sustainability performance. Metrics to measure the "greenness" of the companies' chemistry have been developed by the American Institute of Chemical Engineers' Center for Waste Reduction Technologies (AIChE/CWRT) assessing material intensity, energy intensity, water consumption, toxic release, and pollutant effects. The metrics developed are simple, understandable, easy to reproduce, and comparable. They take into notice also the social aspects of sustainability by considering the health effects the chemicals used/produced have. However, they are developed for companies and are adjusted for existing process improvements rather than for a new process design.

Similarly, the Institution of Chemical Engineers (IChemE) has developed a set of metrics to enable process industry companies to measure and report progress along the path of sustainable development. The Sustainable Development Progress Metrics are intended to help companies to set targets and develop internal standards and to monitor their progress in time. The IChemE metrics are divided into environmental, economic, and social indicators. The environmental indicators are concentrating on resource use by considering how much energy, material, and water are consumed and land is used. Also atmospheric, aquatic impacts, and impacts on land caused by emissions, effluents, and waste are taken into notice. The economic indicators are concentrating on the profit gained, value added and taxes paid, and investments made by the company. The social indicators are considering the employment situation, health and safety at work, and also impacts to society. Not all the metrics proposed are valid in every case and it is up to the companies to decide which of the metrics are relevant for them. Key indicators have to be chosen from each of the aspects of sustainability to give a balanced view of the sustainability performance. Whilst the IChemE metrics account for all three aspects of sustainability, they are meant as a sustainability management tool for companies, aiming at enhancing their sustainability performance, and are not suitable for assessing processes under development.

In terms of sustainability guidance for chemicals and chemical process design, Green Chemistry was developed to reduce or eliminate negative environmental impacts. The 12 Principles of Green Chemistry have been a cornerstone of environmentally conscious chemical process design since the late 1990s. Green Chemistry had been suggested to be used as a pollution prevention tool as it applies innovative scientific solutions to real-world environmental situations. However, the assessment range of Green Chemistry does not cover the full depth of sustainability. As it was its original purpose, its emphasis is on reducing the toxicity of chemical products and driving inherently safer chemistry.

Protection of human health and the environment from chemicals and associated risks is also the goal of the European REACH (The Registration, Evaluation, Authorization and Restriction of Chemicals) regulation, which came into force in 2007. It renewed and upgraded the previous chemicals regulatory framework of the European Union (EU), in order to ensure that there is free circulation of substances on the internal market and to enhance competitiveness and innovation. REACH confirms that industries are responsible for both assessing and managing the risks associated with chemicals, giving suitable safety information of chemicals to users, and promoting alternative testing methods. About 143 000 chemicals marketed in the EU were preregistered by the December 1, 2008, deadline in REACH. The registration document of chemicals under REACH includes general information, safety data sheets (SDS), chemical safety report (CSR), and chemical safety assessment (CSA). Testing for health hazards under REACH includes acute toxicity, skin corrosion and irritation, serious eye damage and irritation, skin or respiratory sensitizer effect, mutagenic or carcinogenic impacts, toxicity for reproduction, specific target organ toxin in single exposure, specific target organ toxin in repeated exposure, and aspiration hazard.

Table below summarizes some of the main evaluation guidelines or indicators used in the mentioned assessment processes. All methods outlined in Table below take into notice some key aspects of sustainability and clearly intend to evaluate triple bottom line success. Indicators that are possible to assess in the design phase and would give a good signal of sustainability performance are bolded. In terms of sustainability assessment of chemical processes in the design phase, Green Chemistry is the most thorough; however, should it be used to assess sustainability, it is recommended to extend it with social and economic indicators of GRI, AIChE, and IChemE to give a comprehensive measure of sustainability.

Table : Main principles/evaluation guidelines of reviewed assessment methods.

	GRI	AIChE	IChemE	Green Chemistry
Environmental performance	Materials Energy Water Biodiversity Emissions Effluents and waste Products and services Compliance Transport Suppliers	(i) Resource use (a) Energy (b) Materials (c) Renewables (d) Water GHG emissions Waste, wastewater Compliance management Value chain management	Resource usage (i) Energy (ii) Material (iii) Water (iv) Land Emissions, effluents, and waste	Prevent waste Use renewable feedstock Avoid chemical derivatives Catalysts Product degradability
Economic performance	Economic performance Market presence Procurement practices	(i) Sustainability innovation (ii) Strategic commitment to sustainability	Profit, value, tax (i) Investments	Maximise atom economy Increase energy efficiency
Social performance	Labor practices (i) Employment (ii) Health and safety (iii) Innovation and knowledge potential (iv) Diversity and equality society (i) Acceptability and social dialogue Human rights	Social responsibility (i) Stakeholder partnership (ii) Social investment (iii) Image in the community Product stewardship (i) assurance system (ii) risk communication (iii) legal proceedings	Workplace (i) Employment (ii) Health and safety society	Less hazardous chemical syntheses Safer chemicals, products, solvents, and reactions Accident prevention and real time analysis

Following the recommendation of IChemE, we selected key indicators from each of the aspects of sustainability to give a balanced view of the sustainability performance. The suggested indicators are as follows: environmental indicators: (a) feedstock renewability, energy intensity, waste generation, CO_2 balance, economic indicators: yield, feedstock price, process costs, social indicators: process conditions, chemicals safety, innovation potential. These indicators were selected as they can be assessed based on reaction routes as well as laboratory scale experiments and thermodynamic simulations. We propose that these 10 indicators are a necessary and sufficient set of meters for screening purposes at the design phase and give a balanced view of chemical process sustainability.

Assessment of DMC Production Routes

Dimethyl carbonate (DMC, $(CH_3)_2CO$) is an important chemical intermediate that can be used as a fuel additive and a polar solvent in the chemical industry. The production of DMC has received increasing attention over the least years. There are several methods for the synthesis of DMC, such as phosgenation of methanol, oxidative carbonylation of methanol, transesterification method, and esterification of carbon dioxide with methanol. In this chapter, three reaction routes for DMC synthesis are evaluated. The reaction routes are outlined in Table below.

Table : The assessed reaction routes for DMC production.

Route A: direct synthesis from carbon dioxide and methanol

$$CO_2 + 2CH_3OH \rightarrow (CH_3O)_2CO + H_2O$$

Route B: transesterification of methanol and propylene carbonate using ionic liquid (IL) as a catalyst

$$C_3H_6O + CO_2\ C_4H_6O_3 \rightarrow C_4H_6O_3 + 2CH_3OH \rightarrow (CH_3O)_2CO + C_3H_8O_2$$

Route C: oxidative carbonylation of methanol (ENiChem)

$$2CH_3OH + 1/2O_2 + CO \rightarrow (CH_3O)_2CO + H_2O$$

All three routes provide a safer alternative for the primary synthesis pathway, the "phosgene route"

$$COCl_2 + 2CH_3OH\ (CH_3O)_2CO + 2HCl.$$

The use of phosgene route is phased out from the commercial processes, as phosgene is one of the most acutely toxic substances used in industrial scale. As this route presents inherent hazards and potential environmental problems in handling and waste disposal , it is crucial that it is replaced by a more sustainable method.

Route A is currently in academic research phase. This route is particularly attractive for being CO_2-based. Generally speaking, carbon dioxide (CO_2) can be considered as an environmentally friendly and widely available feedstock, available as a waste emission of industrial processes. Chemical utilization of CO_2 for DMC manufacture would be a means to turn this waste into a nonwaste, allowing us to view CO_2 as a useful resource. It has been reported earlier that CO_2-based synthesis processes are meeting many of the provisions for environmental, economic, and social sustainability. Therefore, much academic research has concentrated on the search for benign by design synthesis involving CO_2 as a raw material. The synthesis of carbonic esters is one example. The expectation is that the CO_2-based DMC production routes have significant potentials toward sustainable production. However, there are also numerous challenges of CO_2utilization ; therefore, long-term research efforts for acquiring the necessary knowledge in its chemical reactivity are needed.

Route B is also an attractive "carbon-friendly" route, due to using CO_2 as a reaction feedstock. However, the complexity of a two-step process, the use of toxic propylene oxide, and the coproduction of propylene glycol make this process demanding. The challenge in both Routes A and B is that scale-up of the production would not be economically feasible at the moment.

The commercial route (ENiChem), Route C, is based on the catalytic oxidative carbonylation of methanol. It offers operational and environmental advantages, for example, fewer side products, better atom economy, and safer production comparing to the phosgene route but it is not responding to the current demand of DMC.

In terms of "measuring" innovation potential, we performed a literature review using a simple keyword search in Science Direct to evaluate the volume of publications and calculated the percentage of recent publications of the 50 most relevant publications. Our reasoning is that the volume of publications is indicative of the level of knowledge potential, and the high percentage of recent publications indicates intensified academic interest, which will contribute to the renewal of science and is more likely to drive innovation.

Identifying Best Practices for Evaluating Sustainability in the Process Industry

(*Nanowerk News*) The SAMT project of the European Union will work together with leading industrial actors from the cement, oil, metal, water, waste and chemical industries and review the latest scientific developments within the field of sustainability assessment. In the first phase of the project, a total of 90 methods and tools were reviewed. In the second phase, the best performing methods and practices will be tested with real-life case studies.

"The industries can learn from each other by sharing information on their methods and tools to evaluate sustainability. The SAMT project is offering this possibility for process industries operating in different fields by organizing workshops and conducting case studies", says coordinator of the SAMT project Tiina Pajula from VTT Technical Research Centre of Finland.

Most of the process industry sectors are resource and energy intensive by nature. In the European Union's research programme Horizon 2020, process industries share ambitious goals in reduction of fossil-energy intensity, greenhouse gas emissions and intensity in using non-renewable, primary raw materials.

"Sustainability assessment is increasingly important within European projects and society as a whole in order to establish a basis for the environmental impact of manufacturing. As such, it is a core part of the SPIRE vision for the Sustainable Process Industries", emphasizes Søren Bowadt, Programme Officer from European Commission. For the process industries, sustainability assessment methods are crucial for evaluating and communicating the current state and the achievement of the goals related to resource and energy efficiency. The SAMT project will provide an unbiased, independent review of the life cycle based sustainability assessment methods and tools, and the identified best practices will serve as a benchmark. "The evaluation methods and tools are improving all the time. Neste is a leading company in production of renewable diesel and a forerunner in circular economy in production of fuels. Therefore it is important for us to thoroughly know all the impacts of our products during their life cycle - and also the methods in evaluating them", says Pekka Tuovinen, Director in Sustainability from Neste."In addition to consistent and innovative solutions, our customers are looking to contribute to a more sustainable future where being environmentally and socially responsible goes hand in hand with running a profitable business. Therefore, BASF has embedded sustainability in its corporate purpose, and linked this ambition directly to business factors", states Peter Saling, Director in Sustainability Methods and

Sustainability Strategy of BASF. "The SAMT project is an interesting platform to learn and discuss other companies' views on this topic as well as an opportunity to exchange good practices", Saling continues. "The linear growth model linked to linear produce-consume-and-throw consumption practices is close to its end. It is necessary to accelerate the development of a sustainable model, for more long lasting and harmonious growth. For more than ten years, sustainability assessment tools and Life Cycle Assessment approaches are helping SUEZ environnement, its partners and its customers to make moresustainable choices", says Paul-Joël Derian, Senior Vice President Research, Innovation & Performance of SUEZ environnement. The first open SAMT workshop took place on 2nd of June at Wuppertal, Germany, where industrial actors and researchers gathered to discuss current best practices and challenges in applying sustainability assessment methods in practice. The second open workshop will be held next year, on February 17th at Bilbao, Spain, to discuss the findings of the industrial case studies. The project will publish its results including identification of the best practices, cross-sectorial guidance and recommendations for future actions in October 2016.Partners working in the SAMT project are CEMEX Research Group AG, SUEZ environnement, Neste Corporation, Bayer Technology Services GmbH, BASF SE and Norsk Hydro ASA, VTT Technical Research Centre of Finland Ltd, Wuppertal Institut für Klima, Umwelt, Energie GmbH (Germany) and Tecnalia (Spain). Spanish National Standardization Body (AENOR) acts as an expert for standardization. SAMT is one of the three projects funded under the SPIRE Public Private Partnership to coordinate studies related to sustainability assessment in the process industries.

BUILDINGS SUSTAINABILITY ASSESSMENT

The sustainability level assessment tools began to be used primarily for the evaluation of buildings. Numerous assessment tools have been developed for the construction sector, aiming to gather and report information for decision making during the different phases of construction, design and use of a building. The variety of tools is large, with LCA-based tools, rating systems, technical guidelines, assessment frameworks, checklists, and certificates.

The first available environmental assessment tool for buildings was the Building Research Establishment Assessment Method. This method was established in UK in 1990 and together with the following two rating and certification systems provided the basis for the other approaches used throughout the world: SBTool, developed through the collaborative work of representatives from 20 countries , and LEED, developed in the USA.

In Portugal, a building sustainability assessment method has been developed: . Sustainable Building Tool for Portugal () was developed by the Laboratory of Building Physics and Construction Technology of the University of Minho, in coordination with the nonprofit association International Initiative for a Sustainable Built Environment-Portugal (iiSBE-Portugal) and the private consulting company Ecochoice SA.

The wide dissemination of these assessment tools is contributing to understanding the impacts of the building sector. Assessment tools are proven to provide unique opportunities for designers, owners, contractors, and users to make decisions and choices during the project and construction of buildings, in order to increase their sustainability level.

From Building Sustainability Assessment to Urban Planning Assessment

The impacts of the building sector are a very well documented fact, which can be addressed throughout measures that are included in building sustainability assessment tools. However, these impacts can be addressed in a more adequate way if the sustainability measures are implemented in a larger scale such as the urban planning.

Nowadays, the goal is to achieve the Zero Impact Built Environment. Solutions should not only focus on zero energy, materials, water, or food but also on the integrated management of all resources that have a major impact on the built environment. The challenge is to achieve a built environment as much as possible sustainable, that is, to achieve a built environment that has the lowest possible environmental impacts, that provides the best living conditions, and that is affordable to the population.

How to achieve all these objectives? No doubt that the buildings are one of the most important components of the built environment but a "built environment" is much more than the agglomeration of buildings. Systems such as transportation, energy production, resources distribution, and waste management, among others, have high impacts and go out of the scope of buildings. Therefore, it is needed to consider the interaction between the buildings and their surroundings, taking into account the life style of the population. Buildings can be very efficient but hardly sustainable because sustainability is a broader concept that can only be implemented at a larger scale. For example, it is very hard to achieve the goal of net zero energy buildings without considering energy efficiency and clean energy production at the urban scale. The same applies to water, materials, food, and so forth. Furthermore, the current population move from rural to urban environments also stresses that sustainability studies have to be performed at the urban scale.

Additionally, the rapid growth of cities and the urban regeneration of degraded and/or abandoned areas are the current concerns of authorities, both at international and local levels. These concerns have directed the focus on developing assessment frameworks and tools for urban communities, such as BREEAM Communities , LEED-ND (Neighborhood Development) , SCTool (Sustainable Communities Tool, in development) , CASBEE-UD (Urban Development) , EarthCraft Communities or Green Star Communities. The interest in evaluation systems is increasing among authorities, investors, and especially developers , since these systems allow the comparison of municipalities and urban areas, serving to support decision making processes, benefiting authorities, planners, and designers during this process.

These tools were designed to give opportunity for projects to demonstrate their environmental, economic, and social benefits to the local community, in all the planning stages of development processes. These tools' system consists of frameworks with several indicators, which are grouped into categories. These tools, while evaluating and ranking the sustainability of urban developments, are also instruments that guide and encourage the process of design and development of sustainable, smart, and high quality communities throughout the promotion of reference best practices.

Sustainability Assessment Tool for Urban Planning: UP

One example of the evolution of the focus of sustainable assessment tools from buildings to urban planning operations is the Portuguese sustainability assessment method. A sustainability assessment tool for urban planning operations is being developed under the scope of the urban scale, for urban planning. The tool -UP will follow the steps of the overall methodology, which considers a set of indicators related by categories and evaluated by a set of parameters. These indicators along with their categories represent the three dimensions of sustainable development: environmental, social, and economic.

The structure of this methodology is being developed and adapted to the Portuguese context by the authors, based on the work of Salat *et al.* and on the work of the iiSBE working group SBTool Urban, which is also in development. A cooperative effort is being made for the improvement of these methodologies taking into account the latest scientific developments in sustainability at the urban scale.

During the process of developing indicators from -UP methodology care was taken to create a list of indicators that were organized, transparent, objective, and correct as possible. This list was developed based upon the current state-of-the-art of existing methodologies to assess the sustainability of urban projects and urban communities and based upon the indicators of overall system SBTool version. Subsequently the list was harmonized following the discussions inside the working groups (national and international) consisting of civil and environmental engineers, architects, and urban planners, in order to suit the national context. This tool encompasses twelve categories under the scope of the main sustainability dimensions, environment, society, and economy. Additionally, an extra category is considered covering the sustainability of buildings and the information and communication technologies at urban scale. The forty-one indicators included in this tool, as well as the respective categories and dimensions, are presented in Table below.

Table : List of categories and sustainability indicators for the SBTool^{PT}-UP methodology.

Dimension	Categories	ID	Sustainability indicators
Environment	C1. Urban form	I1	Passive solar planning
		I2	Ventilation potential
		I3	Urban network
	C2. Land use and infrastructure	I4	Natural land aptitudes
		I5	Density and flexibility of uses
		I6	Reuse of urban areas
		I7	Built environment rehabilitation
		I8	Technical infrastructures network
	C3. Ecology and biodiversity	I9	Distribution of green spaces
		I10	Connectivity of green spaces
		I11	Indigenous vegetation
		I12	Environmental monitoring
	C4. Energy	I13	Energy efficiency
		I14	Renewable energy
		I15	Centralized management of energy
	C5. Water	I16	Consumption of drinking water
		I17	Centralized management of water
		I18	Management of wastewater
	C6. Materials and wastes	I19	Sustainable materials
		I20	Construction and demolition waste
		I21	Management of urban solid waste
Society	C7. Comfort of outdoor areas	I22	Air quality
		I23	Outdoor thermal comfort
		I24	Acoustic pollution
		I25	Light pollution
	C8. Safety	I26	Safety in the streets
		I27	Natural and technological risk
	C9. Amenities	I28	Proximity to services
		I29	Entertainment equipment
		I30	Local production of food
	C10. Mobility	I31	Public transportation
		I32	Pedestrian accessibility
		I33	Cycle paths network
	C11. Local and cultural identity	I34	Public spaces
		I35	Heritage valuation and landscapes
		I36	Integration and social inclusion
Economy	C12. Employment promotion and investment	I37	Economic viability
		I38	Local economy
		I39	Employability
Extra		I40	Sustainable buildings
		I41	Information and communication technologies

Category 1—Urban Form

The urban form assessed aspects are the following: the passive solar planning, considering the suitability of thermal energy conservation and the production of renewable solar energy; the promotion of natural ventilation of urban spaces, taking into account the dominant winds of the geographic area; and the urban network, to promote connectivity between routes of different hierarchies, to a human scale, reducing distances and travel times in order to facilitate the movement and commuting to foot and bike lanes.

Category 2—Land Use and Infrastructure

It assessed how the urban spaces are defined in conformity with the natural land aptitudes, promoting land use efficiency. There is also a promotion of flexibility of uses of the different areas, as well as an incentive to reuse and rehabilitate or regenerate preexisting urban areas such as abandoned urban centres. The objective is to avoid urban expansion or urban sprawl that has many impacts such as the construction of long networks of technical infrastructures, which encompasses high environmental impacts.

Category 3—Ecology and Biodiversity

The assessment of this category covers the management of green spaces, promoting the protection and increase in local biodiversity by rewarding varied distribution of green spaces within the urban fabric and designing of urban green spaces network articulated with ecological corridors. The selection of indigenous species for plants and vegetation in these green spaces and the development of an environmental management plan are also promoted.

Category 4—Energy

The main aspects of the assessment methods are related to the implementation of measures that improve energy efficiency of the public energy consuming equipment and of systems that produce energy from renewable sources. It also promoted the central management and monitoring of energy consumption in order to identify problems in time and to disclose consumption data so that decision making and problem solving are made based on knowledge of inhabitants, enabling attitude changing by the population.

Category 5—Water

The main aspects in consideration are the assessment of drinking water consumption and the treatment of wastewater, while the central management of water is also promoted. The objective is to reduce the water consumption in public spaces through the use of efficient equipments. This will reduce the production of wastewater and reduce the pressure on the drainage systems. Local treatment of effluents will also be promoted, as well as the implementation of a monitoring system.

Category 6—Materials and Wastes

The indicators related to the materials' life cycle are included here. In this category the responsible selection of sustainable materials and the destiny of construction and demolition wastes, as well as the management of urban solid wastes are accessed. It is promoted the selection of materials with lower life cycle environmental impacts and the reuse and recycling of RCDs, as well as the selective separation of urban wastes. The objective is to lower the landfill of residue and to reduce the need for the extraction of raw materials.

Category 7—Comfort of Outdoor Areas

The social dimension is related to the health and comfort of inhabitants regarding the air quality and thermal, acoustic and visual comfort. In this category the reduction of pollutants and odours in the public spaces (including the use of transportation using alternative fuels), the reduction of the heat island effect and the application of rain protection systems, and the reduction of exterior noise and the reduction of glare and nocturnal light pollution are promoted.

Category 8—Safety

The safety of inhabitants is addressed in this category, where the pedestrian safety is assessed, taking into account public health and crime prevention through urban design, by promoting and providing the correct distribution and orientation of streets as well as adequate nocturnal illumination levels, discouragement of high speed traffic and the safety against natural or technological disasters.

Category 9—Amenities

The proximity between the residential areas and working places to key amenities (basic necessity goods), as well as for entertainment equipments, is accessed. The assessment is related to the promotion of the reduction of travel distances and consequently travelling times, contributing to the easy access of inhabitants to services. The creation of public spaces dedicated to the production of organic food, such as community gardens is also promoted.

Category 10—Mobility

This category assesses the promotion of public transport and thus the reduction of the need for the use of private vehicles. With this objective, the creation of a pedestrian network and cycling networks that not only increase inhabitants' satisfaction but also are good for the environment is also promoted.

Category 11—Local and Cultural Identity

This category addresses the issues related to the cultural identity of the urban spaces, promoting the maintenance of key identified architectural styles in existing heritage locations, and the improvement of the use and stimulation of these public spaces. The existence of housing affordable to a wide spectrum of social classes (also age, religion, race, genre, etc.) and the civil participation in public affairs are also promoted.

Category 12—Employment Promotion and Investment

The economic dimension is assessed in this category. Aspects such as the economic viability of constructions, including the analysis of life cycle costs, the promotion of the local economy, and the creation of local opportunities for employment and professional education, are assessed in this category.

Extra Category

Additionally, another two indicators are assessed in the methodology in an extra category. This category was created in order to promote some measures that are good to the sustainability of urban areas but are more difficult to implement. Achieving a good grade in these indicators will improve the sustainability rate of the urban area. The first indicator in this category is "sustainable buildings" and its goal is to promote the sustainability of buildings through the implementation of sustainable building assessment tools. With the "information and communication technologies" indicator it is intended to promote the integrated management of the functional aspects of the city in order to facilitate the urban functions and to improve the quality of life in cities.

SUSTAINABILITY ASSESSMENT FRAMEWORK

Sustainability Assessment: Towards a Standardised Methodology?

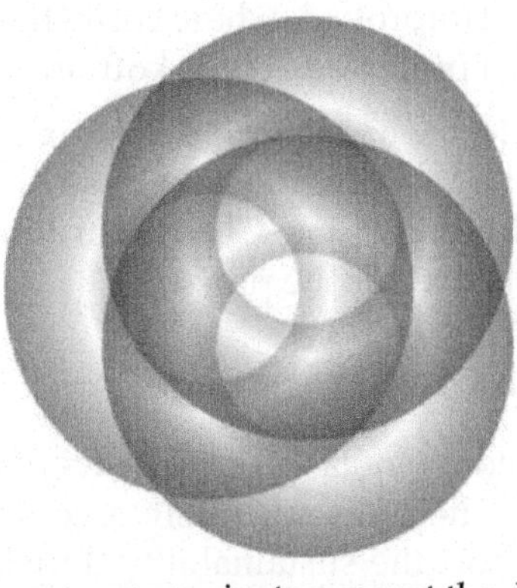

Fig. : SusChem projects and programmes aim to support the development of products and technologies that address societal needs while being environmentally sound and economically viable.

They aim to enable a future in which chemical research and innovation brings new, safe, high performance products and technologies rapidly to market: enabling society to "do more with less". The integration of all aspects of sustainability into the framework of SusChem-inspired projects at an early stage of the development of new processes and products is a key element for their success. However this requires a consistent sustainability framework for the objective assessments of projects. A FP7 project, PROspective SUstaInability assessment of TEchnologies (PROSUITE), looked at the issues involved.

Achieving sustainability requires a balance of economic, environmental and social goals. This triple bottom line, often referred to as 'people, profit, planet', is a key element for Sus Chem.

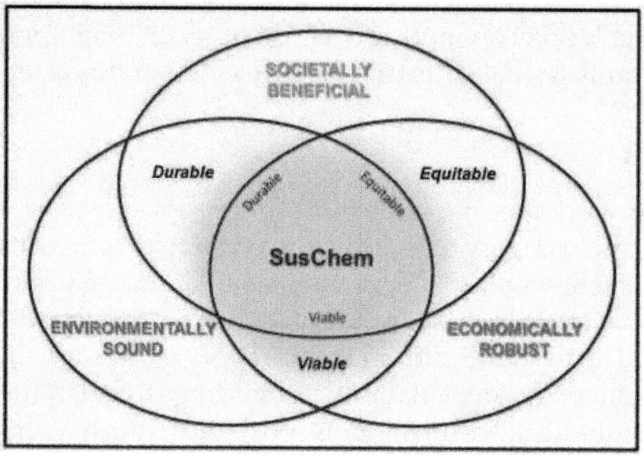

But how can we objectively assess sustainability? To do this requires the development of new tools that can analyse activities and their consequences upon the economy, the environment, and society. And these tools must be adequate for measuring these complex domains, as well as their interactions. Reliably performing sustainability assessments should help point us to better, more sustainable solutions to the challenges that society faces. They should be able to highlight where performance can be improved, where corrections can be made, and where wiser choices can lead to a preferred "triple bottom line" result.

The PROSUITE project was a four-year initiative funded under FP7 that started in 2009 and worked to provide such assessment tools in particular for use in assessing new technologies. SusChem board member Gernot Klotz was a member of the advisory board for the project.

Technologies, Today and Tomorrow

PROSUITE developed its tools through delivering a broad life cycle assessment (LCA) framework. The tools are freely accessible on the project website and are designed to support the sustainability decisions that product developers, policy makers and businesses are facing. The tools have been tested on four technology case studies:

- Biorefineries – producing energy from organic waste
- Nanotechnology - in new textiles
- Multifunctional mobile (telephone) devices - containing rare metals that should be recycled
- Carbon storage and sequestration - to reduce greenhouse gas emissions from power plants and large-scale industrial sources.

Five Pillar Approach

The PROSUITE tools and concepts go beyond the traditional three pillars. In order to deliver even more detailed and meaningful results for decision makers the project adopted a five-pillar framework for assessment, which is supported by a freeware Decision Support System.

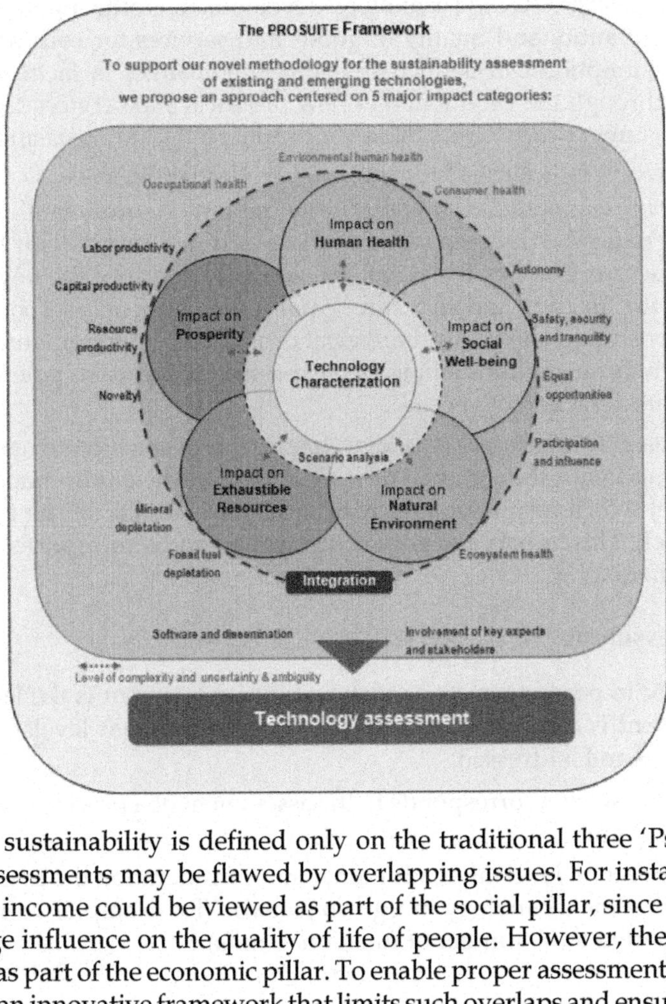

When sustainability is defined only on the traditional three 'Ps' approach, existing assessments may be flawed by overlapping issues. For instance, human health and income could be viewed as part of the social pillar, since both factors have a large influence on the quality of life of people. However, they also could be viewed as part of the economic pillar. To enable proper assessment, PROSUITE developed an innovative framework that limits such overlaps and ensures that each pillar has a unique set of indicators. To achieve this goal, the resulting framework proposed five pillars:

- *Impact on human health:* The impacts on human health of a new technology include all changes in morbidity and mortality that are caused by the introduction of new technologies, through all possible pathways, including environmental, occupational and consumer. These can be quantified using the 'DALY-concept' (Disability Adjusted Life Years).
- *Impact on social well-being:* The social impact includes all impacts on human well-being that are related to inter-human relationships. Impact on human well-being includes everything that affects the quality of life of people both on an individual and collective basis (however, excluding human health and economics). These include impacts on autonomy, safety, security and tranquillity (SST), equal opportunities and participation and influence.
- *Impact on prosperity:* Technology development is often pursued to increase the quantity and quality of goods and services for consumption. Final consumption can be increased through changes in factor productivity or through the production of new products and services satisfying new consumer needs. Gross domestic product (GDP) is a measure of the value of goods and services available for final consumption.
- *Impact on natural environment:* The 'natural environment' encompasses the natural ecosystems around the world in terms of their function and structure. For this endpoint, the aim is to quantify the negative effects on the function and structure of natural ecosystems as a consequence of exposure to chemicals, biological or physical interventions. The impact on natural environment is then expressed in terms of 'potentially disappeared fraction of species'.
- *Impact on exhaustible resources:* Impact on exhaustible resources is concerned with the removal of resources from the environment (and their use) which results in a decrease in the availability of the total resource stock. This impact category comprises abiotic resources: fossil fuels and mineral ores.

Level Assessment

Another important aspect of sustainability assessment is the level at which the assessment is conducted. In PROSUITE three different levels are explicitly differentiated and addressed:

- The first level corresponds to the assessment of a process chain per functional unit.
- The second level takes additional into consideration the market penetration of the technology in a given scenario (for example, expressed in the expected total number of functional units).

- The third level goes beyond the process itself and includes the effect of deploying a technology at the system level by taking into account not only the penetration of the technology but also the implications of such penetration on, for example, trade flows across sectors in the economy.

The work in PROSUITE has resulted in an increased understanding of the complex set of approaches and tools needed to assess the potential impacts of technologies at an early stage of development.

Departmental Sustainability Assessment

The Chemical and Biological Engineering (CHBE) Sustainability Club has developed a methodology to assess the sustainability of a large institution. The framework methodology allows for an assessment of the sustainability of an entire institution, such as UBC, through examination of the institutions sub-divisions or entities including: individual departments, research institutes and/or faculties. With the developed framework, these units will be able to perform their own assessment according to their internal capacity.

The methodology utilizes two levels of indicators. A base level targets individuals in each unit and their personal sustainability. This is measured using an online or paper survey distributed through the department, institute or faculty communities. The second level of measurement is used to measure the sustainability of each entities operation. This would be completed by motivated individuals, such as sustainability coordinators or student groups. These measurements would be conducted by the environmental/sustainability administrative body within the institution, at UBC-Vancouver, this would most likely be conducted by the Campus Sustainability Office.

The series of indicators is based on an number of leading frameworks, such as the Campus Sustainability Assessment Framework (CSAF) and the Sustainability Tracking, Assessment & Rating System (STARS). As well, UBC specific strategies were utilized, such as the Inspiration and Aspiration Guidelines and the Sustainability Academic Strategy.

What is Sustainability?

The concept of sustainability is continually evolving. The famous Brundtl and Commission of the United Nations laid down the foundation of the concept as follows: "sustainable development is development that meets the needs of the present without compromising the ability of future generations to meet their own needs".

The project *Realizing Sustainability Assessment* aims to not only measure sustainability, but also to educate individuals about the various dimensions of sustainability.

The Brundl and definition led to the question of how to achieve this. The 3 pillars, or triple bottom line model was then proposed and is still largely used today.

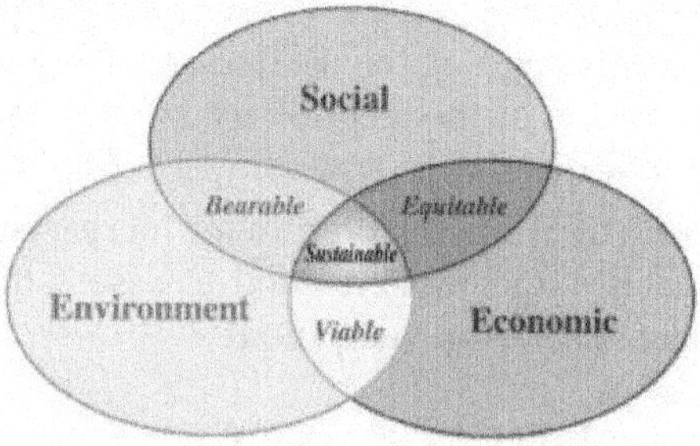

The overlapping circles brought the concept of balance between economic, social and ecological imperatives. But this model can be challenged, because it gives equal importance to each of the three pillars. No economy can exist without people and people cannot live without basic needs of water, air and food provided by the environment. In light of this a nested model seems more appropriate.

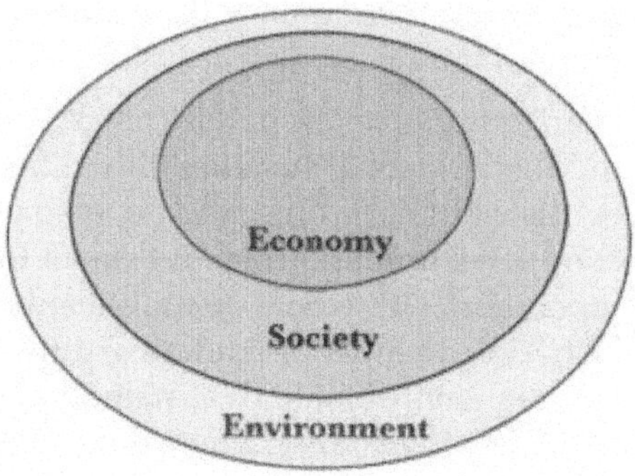

Being sustainable, means to live within the capacity of our environment, maintain a functioning peaceful society and use the economy as a tool to enhance our lives and opportunities. Unfortunately, we as a society are not living within the center balanced region of the model. Many see the economy as the number one priority, followed by the health and social needs of the population with the environment often put last or not considered at all. This actual model of "unsustainablity" is leading us today towards major catastrophes. Examples of these are: the collapse of numerous fisheries, the de-forestation of the tropics and global warming.

How to Measure Sustainability?

Many assessment frameworks have emerged in the past decade according to the type of institutions that they try to assess. The following describes a number of these different frameworks.

Corporations

For large corporations, the Global Reporting Initiative (GRI), is a comprehensive and adaptable sustainability framework for businesses of any size. It has many valuable social, economic and environmental indicators. The GRI offers valuable information around the values of reporting, process and boundary setup. Many large companies are now incorporating the GRI report into their annual fiscal report.

Campuses

Campus Sustainability Assessment Framework (CSAF)

For large campuses, one of the pioneering frameworks for sustainability assessments was the Campus Sustainability Assessment Framework (CSAF), developed by Lindsay Cole (2003). The framework is currently promoted by the Sierra Youth Coalition (SYC) *via* their Sustainable Campus Program.

The CSAF structure is based on the nested model of sustainability. It divides sustainability into 10 dimensions with a total of 170+ indicators. The framework is still considered as one of the most comprehensive assessments for campuses today. However, the CSAF has serious weaknesses. The major one being that many indicators require large time commitments in order to be measured. Also, it does not include a convenient method of presenting the results and comparing with other campuses, or tracking changes over years, as all indicators are measured and reported individually. Finally, the CSAF was written seven years ago and requires updating.

Sustainability Tracking, Assessment & Rating System (STARS)

The **STARS** is currently being developed by the Association of Advancement of Sustainability in Higher Education (AASHE). In September 2009, they released their STARS 1.0 framework. The STARS major advantage is its structure. The framework is divided into operational categories, rather than sustainability dimensions, making it easier to determine where to look to find information. It has a point system that quickly shows strengths, weaknesses, allowing comparison and tracking. It focuses more on outcomes than specific strategies, unlike the CSAF. One of STARS weakness is that it does not have a lot of indicators about social well being. Also, the long list of indicators can still be overwhelming for large campuses, such asUBC.

UBC and Sustainability

UBChas its own set of indicators and sustainability planning documents: TREK 2010, Inspiration and Aspiration, Campus Plan, Climate Action Plan, Focus on People, Aboriginal Strategic Plan, Alternative Energy Committee, and the newly approved Sustainability Academic Strategy. All these strategies and plans developed their own set of indicators and objectives independent of one another. UBC is known for its leadership in sustainability. Our assessment methodology can help to inform the campus community about those strategies and put them into practice at the department level.

How can we Make Assessment a Reality for Large Campuses?

UBC is one of the largest campuses in North America. The Sustainability office participated in a Pilot project of the STARS 0.5 in 2008 and the exercise proved to be quite challenging: very few indicators were measured.

The Sierra Youth Coalition has been promoting CSAF for many years, but no student group atUBChas attempted to use its framework, probably overwhelmed by the size of the task.

The CHBE Sustainability club has developed two assessment tools:

1. A series of indicators adapted to our campus.
2. A method to realize the assessment. The assessment was divided into three sets of questions:
   - Set 1: Indicators measured through a survey, divided for staff, faculty, undergraduate and graduate students; instituted at the department/ institute/faculty level. Surveys can be conducted online, or printed and distributed in classes. They can be promoted *via* the Sustainability Coordinator Program or by any groups interested. (Examples of Set 1 indicators are: transportation habits, awareness of campus sustainability related program, enrolment in sustainability related courses, student debt load.)
   - Set 2: Indicators measured individually in department/institute/faculty consisting in a list of questions. The Sustainability Coordinators or other interested keen individuals can answer them over the range of a full term to a year. This time frame would require spending about 2 hours a month.
   - Set 3: Indicators relevant for the whole campus, where the sustainability office staff paid / honorarium students could determine (Examples of Set 3 indicators: campus GHG's inventories, campus wide water and paper consumption, green spaces).

EMISSION IMPACT INDICATORS

The manufacturing industries are responsible for large emission loads discharged in water. The manufacture of products results in many and varied emissions, depending on the products being manufactured and the processes and chemicals used. Emissions to water include heavy metals (*e.g.* cadmium, mercury), organic micropollutants, suspended solids and organic matter., affecting ecological and/or chemical status of water bodies. Over the last 30 years the urban and industrial wastewater treatment has progressively improved and in many parts of Europe a large proportion of the pollutants are today removed (European waters-assessment of status and pressures). However, pollution caused by inadequately treated wastewater is still in some areas an important source of river pollution and an important source for transitional and coastal waters. Water Framework Directive (WFD) lays down a strategy against pollution of water and requires further specific measures for pollution control and environmental quality standards (EQS). In accordance, Member States should implement the necessary measures with the aim of progressively reducing pollution from priority substances and ceasing or phasing out emissions, discharges and losses of priority hazardous substances.

Consequences of economic activities with regard to water quality and quantity have been analysed under the WFD through the Member States' River Basin Management Plans. The study of the link between water status (quality and quantity), relevant pressures and their economic driving forces provides important basis for the decision making and prioritization of measures with regard to achieving the objectives of the WFD. Moreover, it can help to indicate whether the economic growth in a particular industry/sector is decoupled from its environmental impact and thus whether the sector movers towards higher resource efficiency. Easily understandable indicators will be necessary to provide signals and measure progress in improving resource efficiency.

Decoupling represents a strategic approach for moving forward a global Green Economy –one that "results in improved human well-being and social equity, while significantly reducing environmental risks and ecological scarcities". The emission of pollutants from point sources (industrial facilities) provides an indication of potential environmental impact on water quality. Economic growth is defined as the added (monetary) value of all final goods and services produced within a country (or economic sector) in a given period of time, usually a calendar year.

THE PERFORMANCE OF STEEL USING VEHICLE EMISSIONS IMPACT INDICATOR

Vehicle emissions impact indicator is a sophisticated and interactive tool that allows users to make decisions on the type of engines used in their cars, the materials used to manufacture their cars and judge how these factors affect their overall environmental performance with respect to fuel consumption and total life cycle emissions.

Life Cycle Emissions

Currently, regulations are focused on the emissions that occur while the vehicle is being driven and using fuel, but this is only one part of the vehicles life that generates emissions. Vehicle's lifetime emissions is given more importance, it is possible to make design decisions that can reduce fuel consumption. However, this strategy increases emissions over the total life cycle of the vehicle.

This is attributed to the high emissions that occur during the production and recycling of technologies or materials that are selected. In other words, if more energy is used to make something, then there would be a higher level of emissions.

However, the issue is not solved completely and is transferred to another place in the life cycle. This means, cars are not better in terms of overall environmental performance and may ultimately lead to an even bigger environmental problem. The emission regulation and a life cycle approach to vehicle design can help mitigate the overall environmental impact of vehicles.

Use of Steel on Vehicle Emissions Impact Indicator

Based on solid scientific methodology, vehicle emissions impact indicator is a simple tool that shows technology tradeoffs for a specific set of parameters. Using steel on the indicator may not always be the best option, however the production of other materials such as plastic and aluminium composites can generate up to 20X more emissions. Therefore, a life cycle assessment is required to guide the design decision process which allows automakers to make the right decisions and avoid unintended consequences.

Basic Information on Indicator

Users can compare four different types of engines such as electric, hybrid electric, gasoline, and diesel as well as three different types of materials such as aluminium, advanced high-strength steels, and multi-material.

For example, let us consider that the vehicle is an A/B-Class Sedan, driven for a 200,000km life span of approximately 12 years. In order to better understand the CO_2 emissions saved by recycling the materials from which the vehicle is manufactured. Since CO_2 emissions are often lower for recycling when compared to virgin production, the recyclable product is credited by eliminating the extra emissions, and this credit is deducted from the total emissions of a product.

Fuel fill ups saved refers to the number of times the vehicle owner can avoid going to the fueling station over the course of a year. Such comparisons are quite surprising.

Chapter 6

Integrated Business and SHESE Management Systems

Market orientation is a key driver of company performance. Accordingly, companies should align their core activities with market requirements. Among a company's most important activities is the development and market introduction of new products because this determines sustained company success in many industries. It has been frequently shown that the alignment of all process steps from idea to launch with market requirements is a success factor of new product development. Consequently, research and development (R&D) should be focused on designing new products which satisfy market requirements.

Technological change has been found to have a decisive impact on the competitive structure in many industries. Thus, even if market requirements remain basically unchanged, technological progress can severely effect a company's competitive position. These changes need to be identified and assessed as early as possible in order to align the firm's R&D strategy accordingly. Thus, market requirements and technological capabilities need to be assessed simultaneously in order to formulate effective strategies for new product development.

It has been shown in empirical studies that the integration of market requirements and technological capabilities during new product development increases success. At the same time, integration ought to be achieved also on the strategic level, *i.e.* Marketing and R&D strategy must be aligned. There is empirical evidence that a strategy which balances market requirements and technology capabilities leads to higher new product success.

The strategic management of a firm's R&D portfolio enhances the success of new products. To support strategic planning various portfolio concepts have been developed over the years. In the 1970s, market portfolios became popular to support strategic planning in large corporations. However, a product's underlying technologies were not assessed in market portfolios which posed the danger to overlook technological changes and to make wrong investment decisions.

Consequently, technology and patent portfolios were developed from the 1980s onwards to assess technologies according to their competitive impact. However, technology portfolios lack an explicit market focus. Thus, the need arises to integrate market and technology portfolios based on objective data in order to achieve a better alignment of Marketing and R&D strategies in companies. We first develop such an integrated portfolio approach to support market-oriented R&D planning. Subsequently, we test the proposed portfolio method in a practical application. This chapter concludes with an evaluation of the proposed portfolio method and with suggestions for further applications.

DEVELOPMENT OF AN INTEGRATED PORTFOLIO CONCEPT

The concept of product portfolio analysis is one of the most widely used tools in the field of strategic planning. Its origins date back to the late sixties when diversified companies were facing an increasingly complex strategic planning process. Therefore, companies had to find new ways to assure an effective and efficient management of the company's resources. The Boston Consulting Group developed its market share/market growth matrix for this purpose. It is still the most widely known and implemented approach. It enables managers to easily comprehend and communicate a complex problem, thus having a powerful tool supporting strategic decision making.

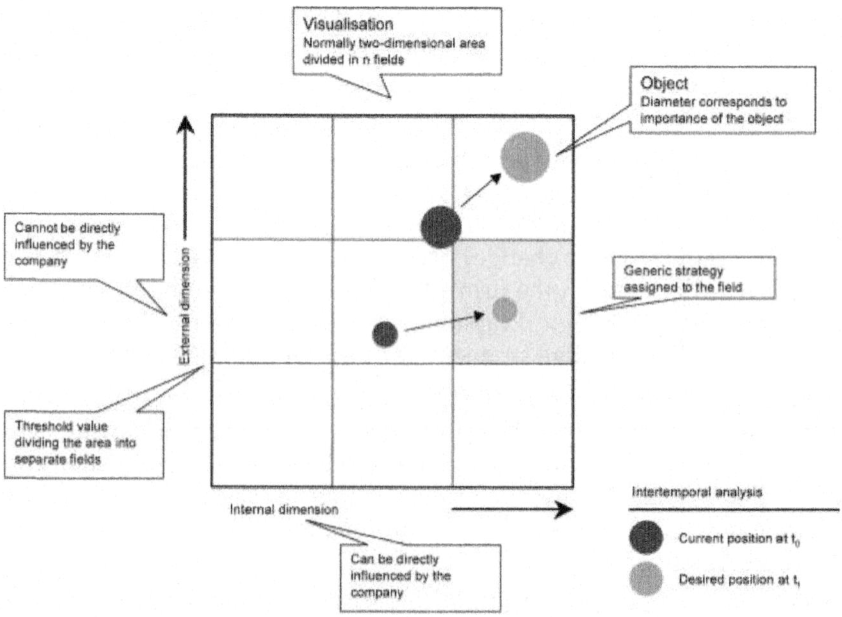

Fig. : A generic portfolio matrix.

Subsequently many variants of this approach were developed, yet all varying the same basic structure. All share a two dimensional framework with one dimension representing the competitive position in the respective market and the other one illustrating market attractiveness. The dimensions are either univariately or multivariately defined. The area spanned up by the dimensions is divided into four or more fields, each of them being assigned a generic strategic recommendation for the products or business units being positioned in the respective area. The size of the circles representing products or strategic business units is usually determined by the respective share a measure of profitability.

The market share/market growth approach is a sound concept regarding required data and underlying assumptions. It has been subject to much empirical research providing supporting evidence. The required data is relatively easy to capture: market share and market growth can be objectively measured, the only problem being the correct definition of the relevant market.

However, critics quickly argued that the underlying assumptions are flawed. Indeed, there is empirical evidence that the relationship between market share and profitability is not as clear-cut as postulated in the generic strategies of the portfolio matrix. Moreover, it is often argued that the data basis is too weak to make profound strategic decisions. This led to the development of product portfolio concepts using multivariate dimensions such as the market attractiveness/ market position matrix developed by McKinsey. In addition to market share, it uses a multitude of factors identified in the PIMS project, several of them qualitative and thus harder to measure. Their estimation may lead to biases due to the subjective nature of the respondent's opinions.

Thus, one faces a trade-off between a simple concept incorporating few variables, which can be objectively measured, or a complex model requiring data, which are hard to capture, thus leaving room for biases. When using the approaches based on few variables, it is important to keep in mind the limited data base of the simpler approaches, on which the generic strategies are recommended.

Apart from these issues being more technical in nature, there is substantial criticism highlighting a central drawback of market portfolios. As they capture the market positioning of the company's products or strategic business units and their respective current competitors, product portfolio analysis is essentially short-term oriented. It fails to capture external turbulences such as new technologies which affect the positioning in the portfolio and thus the recommended strategies.

The Technology Portfolio

The Patent Portfolio

Patent data and patent statistics have long been used by economists as a proxy for technological change and technological process. There is significant empirical evidence supporting the validity of this approach. Patent data have the big advantage that they are widely available and objective in nature. In a further step, entire portfolios assessing a company's patent position were proposed. These assign the company's patents to different technology fields relevant to the company and evaluate them afterwards using different indicators entirely based on patent data. The patent portfolio shares its basic structure with the technology portfolio, having one dimension assessing technological position and the other one assessing technology attractiveness. As patent portfolios will be an important element of the integrated portfolio concepts developed, we will briefly describe the indicators used to assess technological positions.

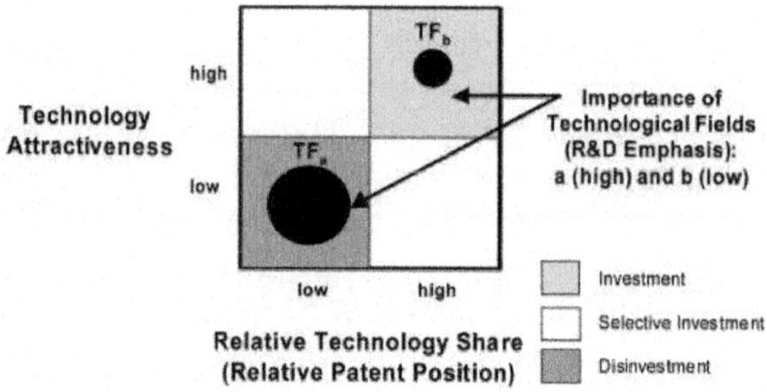

Fig. : The patent portfolio.

In analogy to the market growth/market share portfolio, it has been suggested to use the term 'relative technology share' in order to describe the competitive technological position in analogy to 'relative market share' commonly used to measure a firm's competitive position in the market.

As this is a measure only incorporating patenting activity, but not accounting for the quality of patents, other operational definitions of the internal dimension have been proposed. Accounting for patent quality is an important link to establish an analogy between patent and market portfolios: both relative market share and relative technology share, incorporating patent quality, have a positive impact on a company's business performance. Therefore, a construct for patent quality consisting of different quality measures is used in newer approaches to assess the technological impact :

1. *Rate of patents granted:* A patent will be granted only if the invention consists of new and non-obvious technological elements. Thus, a patent is believed to have a higher technological value than the patent application. The rate of patents granted substantially differ between companies and can thus serve as one indicator for the quality of research.
2. *International scope of patent applications:* As the cost for obtaining valid patent for a number of countries is significantly higher than for a single national patent, the former have been empirically found to be of higher value than national applications only.
3. *Patent Citation Ratio:* Patents are used by examiners at the patent office to document the state of technology when they check if a patent application contains new and inventive features which go beyond what has been known so far. This procedure leads to patent citations. The number of citations received by a patent in subsequent patent documents is often interpreted as a proxy for the economic importance of the invention.

The use of different measures allows to achieve a stable assessment of a company's patent position.

The external dimension 'technology attractiveness', which is displayed on the ordinate, is measured by the growth rate, either absolute or relative, of patent applications in the respective technology field. Using patent growth as a proxy for a technology's attractiveness is supported by empirical studies showing a strong correlation between patent and market growth in various industries.

The third dimension represented by the circle size of the technology fields in the patent portfolio is called 'R&D emphasis'. It is calculated as the total number of patent applications of the company in one technology field divided by the total number of all patent applications of the respective company. Thus, "R&D emphasis" indicates a technology's importance in the company's total R&D portfolio.

Generic strategies largely correspond to the generic strategies of the market growth/market share portfolio, *e.g.* to aggressively invest in technology fields with high growth and strong positions, *i.e.* high impact pacing and key technologies. Thus, the patent portfolio is the only technology portfolio grounded on a sound empirical basis.

Despite this fact, patent portfolios share the most important drawback with classic technology portfolios. R&D planning is only one aspect of strategic planning. It necessarily leads to the misappropriation of resources if only technological aspects are considered in the planning process. Therefore, they need to be aligned with other strategic planning tools in order to avoid one-sided misconceptions.

The Integrated Portfolio

Pure technology or market portfolios have a one-sided focus on either technology or product market. We already elaborated on the importance of market orientation in the R&D process. New technologies have to fulfill market needs. Otherwise products based on them will fail in the market. The integrated portfolio concepts trying to overcome this shortfall by combining market and technological analysis can be classified into three distinct groups:

The first group builds upon an existing technology or market portfolio and adds generic strategies for the missing perspective. Although this approach is intriguingly simple, it does not solve the central problem of incorporating the interdependencies. Applying similar strategic recommendations will only be valid if both technology and product are positioned in the same field of the portfolio. In all other cases, this will lead to a misallocation of resources and could even result in eroding the company's competitive base.

A second class of concepts develops entirely new portfolio approaches. Some of them use the classic matrix-type visualisation, while others follow different approaches for the formulation of the optimisation problem. Some approaches in the latter category use a mathematical formulation of the portfolio problem and usually have a very stringent general formulation. Being often rather complex and thus difficult to communicate and implement, these approaches lack practical relevance as well.

The third group of concepts tries to aggregate two or more single portfolios into one holistic portfolio. This could either be a new single matrix-type portfolio, *i.e.* a condensation of four dimension into only two. This procedure leads to a visualisation which is easy to understand and to communicate, but encounters several methodological problems. Another approach is the integration using a common axis. Generally, this type of integration is theoretically sound as no data are omitted and the original portfolios remain largely unchanged. Their added value is the development of new integrated generic strategies.

Therefore, we will develop an integrated portfolio using this design, but following a more formalised approach to avoid the pitfalls of earlier concepts. Unlike other approaches, we will keep the amount of required data within reasonable limits in order to achieve an easy implementation while simultaneously building onto a theoretically sound basis.

In the following, we introduce a framework for the creation of concepts for integrated portfolios and simultaneously develop a new portfolio. It draws on elements of a framework by Wind/Mahajan, however incorporates necessary changes for the formulation of an integrated portfolio concept. The process of developing our integrated portfolio concept follows these steps. First, the portfolios which are subsequently integrated into a single portfolio are identified. We already made a case for the use of patent portfolios instead of other technology portfolios, the biggest advantage being the sound empirical foundation and the objective data used by them. Therefore, a matching market portfolio, which captures the advantages of the patent portfolio, needs to be identified. The only market portfolio entirely relying on objective data is the market share/market growth matrix.

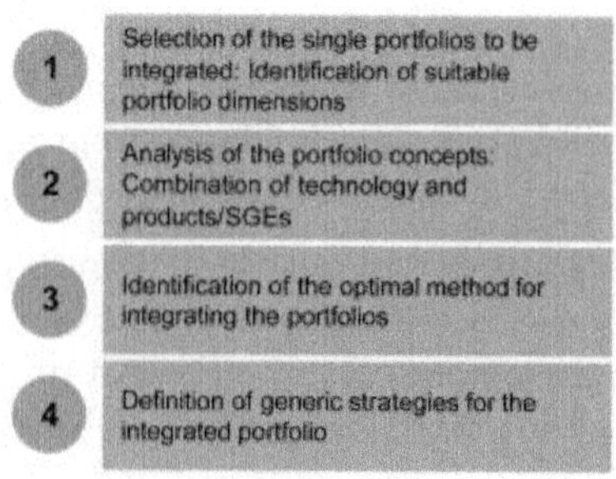

Fig. : Schematic process for designing an integrated portfolio.

In a second step, a suitable integration mechanism has to be found. In order to come up with an integrated portfolio which can be easily communicated, the integration using a common dimension is desirable. In order to emphasise the importance of market orientation in the R&D process, we suggest using market growth in the integrated portfolio. This approach clearly follows the notion that the attractiveness of a technology has to be mainly judged on the basis of its market impact. The growth/share matrix is now easy to integrate as both portfolios share a common dimension. In case a technology has not yet been integrated into products, we suggest to use the measure "patent growth" as described in the initial patent portfolio in order to capture the attractiveness of the technology.

The third step is the inclusion of possible interdependencies between technologies and products into the analysis. The attractiveness of a technology field could be determined by the weighted average of the market growth rate of the products in which the technology field is applied. Weights could be defined as the share of one product's contribution to the total sales generated by all products incorporating the respective technology.

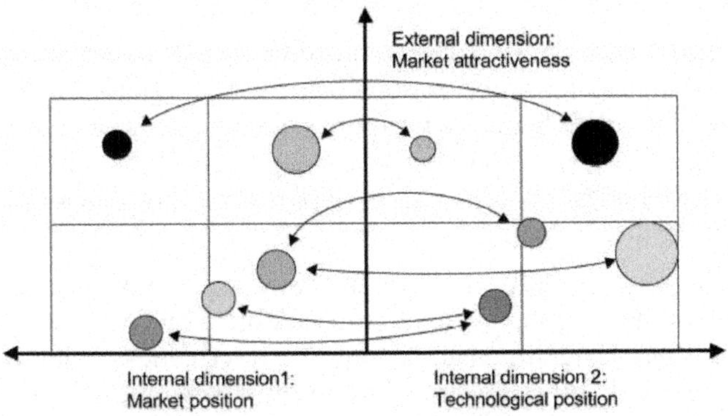

Fig. : The integrated portfolio.

Alternatively, a similar measure could be based on profit data. This way, all interdependencies between technologies and products can be captured. To assess new technologies not yet incorporated in existing products, we suggest using expert assessments to compensate for eventually missing quantitative market data.

Each half of the proposed integrated portfolio is split into four quadrants: the dimension "market attractiveness" is divided by a hurdle growth rate, *e.g.* average sales growth or a sales growth target, the dimension "technology share" at a value of the relative technology share of 1, the "market share" dimension equally at a value of the relative market share of 1.

In the case of a positioning in the same quadrant in both market and patent portfolio, the combined generic strategies remain appropriate: for a product with a star/star positioning, increased marketing and R&D expenditures are necessary to reinforce market and technological positions and to counter attacks from competitors.

However, the generic strategies cannot be simply combined in case of disparate positionings. These could be interpreted as evidence for a misappropriation of R&D and/or marketing expenditures, but also simply result from industry specificities and thus not require any strategic action.

In all cases, the underlying situation has to be carefully evaluated in order to derive effective strategic implications from the portfolio analysis.

An Application of the Integrated Portfolio in the Chemical Industry

Data Collection and Measurements

A first implementation of the proposed concept was carried out for two business units of a company operating in the specialty chemicals industry. For each unit, distinct product areas were defined. The entire analysis incorporates seven product fields and 22 products of Business Unit 1 and three product fields and 15 products for Business Unit 2. Here, we can only report on some characteristic results. The required patent data were supplied by the firm's patent department. Patent applications were assigned to product fields and individual products in a workshop with senior patenting experts from the company.

The indicators described earlier required to draw patent portfolios had to be calculated. The share of granted patents was measured based on granted patents at the German Patent Office and international scope was calculated as the share of triad patents, filed in Europe, the US and Japan simultaneously, because it was believed by the experts from the patent department that this would best capture patent quality in this industry.

Due to the very specialised nature of the business units' products, no market growth rates and market share data were available. Thus, we had to rely on expert evaluations instead. The dimensions market attractiveness, relative market share and revenue share driving the circle size in the market portfolio were evaluated using a questionnaire sent out to marketing managers asking for their estimation of relative market share in relation to the strongest external competitor, product market growth relative to the average growth in the business unit and product sales relative to the average sales per product in the business unit.

Results of the Portfolio Analyses

Based on these raw data, we constructed multiple portfolio visualisations on the business unit and product field levels. The visualisations were discussed in a joint meeting with marketing, R&D and patent managers from the company. Some of the discussion helps to better understand and interpret the portfolios.

The analysis on the product area level draws a picture of a very attractive area, *i.e.*, each product growing at least as fast as the business unit. In most cases, the products are positioned in corresponding quadrants of the portfolio. Thus, R&D and Marketing strategies are well aligned in these product areas.

In some cases, however, there are obvious differences. Product 6 enjoys a strong market position in a high growth market. In contrast, its technology position is fairly weak.

This disparate positioning could be attributed to the fact that the company was still in the process of reinforcing its technological position and thus still intended to file for more patents. The need for further R&D investment in this product is clearly highlighted by the portfolio analysis and will be further emphasised by the company.

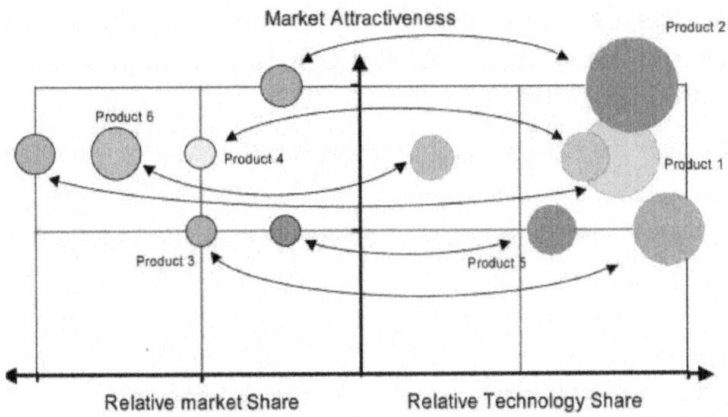

Fig. : The integrated portfolio: An application for a product area with six products (disguised).

Product 3 is characterised by a strong technological position and average numbers for market share and market growth. The company under consideration has a strong R&D emphasis on this product, however, the contribution to sales is rather modest. The discussion revealed that Marketing and R&D perceive the potential of the recently introduced product quite differently. It was decided to conduct a workshop with multiple company experts in order to reassess the attractiveness of this product. Based on the outcome a decision ought to be made whether R&D investments in this product should be realigned. The analysis at the business unit level follows a similar logic.

In sum, the integrated portfolio analyses have shown that market and technology positions in the company under consideration are not always aligned. These results have led to a first discussion among company managers on the root causes for the observed inconsistencies of portfolio positions. This already highlights the need for intensive discussion based on the portfolio presentations before final conclusions are made. All managers perceived this to be very helpful to gain a common understanding of Marketing and R&D strategies. To get a more details picture and in order to incorporate further knowledgeable experts and decisions makers in the company, it was decided to carefully review the marketing, patenting and R&D strategies for some of the concerned areas.

The integrated portfolio can be a powerful tool for strategic planning purposes because it offers an efficient and effective way to better align Marketing and R&D strategies. The integrated portfolio combines a widely known and used market portfolio concept with a patent portfolio capturing technological aspects. Both portfolios' dimensions, especially the abscissa, have been shown to impact business performance, thus making the portfolio illustrations especially meaningful for decision makers in firms. The clear focus on market orientation in the integrated portfolio as expressed on the ordinate where the integration of both portfolios is achieved by using the market portfolio's initial dimension "market attractiveness". It is believed that a technology's attractiveness has to finally show on the market and that R&D strategies ought to be consequently aligned with market requirements. In cases where market data is not yet available for new and future products, one should rely on accurate market forecasts derived from market research or other sources. The strategic implications which can be derived from the integrated portfolio.

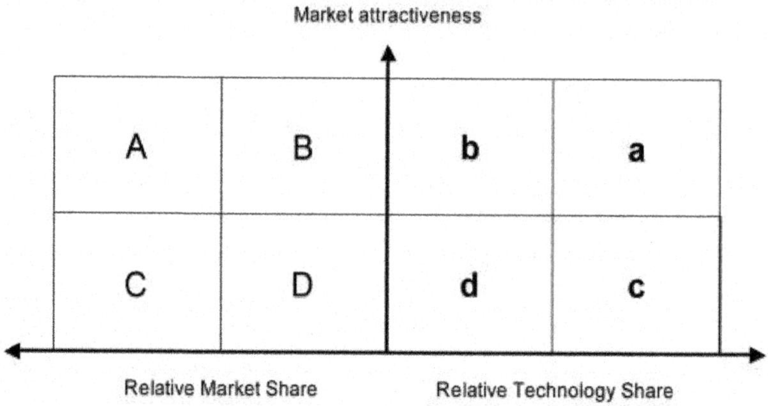

Fig. : Overview about partial generic strategies in the integrated portfolio.

If both technology and products are placed in the same field of the market and technology portfolios represented by the combinations Aa, Bb, Cc and Dd, the generic strategies from the separate portfolios can simply be added up and remain valid. The main advantage of our portfolio concept is the detection of disparate positionings. In the case example, we elaborated on the combinations Ab, Ba, Cd and Dc. In these cases, different generic strategies have to be applied and those from the separate portfolios might become invalid as demonstrated by our example. Several hypotheses can explain the differences and we tried to find some evidence in expert discussions. For a disparate positioning in the upper half of the portfolio, *i.e.* a high growth environment, a likely explanation for a strong technological position, but weak market position would be an insufficient satisfaction of market needs. In this case, R&D expenditure should be cut back as it cannot be earned back in the future. In contrast, it could also be possible that the weak market position is due to strong competition in the product market. This would call for larger investments in marketing, R&D investments simultaneously held constant or scaled up. In a low growth environment, our hypothesis explaining a differing position would be a misappropriation of R&D investments. As market "cash cows" usually require massive economies of scale to be profitable, aggressively building market share would be a loss-making strategy. The sustainability of a market "cash cow" could also be threatened by a weak technological position. However, this need not be the case: if there are high barriers to entry due to economies of scale, the company does not need to protect its competitive position by property rights any more to prevent new firms from entering the market. We provided first evidence for possible strategic actions, which might be applied in these situations. However, it has to be stressed that these always need to be checked for appropriateness in the specific situation under evaluation.

Another strength of the portfolio method is the visualisation of complex decision problems. It lays the ground for an intensive discussion of portfolio positions, their root causes and their implications. The case example clearly shows that the integrated portfolio establishes a framework for joint strategy discussions between technical and non-technical functions like R&D, Marketing and the Patent Department.

We believe that the main benefit of portfolio techniques in strategic planning is the stimulation of cross-functional reviews of R&D and market strategy: there are hardly any combined generic strategies for disparate positionings which can be applied across industries and in different contexts. This opinion was confirmed by managers from the company, which found the portfolio very useful as a tool supporting strategic decision making.

A further important benefit of the integrated concept is that further knowledge is created during the conceptualisation and implementation phases. The systematic condensation of patent data and the simultaneous visualisation of technology and products create a significant amount of knowledge. The analyses create a significant amount of important knowledge which separate portfolio analyses cannot provide.

The limitations of this work can first be attributed to the case study approach, which is in a first step justified when new concepts are developed and preliminarily tested. A second limitation results from the data needed for the patent portfolio: if a company does not patent innovations, the method cannot be applied. However, patenting product innovations is a dominant strategy in many industries. Thus, we do not believe that this severely limits the applicability of our approach in most cases. Finally, the proposed portfolio concept does not capture future technological developments which have not led to any patent applications yet. Here, traditional forecasting techniques, *e.g.* Delphi studies or scenario analysis, can be used which, however, have their own shortcomings.

INTELLECTUAL CAPITAL AND COMPETITIVE ADVANTAGES: THE CASE OF TTY BIOPHARM COMPANY

Intellectual capital includes human capital and structural capital wrapped up in customers, processes, databases, brands, and systems, and has been playing an increasingly important role in creating corporate wealth and growth. The growing divergence between firms' market value and book value shows that much of the source of economic value is no longer the production of material goods, but the creation of intellectual capital. Cañibano *et al.* (2000) summarized an extensive body of empirical evidence on the value relevance of investments in intellectual capital such as R&D, advertising, patents, brands and trademarks, customer satisfaction, human resources.

Pharmaceutical industry is heavily characterized as one of the most innovative, knowledge-based, and R&D-intensive industries. The innovation in the industry has been in a large measure driven by the persistent and successful pharmaceutical R&D effort. As innovation becomes key to success in the pharmaceutical industry, pharmaceutical companies usually needs vast investment in a new drug R&D and takes lengthy processes to launch a new drug to the market. Consequently, it demands a sizable product market to support the mega R&D costs.

Given the small home market size, pharmaceutical companies in Taiwan may not afford the vast costs of a whole-range new drug R&D, covering from discovering new chemicals to developing new drugs. Therefore, most pharmaceutical companies in Taiwan are essentially generic drug manufacturers. The competition for this type companies mainly focuses on low costs and low prices.

Established in 1960, TTY Biopharm Company (hereafter, TTY) used to be a traditional, generic drug pharmaceutical company that focused on production and sales. Due to the keen price-cutting competition and deteriorating in generic product margins, TTY was once close to being bankrupt. However, in 1996 a new management team came in TTY and started reforming and repositioning the company. It initiated the strategy of branding generic drugs to differentiate from other generic manufacturers. To carry out this strategy, TTY began to build up new drug development capabilities to enhance product brand and quality. The approach achieved great success in increasing sales margins and promoting the company's image, helping TTY to leave out the endless price-cutting competition in the generic drug market and, thus, can channel more cash into long-term new drug development. The transition runs fairly well. In 2001, the company successfully launched IPO in Taiwan OTC technology listings and its registered capital increased triply from (US) $4.29 million in 1997 to $14.29 million in 2003.

Using the case study methods, this chapter is to investigate TTY's intellectual capital and depict the role of TTY's intellectual capital in enhancing its key successful factors to achieve corporate strategies. In their conceptual intellectual capital model, Roos *et al.* (1998, p.63) also emphasized that in cultivating and managing intellectual capital, selected measures for intellectual capital should link to firms' key successful factors derived from corporate strategies. Therefore, this study is expected to have contribution to illustrating Roos *et al.*'s conceptual model by analyzing the case company's transformation and intellectual capital management.

TTY's intellectual capital and competitive advantages

The role of TTY's intellectual capital in supporting the achievement of corporate visions, strategies, and competitive advantages. Having successfully transformed from a traditional generic manufacturer into a new drug development company, TTY's vision now is seeking to be an innovative biopharmaceutical company specializing in anti-cancer drugs in Asia and focusing on new drugs for prevalent Chinese diseases.

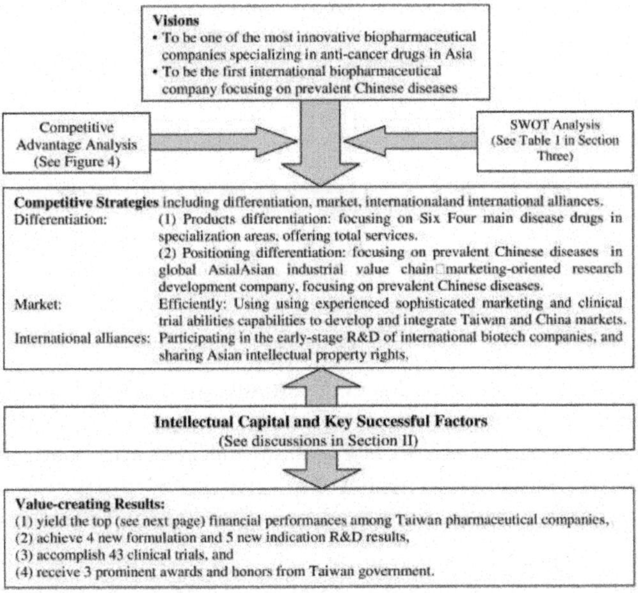

Fig. : An Illustration of the relationship between TTY's intellectual capital and corporate visions, strategies, and competitive advantages.

In light of the increasing competition in pharmaceutical industry, TTY identifies five key successful factors (KSFs), including:

(1) bring corporate strategies into operation,

(2) recruiting and training right people,

(3) effective manage-ment of pipeline and portfolio,

(4) becoming a preferred partner for strategic alliances, and

(5) reducing time to reach the highest sales.

TTY's intellectual capital plays an important role in enhancing those key successful factors.

The first KSF is related to the execution of corporate strategies. TTY positions itself as a marketing-oriented, new drug development company. In order to demand high prices for its branded new drugs, TTY focuses its major relational capital on large hospitals. In developing innovative capital, TTY's R&D strategy is not to conduct a whole-range new drug R&D. Instead, it focuses on the phase III new drug development, which includes developing new indication and new formulation drugs. Consistent with the company's self-positioning, the phase III development is the closet stage to the product market; therefore, focusing on this stage can limit TTY's R&D risks and expedite the payback periods. TTY's R&D intensity, gross margins, and growth rates with other pharmaceutical companies listed in Taiwan stock market. Although TTY invested in the highest R&D intensity, it remains to be the most profitable and fast growing company.

Table : A Comparison of R&D, gross profits, and growth rates among Taiwan listed pharmaceutical companies in 2002.

Company	Chinese Chemical	Yung Shin	Standard Chemical	Sin-pharma	Yung Ri	Chi Cheng	TTY
Total assets (in millions)	178	151	74	45	16	39	48
Net Sales (in millions)	69	68	36	24	7	11	36
R&D %*	5,5	8,7	5,3	8,4	5,7	1,4	11,5
Cross profit%	31,9	65	51	45	38,7	39	67,7
1998-2002 growth rate	15,5	-3,5	31,6	44,1	25,1	29,2	121

* R&D% is measured by R&D expenditures divided by net sales.

In supporting of the KSF of recruiting and maintaining quality work force, TTY stresses on investment in human resources. Its average employee training and education costs are one of the highest among Taiwan's listed companies. The payoff for TTY's investment in human capital is quite fruitful. The company's average employee productivity increased from $108,000 in 2000 to $148,200 in 2002.

To reduce time to reach the highest sales, TTY's strategy is to invite its major customers to participate in the early stage of new drug R&D, thereby developing products better fit customers' needs. Co-developing new drug with physicians not only expands the company's innovative capital, but enhances relational capital with major customers as well. TTY's optimalization capability of the marketing-oriented development strategy in building its core competitive edges. Based on its strategic position, TTY devotes substantial resources to building solid distribution channels with large hospitals and medical centers.

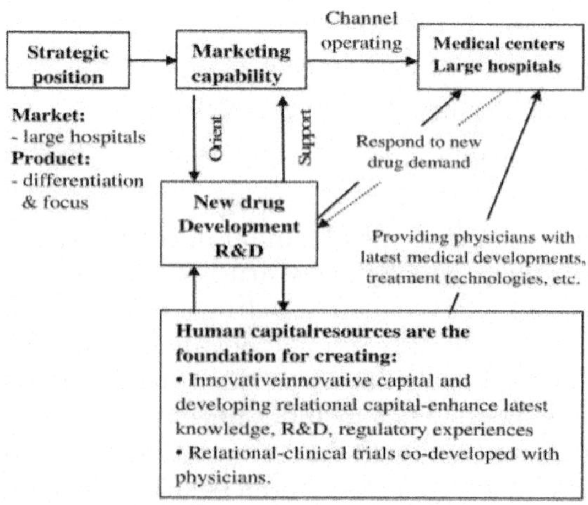

Benefits:

Developing a close relation with physicians Providing better products fit physicians needs Forming strategic alliance with physicians

Fig. : TTY's core competitive advantage: Marketing-oriented development strategy and optimalization capability.

Physicians of large hospitals and medical centers are more willing to use new treatments and new drugs for patients. Feedbacks from them often provide the direction for future new drug development. Finally, TTY's innovative capital and relational capital also bring the advantages in seeking international strategic alliances. By establishing expertise and clinical experiences in prevalent Chinese diseases as well as distribution channels in Chinese drug market, TTY is well equipped as a preferred partner for international pharmaceutical giants seeking to enter the Chinese market.

Value-creating Results

TTY's successful transformation brings fruitful results in both financial and nonfinancial performances. TTY's financial performances yielded leaped growth, with sales growing from $22.6 million in 2000 to $35.6 million in 2002 and EPS growing from $0.05 in 2000 to $0.17 in 2002, on a retrospective basis.

TTY's financial performances come from its advancements in R&D results. It is the first company in Taiwan and the third in the world to launch the liposomal (nanotechnology) formulations. In addition, the company's new indication development has successfully launched several anti-cancer drugs in both Taiwan and overseas markets. During 2001-2003, TTY received three prominent awards and honors from Taiwan government in recognition of its technological innovation and achievements in the pharmaceutical industry.

Conclusions and Discussions

Given Taiwan's small pharmaceutical market size, it seems economically infeasible to support the whole phases of a new drug R&D, covering from discovering new chemicals to developing new drugs. However, TTY's success illustrates that even in a relatively small home market size, adopting adequate R&D strategies and investing in intellectual capital can bear fruitful results. By developing intellectual capital, the company is able to maintain quality workforce, enhance relations with stakeholders, fulfill fit R&D strategies, and establish a supportive organization.

Despite recognizing the importance of intellectual capital and considerably investing in it, TTY, like many other companies, still has not developed appropriate measures for intellectual capital.

As executives and employees paid attention to what they measured and could not manage well what they were not measuring, the lack of measures makes TTY difficult to objectively evaluate and monitor the changes in intellectual capital. Further, Kaplan and Norton indicate that the value of intangible assets derives from their ability to help the organization implement its strategy. Therefore, the value of intellectual capital depends on how it helps achieve corporate strategies. In developing intellectual capital that can effectively help achieve corporate strategies, it is essential to identify and focus on those intellectual capital indicators that closely link to firms' key successful factors for achieving corporate strategies.

Delineating strategy-related intellectual capital indicators needs deliberate effort and is a difficult task. However, it is crucial for companies to effectively link intellectual capital to strategy-implementation and, more importantly, to convey the value of intellectual capital to outside stakeholders.

Finally, the success of TTY's transformation also has implications for pharmaceutical companies in developing countries. Cost-efficiency is, in general, an important competitive edge for developing countries. However, the long product life cycle and low production costs in the pharmaceutical industry suggest that the competition in the pharmaceutical market does not lie in cost-efficiency, but in innovation and product quality. Creating own high-value core products is crucial for pharmaceutical companies to build sustainable competitive advantages, and, thus, risk-taking is inescapable for the executives of pharmaceutical companies.

However, it might be economically infeasible for most pharmaceutical companies in developing countries to compete with those global pharmaceutical giants in new drug R&D. Therefore, to have a role in the global market, it is essential for pharmaceutical companies in developing countries to define their niche position in the industrial value chain and then develop strategy-related intellectual capital to build up sustainable competitive capability.

DESIGN & THE IMPLEMENTATION OF KNOWLEDGE MANAGEMENT SYSTEM

Business enterprises typically are valued at the net tangible assets recorded on their books. When the market value of a firm succeeds its book value, conventional stock market theory regards the premium as the market's assessment of intangible assets or intellectual capital of the firm. Svieby classified intellectual capital into three categories: employee competence, internal structure and external structure. The managing and measuring of intellectual capital are widely discussed in commercial applications. The implementation of key factors in intangible assets using indicators in day-to-day operations highlights the company's value in creating processes. Many other firms apply similar frameworks in managing their intellectual capital, for example: Exxon Chemical's, "Learning Forums"; World Bank's, "Communities of Practice"; Toyota's, "Alignment Rooms"; Ryder Trucks', "Knowledge Center"; Allied Signal's, "Mental Model Learn"; and Scott Paper's "Real Bottom-Line Customer Service". The purpose of this paper is to explore how an organization can adopt the concept of knowledge management and subsequently measure the intellectual capital - specifically in our company.

Knowledge management is the practice of harnessing and exploiting intellectual capital in order to gain competitive advantage and customer commitment through efficiency, innovation and effective decision-making. Both the ideas of knowledge worker and expert labor think that knowledge management is important to any entity.

The knowledge management system is the framework of an integration of organizational elements in organizational culture, organizational information technology infrastructure and the organization's store of individual and collective experiences, learning, insights, values, etc. Members can effectively accomplish organizational goals through knowledge management processes and procedures.A firm that effectively manages knowledge is likely to be considered a learning organization. Knowledge dissemination and responsiveness to knowledge are cited repeatedly as the most effective way to a competitive advantage While the need for effective managing of knowledge is accepted, much of the literature continues to explore measurement and its effect on outcomes.

KMS (Knowledge Management Strategies) Conceptual Model in Industries

There are two types of knowledge involved in industries settings: academic knowledge and organizational knowledge. Academic knowledge is the primary purpose of universities and colleges. Organizational knowledge refers to knowledge of the overall business of an institution: its strength and weaknesses, the markets it serves, and the factors critical to organizational success. This chapter suggests strategies for the formation of KM in an academic knowledge framework and organizational knowledge framework. It is believed that knowledge management can be used to support Company, which in turn supports teaching and learning

Academic Knowledge Framework

Huang (1998) suggested four major processes to form a culture of knowledge sharing and collaboration. They are:

(1) making knowledge visible,

(2) increasing knowledge intensity,

(3) building knowledge infrastructure, and

(4) developing a knowledge culture.

From an academic knowledge perspective, the learning community should start at the individual level, create departmental knowledge, create domains of knowledge across departments that share academic interests or disciplines, create institutional knowledge networks and networks with other institutions and corporations. This research presents three strategies to establish knowledge ecologies within the academic framework: individual strategy, institutional strategy and network strategy.

Organizational Knowledge Framework

The most generally recognized four organizational knowledge management strategies are culture, leadership, technology, and measurement. The examples of culture strategies in industry include staff development and training, communities of practice, and promotion of learning organizations. KM leadership strategies in industries include KM strategic planning in alignment with mission/vision, hiring knowledgeable employees, and evaluating employees for knowledgeable contribution. Company is responsible for providing infrastructure of tools, systems (intranets, web pages, electronic repositories, and data base, etc.), platforms, and automated solutions that centralized the development, application, and distribution of organizational knowledge. Measurement strategies may include benchmarking against other industries, allocating resources toward efforts that measurably increase the knowledge base, and linking and accessing impact of KM to the strategic plan.

Many KM application experts recommend the multi-perspective modeling approach. Models have been used previously from business management, such as SWOT (Strengths, Weaknesses, Opportunities, and Threats) analysis, balanced scorecards and benchmarking as good sources of techniques.

Despite the issues is better known at the end of this stage of the process knowledge management course,we review the main factors of knowledge management projects fail. The main reasons are as follows:

1. Hasty planning and command and to desire to extract the knowledge of experts.
2. Sheer care on expository aspect of design and sacrificing accuracy for speed.
3. Lack of primary study and evaluating knowledge requirements.
4. Outsourcing justify irreversible renal extraction process to consultants outside the organization and Disclaimer.
5. Lack of proficiency and familiarity of advisors.
6. Lack of employer and advisor's care to necessity of make culture and planning in order to making physical and spiritual incentives for experts.
7. Mere reliance on a very inefficient and display software to display in bringing the so-called sciences extracted.
8. Designating a too little time in order to extracting skillful and experienced experts.
9. Lack of stress on educating particular courses of knowledge management to experts.
10. Releasing the extractions after interviews and lack of proper evaluating for completing, editing and filtering the experts.
11. Lack of predicting of strategies in order to accretion of practical knowledge and creating collective knowledge.
12. Lack of planning for participating before, during and after the interviews with experts.

Obstacles in making knowledge management in organizations:

1. The shortage of human relations in organizations which causes increase of knowledge.
2. Lack of proper relations between knowledge management and organizational guidelines.

3. Non-clearance of knowledge and knowledge management value and inaccurate assessment of knowledge assistant which can be given to the organization.
4. Lack of complete value in knowledge management practices.
5. Oral skills problems in organization.

Obstacles in making knowledge management according to Davenport's viewpoint:

A-Organizational viewpoint:

1. Time destruction, necessity of heavy work and high expenses .
2. Imposing extra work from knowledge management.
3. Limitation in knowledge based systems technology.
4. Problems of data extraction from the text.
5. Increase over information.
6. Coding difficulty tacit knowledge.
7. Necessity of powerful culture for creating observation justification.
8. Possibility of sending incorrect message from knowledge staff and information manager.

B-Knowledge committee and members' viewpoint:

1. Personal effort to perceive knowledge with expecting reward.
2. Scare of criticism and be charged from management and other chairman.
3. Lack of respect to the other fields.
4. If lack of respect, trust and shared goals, efforts to reverse.
5. Extra works which imposed to other parts of staff.

C-Personal viewpoint:

1. Distaste to joint information.
2. Knowledge is power source, advantage, reward and penalty.
3. Challenge among professionals.
4. Sense of worth and dignity due to skills.

Preventing obstacles, knowledge management usage and creation new knowledge:

1. Assure employees to declaring creative thoughts without fear of interested critics.
2. Encourage the employees to have critical viewpoints to accepted procedures and ideas about organization products and strategies.
3. Making epistemology knowledge.
4. Promotion of meritocracy.
5. Designating part of a time to some people in order to do their favorite activities.
6. Encourage to relaxation and recreation in desirable workplace.

Case Background

Saveh rolling and Profile Mills Company first started its operation in 1354 (1975) to produce various types of pipes and steel profiles with the production capacity of 50000 tons a year. In order to fulfill development strategy and increasing the variety of products and entering exporting markets the management of this company has started its precise investment planning to promote the quality and the quantity of products. Currently we are producing 2000000 tons of different kinds of steel products in various sizes which has been regarded the outcome of this effort and precise management planning.

Knowledge Management Application Framework in Saveh Rolling and Profile Mills Company

Leaders of our company recognized the need for organizational change to reposition strategically in the competitive industry.

Organization culture can inhibit or enhance organizational change efforts in knowledge management initiatives. Proper procedures are taken to ensure the concept of KM is correctly understood and thus creates a culture of sharing using organization-wide vocabulary. An outside KM specialty team is brought in as mediator for KM implementation. To ensure successful implementation of KMS, the Institute selects 50 representatives from faculty and staff of various departments to attend 14 brainstorming sessions.

The design of an implementation process should focus on the user-How KM can improve a worker's daily work? At the strategic level the organization analyzes and plans its business in terms of the knowledge it currently has and the knowledge it needs for future business processes. Road Map serves as a living document regularly updated and a framework for the monitoring of the knowledge management program. This document reflects the current state of the interrelationship between work in progress and proposed for the future and the overall milestones and aims of the program.

KM Implementation Model in Saveh Rolling and Profile Mills Company

Core Team is formed as center for KMS for each community, including Router for initial evaluation of knowledge proposal, Reviewer for verification, Structures for categorizing, Editor for formalizing, Category Owner for maintenance and Communicator for knowledge transfer and sharing. For each community, specific KM activities are designed for acquiring and sharing knowledge. Major categories in each community are clearly defined.

The success of KM implementation should be linked to economic performance or industry value. The performance of a COMPANY ties to rating, recruiting and financial stability.

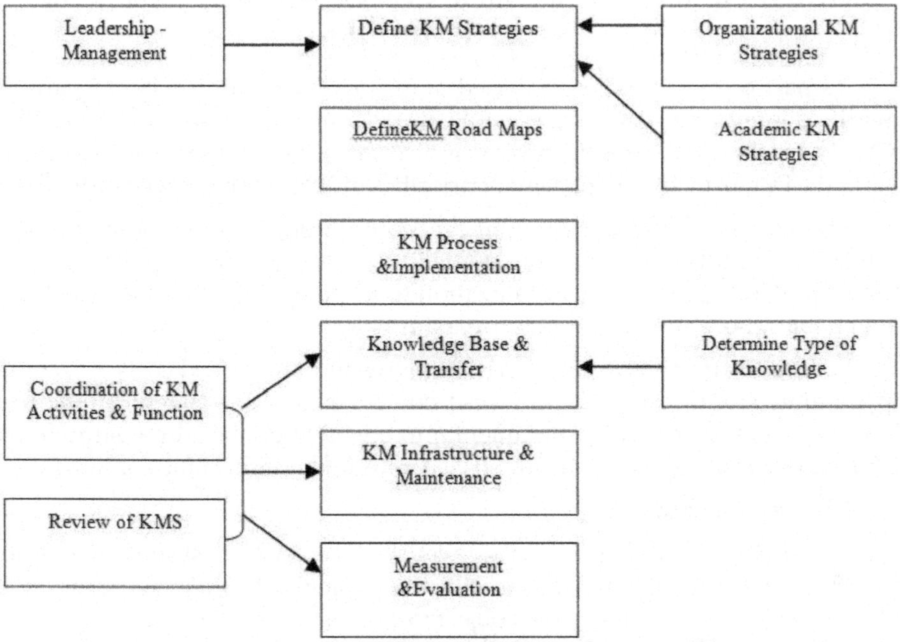

Fig. : Implementation model.

Key success factors for Saveh rolling and profile mills company:

1. Knowledge of consulting field
2. Experience
3. Professional knowledge of consultant
4. Know-how and Experience
5. Time for research and investigation
6. Knowledge
7. Quality of consulting Service
8. Control of service Quality
9. Ability of listening, surveying, and analyzing
10. Capability
11. Company reputation
12. Good public relation
13. Feeling
14. Sense of humor
15. Mastering of public Relation
16. Innovation capability

THE CANADIAN INTEGRATED RISK MANAGEMENT FRAMEWORK (IRM FRAMEWORK)

As part of the annual risks-based audit planning process, the Department identified integrated risk management (IRM) as an area that should be audited in 2013. The rationale for this decision was also supported as IRM was identified as one of the Department's management priorities in 2013. PS has never audited IRM.

Risk management makes a significant contribution in strengthening capacity to recognize, understand, accommodate and capitalize on new challenges and opportunities. It prepares an organization to respond to change and uncertainty and enables more effective decision-making.

Failure to effectively manage risks can result in increased costs and missed opportunities, which can compromise outcomes, and ultimately public trust. Sound risk management is fundamental to effective public administration as it can lead to a more effective, results-based, and high performance organization.

Risks can include:

- Strategic risks: Loss or damage caused by external or systemic conditions or events which may negatively and strategically affect the ability of the Department to achieve its objectives.
- Operational risks: Loss or damage caused by people, processes or technology.
- Hazards: Loss or damage caused by natural, accidental or pre-meditated actions.

IRM is characterized by several key dimensions:

Table : IRM is characterized by several key dimensions.

Dimensions	Description
Horizontal Integration	This refers to the harmonization and alignment of risk management practices horizontally across departmental Branches and programs. Horizontal integration is needed to ensure that consistent approaches for similar risk-based decisions are used, so that overlap and duplication is avoided and that best practices can be shared and leveraged. It is also critical to ensure the appropriate sharing of risk information across the organization, which ultimately enables more informed and robust decision-making.
Vertical Integration	Vertical integration exists when risk management practices at various levels in the hierarchy of the Department are aligned and integrated with one another such that risk information generated at lower levels of the organization can, as appropriate, be consistently aggregated and escalated up to senior management. Vertical integration is needed to ensure that risk information is appropriately shared with senior decision-makers in a way that balances their need to know about, and act on, certain types of risk with lower levels of management's authority to manage.

Functional Integration	Functional integration exists when risk management practices are incorporated and integrated directly into existing business functions and decision-making processes. It is vital to ensuring that decisions and managerial functions are, where applicable, appropriately risk-informed and that consistent, efficient and regularized processes exist to ensure that these risk-informed processes are robust and reliable.

International standards and government-wide expectations of good governance emphasize that all types of risks should be considered and integrated into all planning and delivery activities, at the strategic and operational levels. The Treasury Board (TB) *Framework on the Management of Risk* provides guidance to Deputy Heads on the implementation of effective risk management practices at all levels of their organization. This supports strategic priority setting and resource allocation, informed decisions with respect to risk tolerance, and improved results.

The *Management Accountability Framework* (MAF), which is a key performance management tool for the Federal Government, also expects departments to incorporate risk management practices and principles into the organization's strategic, operational and functional activities.

In 2010 Public Safety Canada (PS) approved an *Integrated Risk Management Policy* (IRM Policy) and in 2013, subsequent to the audit period, updated and formally approved the *Integrated Risk Management Framework* (IRM Framework). "This Framework is designed to communicate PS's risk management strategy to all levels of staff, and supply the means to build and maintain a strong risk management capacity. Moreover, it establishes the importance of integrating risk management into the Department's policy, planning, resource allocation and decision-making processes, and creates a link with the organization's strategic documents." The Framework also outlines the key activities in the risk management lifecycle including risk identification, assessment, response, and monitoring.

One of the processes identified within the IRM Framework is the Corporate Risk Profile (CRP) which presents a departmental perspective of the top risks and opportunities by Program at the Program Alignment Architecture (PAA) level. Developed in accordance with the departmental Performance Measurement Framework (PMF), this document, which is now in its third year of implementation, creates a direct link between identified risks and opportunities and expected results identified in the PMF.

Public Safety Canada's Objectives related to Integrated Risk Management

The Department's objectives as they relate to Integrated Risk Management are stated in the Department's IRM Policy and are summarized as follows:

- Implement a process for risk identification and assessment that is comprehensive and systematic;
- Create a management environment that effectively controls and manages risks and allows for appropriate risk-taking within defined parameters;

- Integrate risk management within PS's ongoing activities and management functions, with an emphasis on risk management across Branches and program activities;
- Foster a culture of continuous improvement in risk management through education, training, monitoring and evaluation;
- Promote clear communication and awareness of risk; and,
- Fulfill the requirements outlined in the TB Framework for the Management of Risk.

Public Safety Canada's Roles and Responsibilities related to Integrated Risk Management

The ADM, Strategic Policy Branch as the Chief Risk Office (CRO) is responsible for:

- Positioning risk management as an integral component of the Department`s ongoing activities and management functions;
- Determining departmental risk tolerance in consultation with relevant stakeholders;
- Establishing risk management objectives and strategies that align with PS priorities and objectives;
- Developing and monitoring the implementation of risk management guidelines;
- Informing and supporting the Deputy Minister and Management Committee on all identifiable and relevant departmental risks; and,
- Facilitating the opportunity for training on risk management.

The Departmental Management Committee (DMC) is responsible to:

- Promoting the importance of implementing Integrated Risk Management within the Department;
- Determining the most effective way to implement Integrated Risk Management within the Department;
- Incorporating risk management into decision-making;
- Ensuring that corporate risks are properly identified, assessed, managed and discussed at least twice a year;
- Ensuring the capacity to report on the performance of the Department's risk management function; and,
- Ensuring the existence of a supportive learning environment and appropriate communications related to the management of risk.

The Assistant Deputy Ministers are responsible for:

- Integrating risk management into branch management practices;

- Ensuring that branch risks are properly identified, assessed and managed; ensuring that strategic risks are clearly identified through the departmental planning framework and that they are continuously being brought forward to the relevant committees and/or the Deputy Minister, as deemed appropriate;
- Providing clear direction with regard to the appropriate level of response to various types of risk; promoting a supportive environment in which effective risk management and sensible risk taking are encouraged; and,
- Managing levels of risk associated with branch programs, plans and policies.

All employees are responsible for:

- Assisting managers with the identification and assessment of risk and developing efficient measures to manage these risks;
- Being proactive about identifying and acting upon risks; and,
- Incorporating risk management as an integral part of business decision-making.

Audit Objective

The objective of this audit was to provide reasonable assurance that the Department's approach to IRM was adequate and effective to ensure a consistent approach was used and that risk information was appropriately integrated into decision making.

Scope and Approach

The audit covered the period of September 1, 2011 to December 31, 2012 as this period ensured that findings were current and a complete business management cycle was included. Some of the testing was done on an "under development" basis, examining the processes as they were evolving to provide management with timely advice and insight.

The scope of the audit included all PS Branches and focused on their integrated risk management processes.

The audit reviewed the appropriateness of the Department's fundamental risk management processes, their horizontal and vertical integration into department's decision making processes, and their regular monitoring assuring their ongoing effectiveness.

The audit's focus was on the activities the Department undertook to identify risks within its operational and management activities, however it did not include an examination of the appropriateness or applicability of the individual risk management tools, given their complexity and required subject matter expertise.

Risk Analysis

PS exerts its leadership and oversight role against the backdrop of internal and external business conditions that expose the Department to a range of operational and strategic risks. Specifically:

- The management of risk in the context of PS is inherently complex. The multitude of players (both within and beyond the Department) that create and use risk information, coupled with varying conceptions of risk and risk tolerance makes the horizontal and vertical sharing of risk information more challenging. Of note is the increasing inter-connectedness of Public Safety programs and priorities, as well as the increasing horizontality of files and issues across government. This inherent horizontality not only increases the need for horizontal sharing of risk information but at the same time, makes it more challenging due to factors such as organizational barriers (silos) and different approaches to risk management throughout the Department and the Government. This is further compounded by an organizational culture and practice that is more naturally prone to the safeguarding, not disclosure or sharing of risk information, simply because of the business of the Department.
- The nature of PS's mandate is such that managers intuitively manage risk every day, in the course of their operations. This, coupled with the relative newness of formalized risk assessment and risk management practices, makes it more challenging to implement and regularize risk management processes. Efforts to deploy risk management have been positive; however, as with all change management initiatives (particularly during times of resource constraint), there is a risk that IRM may be seen as a mere compliance requirement, rather than a value added function, and that managers revert to past behaviors.
- The impact of resource constraints and the Deficit Reduction Action Plan must also be borne in mind. Reduced resources make it more challenging to invest in new processes; however, risk-based decision-making is an integral tool to assist managers in making tough policy, programming, and resource use decisions.

The detailed risks to which the Department is exposed as a result of this part of its business. These risks were identified during the planning phase of the audit and were based on extensive interviews and documentation review. The goal of risk-based auditing is to focus the audit examination on the areas that are characterized by the greatest degree of inherent risk. In this way, resources are used efficiently and value to management is optimized. Guided by these risks, the audit's audit objectives, scope and lines of enquiry were identified.

Audit Opinion

In my opinion the Department's approach to IRM at the strategic level was generally adequate and effective to ensure a consistent approach was used and that risk information was appropriately integrated into decision making. Opportunity exists to further strengthen the approach of IRM at branch and directorate levels to ensure the appropriate integration of risk into decision-making.

Statement of Conformance and Assurance

The audit conforms with the Internal Auditing Standards for the Government of Canada, as supported by the results of the quality assurance and improvement program.

In my professional judgment as Chief Audit Executive, sufficient and appropriate audit procedures have been conducted and evidence gathered to support the accuracy of the opinion provided and contained in this report. The opinion is based on a comparison of the conditions, as they existed at the time, against pre-established audit criteria that were agreed upon with management. The opinion is applicable only to the entity examined.

FINDINGS, RECOMMENDATIONS AND MANAGEMENT RESPONSES

Policies and Directives

The audit expected to find a clear and appropriately communicated policy, directives, and/or guidelines relating to IRM. It was also expected that employees understand the following:

- the objectives of the Policy;
- their roles, responsibilities and accountabilities;
- how often risks and opportunities should be identified and assessed;
- how risks should be integrated into business-planning, decision-making and operational processes; and,
- how compliance to the Policy is monitored to ensure a continuous learning cycle.

The audit found a well-defined and appropriately communicated PS *Integrated Risk Management Policy* (IRM Policy), and subsequent to the audit examination period, a PS*Integrated Risk Management Framework* (IRM Framework) was approved by senior management. Both documents had identifiable objectives, clear roles and responsibilities, and reporting timeframes which were aligned with the TB *Framework for the Management of Risk* and TB *Guide to Integrated Risk Management.*

The IRM Framework that existed at the time of the audit for all intents and purposes was similar to the recently approved version. It provides guidance on the key activities of a complete risk management lifecycle. A separate quick guide provides a summary of similar risk management notions and approaches, however the audit found that the distribution of the quick guide was limited to the Branch Heads and the participants. The audit also found that the guidance concentrated on how risks and opportunities should be identified, assessed and to some extent integrated into planning instruments, however, neither one offered guidance on how to integrate risk management into decision-making or work processes. While the existing guidelines are a good foundation, the TB *Guide to Integrated Risk Management*, articulates a risk-informed approach as "building risk management into existing governance and organizational structures, including business planning, decision-making and operational processes."

The IRM Policy outlines the following processes to monitor the implementation of the Policy:

- Departmental compliance with this Policy may be assessed by Internal Audit, as part of its approved risk-based audit plan. Results of the internal audits will be reported to the Deputy Minister and the Departmental Audit Committee.
- Select monitoring and reviews may also be requested by the Chief Risk Officer to assess changes required to policies, directives, procedures, guidelines and profiles in order to effectively manage risks; and/or to provide reports to the Deputy Minister, as required.

The previous and current version of the IRM Framework stipulates that in order to ensure that integrated risk management is implemented throughout the Department, the following three key mechanisms will be used:

- The Strategic Planning Division will promote the application of the IRM Policy in decision-making and planning through the development of tools and mechanisms to facilitate the use of risk information in the Department;
- The TBS Management Accountability Framework (MAF) assessment will highlight best practices and areas of improvement; and
- The Departmental Audit Committee's recommendations will be used by the Strategic Planning Division to improve IRM processes and ensure they are responsive to departmental realities.

While these monitoring mechanisms are positive and will provide some evidence of effectiveness, they remain periodic and at a high-level. There was limited documentation or evidence to support the select monitoring and review of the implementation of the policy. There are no mechanisms that enable timely information on the state of implementation. The audit recognizes that the Strategic Policy Branch (SPB) has succeeded in putting in place many foundational elements with very few resources however a few select indicators would ensure that the momentum gained is not lost.

Risk Management Tools

The audit expected to find tools that would facilitate the full risk management lifecycle. This would include tools that support the management of:

- Strategic level risks and opportunities;
- Individual specific risk applications, such as the management of grants and contributions; and,
- Operational level risks and opportunities.

Strategic Risk Management Tools:

The Corporate Risk Profile (CRP) is the departmental tool which captures the high-level strategic risks. PS has received strong MAF ratings for its CRP process. Additionally, there is positive evidence of the use of the CRP outputs in support of senior management decision-making.

On an annual basis each Branch is required to identify their risks and opportunities and bring this information forward for discussion and prioritization at horizontal departmental integration sessions. Each of the identified risks and opportunities were then assessed on the likelihood of occurrence and their impacts. The purpose of these integration sessions is to allow a "challenge" function and more importantly, to ensure that the cross-cutting impacts of these risks and opportunities were appropriately identified and understood departmentally. Following these sessions the top three risks and opportunities by Program level of the Program Alignment Architecture (PAA) formed the CRP. Responsible Directorates and mitigation plans were then identified for each strategic risk or opportunity.

Identifying Strategic Risks

Generally all interviewees indicated that branch meetings were held to identify potential risks related to their areas of responsibility in preparation for the CRP integration sessions. They also indicated that these risks were informed by other inputs such as outputs from their specific use risk management tools, stakeholder consultations, and inter-departmental committees. The audit found that the approaches to inform and assess the risks were not documented. Consequently, it was not possible to determine the completeness and sufficiency of this risk identification step and whether some operational and strategic risks were overlooked. The absence of documentation may also contribute to a misunderstanding and/or misinterpretation of the basis of these risks.

The audit noted that in some cases the input of risks for certain PAA level Programs was limited in quantity; for example one Program identified only four high-level risks. Given the complexity and sensitivity of each Program, the audit expected more detailed lists. The audit did not attempt to assess the sufficiency, validity, and relevancy of these risks. The limited number did raise concerns as to whether the Department has overlooked any risks and whether the process benefited from a robust horizontal "assessment and challenge".

Assessing Strategic Risks

Interviewees indicated that the integration sessions were important in providing an opportunity for all participants to understand the departmental risk environment as well as the proposed mitigation plans and acceptable tolerances. Although there was no evidence to support the deliberation of these sessions, it was noted by participants that their ability to report back to their colleagues on a department-wide risk perspective contributed to a heightened risk awareness and knowledge.

The CRP presents the top three risks and opportunities for each Program level of the PAA. As a result it is possible that risks ranking lower than the top three within one specific Program, but higher than the top three risks of other Programs will not be included in the CRP. Consequently, there is no mechanism in place that ensures that the Department's highest ranking strategic risks are managed corporately and monitored by the most senior governance committees.

Developing Risk Responses for Strategic Risks

CRP risks and mitigation plans are developed in the fall as part of the regular departmental planning and reporting cycle. This information is then prepopulated by the Strategic Policy Branch into the Branch Business Plans (BBP) to ensure continuity. It is then expected that branches appropriately update the mitigation plans into the respective directorate workplans. This exercise is positive in that it ensures that all CRP risks and mitigation plans are assigned to the appropriate branches.

To better understand the alignment of the risks and the associated mitigation plans, the auditors developed a cross-walk between the risks identified in the CRP to the risks identified in the business plans to the actions identified in the directorate work plans. The audit was able to trace the CRP outputs into the BBPs.

As expected, the directorate workplans contained detailed actions that were intended to address the high level CRP mitigation plans. However it was not always clear whether:

- The mitigation plans identified responded to the original high level CRP mitigation plans,
- The mitigation plans cascaded into the individual directorate workplans as appropriate, and,
- Changes to the mitigation plans were reassessed to ensure they still sufficiently mitigated the risks.

There was also no evidence that DMC was made aware of, or approved, modifications to CRP mitigation plans.

An example that highlighted the potential misalignment between the CRP mitigation plans and the directorate work plans was observed in one of the 2011-12 CRP risk statements: "That the Government Operations Centre (GOC) infrastructure may not support a coordinated response to an event affecting the national interest – space, security, survivability, and sustainability requirements." This risk statement and the related mitigation plans remained unchanged through to 2013-14. It was uncertain whether the CRP mitigation plans and the associated directorate work plans were either not addressed or not sufficient, or whether the Department chose to accept the risk without mitigating it.

Specific Use Risk Tools

As part of the audit the following program areas specific use risk management tools were examined:

- Public Safety Information Management System (PSIMS) which captures risk information for individual recipients for departmental grants and contribution programs;
- All Hazards Risk Assessment tool (AHRA) which, led by PS, in close partnership with Defense Research Development Canada - Centre for Security Science, supports all federal government institutions in fulfilling their legislative responsibility to conduct mandate-specific risk assessments as the basis for Emergency Management planning; and,
- Critical Infrastructure tool which was developed to support the goal of the National Strategy for Critical Infrastructure in partnership with federal, provincial and territorial governments and critical infrastructure sectors for the purpose of improving information sharing and protection and sustaining a commitment to all-hazards risk management. This all-hazards approach considers: natural hazards; accidental hazards, and intentional threats.

The audit found the development of these specific use risk management tools to be a very positive step in fulfilling risk management and providing outputs to inform decision-making processes in support of these program areas.

It was indicated that these tools were in the early stages of implementation and therefore did not yet provide consistent and systematic outputs. The audit also noted that there was limited documented guidance to support how these tools were to be integrated into other risk-informed operations such as:

- How inter-dependencies and information should be shared;
- How outputs should be integrated into directorate, branch and departmental risk management activities; and,
- How these tools should support the complete risk management lifecycle including response and monitoring.

While the tools supported risk identification, assessment, and some aspects of response, none of the tools had defined approval, reporting and monitoring mechanisms or processes.

Operational Risk Management Tools

Branch Risk Profiles (BRP) as identified in the PS *Integrated Risk Management Framework* are intended to inform each Branch's decision on plans, priorities and resource allocation by capturing, assessing and summarizing key risks that could most impact the achievement of branch objectives. Once finalized the BRPs are used to drive discussions in the development of the CRP.

The audit found that BRPs were generally not completed or consistently maintained nor did branches have any other consolidated risk document. In the instances where branches had a documented BRP, its content was essentially the same as the CRP risks with little additional information. Employees did not use any specific use or operational risk management tools to identify, assess, respond and monitor risks associated to policy development or research activities. Rather the audit was told that these risks were identified through undocumented intuitive means or through stakeholder engagement and committees.

Without having a consolidated risk document, risk management is fragmented and disconnected and consequently the department cannot easily be assured of the completeness and consistent interpretation of risk information and that all risks are appropriately considered throughout the planning and reporting stages. This may also impact the integrity of the planning process and may jeopardize the achievement of better performance.

Governance

The audit expected to find governance structures that promote a risk informed culture and management practices throughout the Department. It was also expected that departmental governance structures would monitor the management of risks, and the implementation of the IRM Policy and Framework. The audit examined the following departmental committees:

- Departmental Management Committee;
- Departmental Audit Committee; and,
- Branch and Directorate Management Committees.

Departmental Management Committee (DMC)

The DMC terms of reference states that the committee`s objectives are to review and approve policies, projects, plans, performance, and reports relating to a broad and diverse suite of corporate management programs and services for the Department including risk management. Further the IRM Policy requires that the DMC identify, assess, manage, and discuss corporate risks at least twice per year.

The audit found:

- In 2012-13 DMC did not review the initial CRP update as it was approved bi-laterally with the DM. The audit was informed that there were opportunities for all members to review and provide their input individually. However, members did not benefit from a collective discussion allowing for the understanding of the Department`s risk tolerances.
- An update on the risks and their respective mitigation plans was presented during the mid-year review process. Interviewees noted that there was limited discussion, only twenty minutes were allotted on the agenda for the entire mid-year review of performance and risk; this is not conducive to a fulsome discussion. Further the documentation supporting the mid-

year discussion only included a "symbol" indicating risk status. The onus was on each individual ADM to determine what to present. The audit expected a more deliberate discussion and consistent presentation of strategic risks and their mitigation by the senior management team.

- Finally the audit noted that the onus was on individual presenters to the DMC to determine the criteria for the disclosure of risk information.

The Departmental Audit Committee (DAC)

DAC, which provides advice on such areas as risk management, was informed of "risks" and risk management processes on a regular basis throughout the audit period. DAC commended the progress made in the implementation of IRM. It was noted that this committee actively uses the outputs from the CRP to challenge departmental activities such as resource allocations. Advice from the Committee members focused on the importance of not losing track of risks when moving on to the next issue; and the need to use risk information more strategically for such activities as linking resources to the areas of risk and priority.

Branch and Directorate Management Committees

Branch and Directorate management committees were identified as fundamental governance structures. However none of these meetings had specific terms of reference or other agreed upon processes defining "when" or "what" risk information should be presented. Nor were any of the meeting discussions documented, thereby making it challenging for audit to conclude whether these committees were providing any form of risk oversight. Compounded by the absence of BRPs- Operational Risk Management Tools, and the lack of directorate risk information within the BBP, the audit could not conclude whether risk information was being used to inform decision-making.

Without defined information requirements, documented discussions, and rigor in ensuring compliance, sufficient and appropriate risk information including risk tolerances and mitigation status, risks may not be effectively reported and monitored. Organizations may be keeping information that prevents them for achieving objectives and the highest of performance.

Integration – Risk Informed Decision-making and Culture

The audit expected to find risk-informed approaches to business planning, decision-making, and operational processes. As stated in the PS *IRM Framework*, "the objective is to create an environment where all levels of the organization will instinctively look for risk and opportunities and take into account their impacts on departmental outcomes when making decisions at both the operational and the corporate levels."

Departmental Level

The audit found that senior managers have started to integrate risk information into key decision-making activities as "risk" has become more engrained. The inclusion of risk information has largely been done through the use of the CRP outputs and specific risk information presented to DMC. The audit noted the following examples:

- The Banking Day exercise which re-allocates resources to new funding pressures now prioritizes the status of implementation of the CRP risk mitigation plans before considering other pressures.
- The Risk and Results Based Approach to Staffing developed by the Human Resources Directorate, informs delegated managers on the risk tolerances of senior management for different types of staffing actions and categorizes them such that controls and reporting requirements are commensurate to the different levels of risks.
- During the development of the departmental priorities, SPB incorporated the CRP outputs into the planning process ensuring the visibility of risks and opportunities and their alignment to each priority. The CRP outputs were further used in the development of the Report on Plans and Priorities. The integration of risks in this manner was a positive step in guiding the focus of the Department.

Even so, there continues to be gaps in regard to integrating risk information in decision-making. For example, there was no systematic requirement for risk management information to be included in briefing notes, memos, status reports, or financial and non-financial performance reports.

The audit did note, the Strategic Planning Division, Strategic Policy Branch has started a *Risk and Threat Assessment Community of Practice,* which will provide a valuable means of tapping into departmental strengths, and to share tools, information, and expertise both horizontally and vertically on an ongoing basis. This forum will provide an opportunity to learn and collaborate where possible and possibly reduce duplication of efforts.

Departmentally the audit found senior management generally supportive of risk management and that tolerances for certain risks were communicated to lower levels. Examples of the mechanisms used to establish a risk management environment included the official approval and communication of the IRM Policy, and the completion and communication of the annual CRP.

However, as noted in previous audits a culture of silos still persists in some areas which impede broad and transparent departmental risk discussions. Without building the transparency and trust needed for a comprehensive risk understanding, departmental risks strategies and tolerances may not be clearly communicated leading, to sub-optimal decisions and use of resources.

Branch/Directorate Level

As noted earlier there were generally no BRPs and few operational risk management tools to facilitate risk awareness and understanding. Generally, at the Branch and Directorate level the use of risk was more intuitive. Risk management was less structured and often only formalized for the purposes of contributing to the required high-level CRP process or applied for a specific program need such as the management of contribution recipients. A number of interviewees noted that the CRP outputs were of limited use and perceived it to be an administrative task with minimal value-add. Therefore in some cases it did not guide their decision-making activities.

Generally interviewees noted that information related to finances, HR capacity, and stakeholder concerns were the key sources that informed decision-making. Risks per say were not the focus of discussions. These indicators in and of themselves are not sufficient to enable robust integrated risk management.

While there were positive signals that some decisions were informed by risk, without a more structured and deliberate risk-informed approach, there is a risk that sub-optimal decisions may be made.

Recommendations :

- Branch Heads should develop annual Branch Risk Profiles or similar tools which ensure risks and opportunities are appropriately identified, assessed, mitigated, and monitored. These Branch Risk Profiles should inform the branch and departmental planning, decision-making, and operational processes.
- The ADM, Strategic Policy Branch, should develop indicators that inform on the state of implementation of the IRM Policy and Framework.
- The ADM, Strategic Policy Branch should strengthen the processes surrounding the development, modification, and alignment of CRP mitigation plans into the departmental and branch planning and reporting cycles.
- The ADM, Strategic Policy Branch, as Chief Risk Officer (CRO) should ensure that DMC has the opportunity to conduct a fulsome review of the departmental strategic risks at minimum twice per year. The CRO should also facilitate the inclusion of appropriate risk information, including those risks identified in the Corporate Risk Profile and the Branch Risk Profiles, into key DMC decision-making activities.

Chapter 7

Supporting Process Design by a Sustainability

SUSTAINABLE DESIGN

Sustainable design seeks to reduce negative impacts on the environment, and the health and comfort of building occupants, thereby improving building performance. The basic objectives of sustainability are to reduce consumption of non-renewable resources, minimize waste, and create healthy, productive environments.

Sustainable design principles include the ability to:

- Optimize site potential;
- Minimize non-renewable energy consumption;
- Use environmentally preferable products;
- Protect and conserve water;
- Enhance indoor environmental quality; and
- Optimize operational and maintenance practices.

Utilizing a sustainable design philosophy encourages decisions at each phase of the design process that will reduce negative impacts on the environment and the health of the occupants, without compromising the bottom line. It is an integrated, holistic approach that encourages compromise and tradeoffs. Such an integrated approach positively impacts all phases of a building's life-cycle, including design, construction, operation and decommissioning.

Theory

The intention of sustainable design is to "eliminate negative environmental impact completely through skillful, sensitive design". Manifestations of sustainable design require renewable resources, impact the environment minimally, and connect people with the natural environment.

Beyond the "elimination of negative environmental impact", sustainable design must create projects that are meaningful innovations that can shift behaviour. A dynamic balance between economy and society, intended to generate long-term relationships between user and object/service and finally to be respectful and mindful of the environmental and social differences.

CONCEPTUAL PROBLEMS

Diminishing Returns

The principle that all directions of progress run out, ending with diminishing returns, is evident in the typical 'S' curve of the technology life cycle and in the useful life of any system as discussed in industrial ecology and life cycle assessment. Diminishing returns are the result of reaching natural limits. Common business management practice is to read diminishing returns in any direction of effort as an indication of diminishing opportunity, the potential for accelerating decline and a signal to seek new opportunities elsewhere.

Unsustainable Investment

A problem arises when the limits of a resource are hard to see, so increasing investment in response to diminishing returns may seem profitable as in the Tragedy of the Commons, but may lead to a collapse. This problem of increasing investment in diminishing resources has also been studied in relation to the causes of civilization collapse by Joseph Tainter among others. This natural error in investment policy contributed to the collapse of both the Roman and Mayan, among others. Relieving over-stressed resources requires reducing pressure on them, not continually increasing it whether more efficiently or not

WASTE PREVENTION

Negative Effects of Waste

About 80 million tonnes of waste in total are generated in the U.K. alone, for example, each year. And with reference to only household waste, between 1991/92 and 2007/08, each person in England generated an average of 1.35 pounds of waste per day.

Experience has now shown that there is no completely safe method of waste disposal. All forms of disposal have negative impacts on the environment, public health, and local economies. Landfills have contaminated drinking water. Garbage burned in incinerators has poisoned air, soil, and water. The majority of water treatment systems change the local ecology. Attempts to control or manage wastes after they are produced fail to eliminate environmental impacts.

The toxic components of household products pose serious health risks and aggravate the trash problem. In the U.S., about eight pounds in every ton of household garbage contains toxic materials, such as heavy metals like nickel, lead, cadmium, and mercury from batteries, and organic compounds found in pesticides and consumer products, such as air freshener sprays, nail polish, cleaners, and other products. When burned or buried, toxic materials also pose a serious threat to public health and the environment.

The only way to avoid environmental harm from waste is to prevent its generation. Pollution prevention means changing the way activities are conducted and eliminating the source of the problem. It does not mean doing without, but doing differently. For example, preventing waste pollution from litter caused by disposable beverage containers does not mean doing without beverages; it just means using refillable bottles.

Waste prevention strategies In planning for facilities, a comprehensive design strategy is needed for preventing generation of solid waste. A good garbage prevention strategy would require that everything brought into a facility be recycled for reuse or recycled back into the environment through biodegradation. This would mean a greater reliance on natural materials or products that are compatible with the environment.

Any resource-related development is going to have two basic sources of solid waste – materials purchased and used by the facility and those brought into the facility by visitors. The following waste prevention strategies apply to both, although different approaches will be needed for implementation:

- Use products that minimize waste and are nontoxic
- Compost or anaerobically digest biodegradable wastes
- Reuse materials onsite or collect suitable materials for offsite recycling

Sustainable Design Principles

While the practical application varies among disciplines, some common principles are as follows:

- Low-impact materials: choose non-toxic, sustainably produced or recycled materials which require little energy to process
- Energy efficiency: use manufacturing processes and produce products which require less energy
- Emotionally durable design: reducing consumption and waste of resources by increasing the durability of relationships between people and products, through design
- Design for reuse and recycling: "Products, processes, and systems should be designed for performance in a commercial 'afterlife'."
- Design impact measures for total carbon footprint and life-cycle assessment for any resource used are increasingly required and available. Many

are complex, but some give quick and accurate whole-earth estimates of impacts. One measure estimates any spending as consuming an average economic share of global energy use of 8,000 BTU (8,400 kJ) per dollar and producing CO_2 at the average rate of 0.57 kg of CO_2 per dollar (1995 dollars US) from DOE figures.

- Sustainable design standards and project design guides are also increasingly available and are vigorously being developed by a wide array of private organizations and individuals. There is also a large body of new methods emerging from the rapid development of what has become known as 'sustainability science' promoted by a wide variety of educational and governmental institutions.
- Biomimicry: "redesigning industrial systems on biological lines... enabling the constant reuse of materials in continuous closed cycles..."
- Service substitution: shifting the mode of consumption from personal ownership of products to provision of services which provide similar functions, *e.g.*, from a private automobile to a carsharing service. Such a system promotes minimal resource use per unit of consumption (*e.g.*, per trip driven).
- Renewability: materials should come from nearby (local or bioregional), sustainably managed renewable sources that can be composted when their usefulness has been exhausted.
- Robust eco-design: robust design principles are applied to the design of a pollution sources.

Bill of Rights for the Planet

A model of the new design principles necessary for sustainability is exemplified by the "Bill of Rights for the Planet" or "Hannover Principles" - developed by William McDonough Architects for EXPO 2000 that was held in Hannover, Germany.

The Bill of Rights:

1. Insist on the right of humanity and nature to co-exist in a healthy, supportive, diverse, and sustainable condition.
2. Recognize Interdependence. The elements of human design interact with and depend on the natural world, with broad and diverse implications at every scale. Expand design considerations to recognizing even distant effects.
3. Respect relationships between spirit and matter. Consider all aspects of human settlement including community, dwelling, industry, and trade in terms of existing and evolving connections between spiritual and material consciousness.
4. Accept responsibility for the consequences of design decisions upon human well-being, the viability of natural systems, and their right to co-exist.

5. Create safe objects of long-term value. Do not burden future generations with requirements for maintenance or vigilant administration of potential danger due to the careless creations of products, processes, or standards.
6. Eliminate the concept of waste. Evaluate and optimize the full life-cycle of products and processes, to approach the state of natural systems in which there is no waste.
7. Rely on natural energy flows. Human designs should, like the living world, derive their creative forces from perpetual solar income. Incorporating this energy efficiently and safely for responsible use.
8. Understand the limitations of design. No human creation lasts forever and design does not solve all problems. Those who create and plan should practice humility in the face of nature. Treat nature as a model and mentor, not an inconvenience to be evaded or controlled.
9. Seek constant improvement by the sharing of knowledge. Encourage direct and open communication between colleagues, patrons, manufacturers and users to link long term sustainable considerations with ethical responsibility, and re-establish the integral relationship between natural processes and human activity.

These principles were adopted by the World Congress of the International Union of Architects (UIA) in June 1993 at the American Institute of Architects' (AIA) Expo 93 in Chicago. Further, the AIA and UIA signed a "Declaration of Interdependence for a Sustainable Future." In summary, the declaration states that today's society is degrading its environment and that the AIA, UIA, and their members are committed to:

- Placing environmental and social sustainability at the core of practices and professional responsibilities
- Developing and continually improving practices, procedures, products, services, and standards for sustainable design
- Educating the building industry, clients, and the general public about the importance of sustainable design
- Working to change policies, regulations, and standards in government and business so that sustainable design will become the fully supported standard practice
- Bringing the existing built environment up to sustainable design standards.

In addition, the Inter professional Council on Environmental Design (ICED), a coalition of architectural, landscape architectural, and engineering organizations, developed a vision statement in an attempt to foster a team approach to sustainable design. ICED states: The ethics, education and practices of our professions will be directed to shape a sustainable future.... To achieve this vision we will join... as a multidisciplinary partnership."

These activities are an indication that the concept of sustainable design is being supported on a global and inter professional scale and that the ultimate goal is to become more environmentally responsive. The world needs facilities that are more energy efficient and that promote conservation and recycling of natural and economic resources.

Applications

Applications of this philosophy range from the microcosm – small objects for everyday use, through to the macrocosm – buildings, cities, and the Earth's physical surface. It is a philosophy that can be applied in the fields of architecture, landscape architecture, urban design, urban planning, engineering, graphic design, industrial design, interior design, fashion design and human-computer interaction.

Sustainable design is mostly a general reaction to global environmental crises, the rapid growth of economic activity and human population, depletion of natural resources, damage to ecosystems, and loss of biodiversity. In 2013, eco architecture writer Bridgette Meinhold surveyed emergency and long-term sustainable housing projects that were developed in response to these crises in her book, "Urgent Architecture: 40 Sustainable Housing Solutions for a Changing World." Featured projects focus on green building, sustainable design, eco-friendly materials, affordability, material reuse, and humanitarian relief. Construction methods and materials include repurposed shipping containers, straw bale construction, sandbag homes, and floating homes.

The limits of sustainable design are reducing. Whole earth impacts are beginning to be considered because growth in goods and services is consistently outpacing gains in efficiency. As a result, the net effect of sustainable design to date has been to simply improve the efficiency of rapidly increasing impacts. The present approach, which focuses on the efficiency of delivering individual goods and services, does not solve this problem. The basic dilemmas include: the increasing complexity of efficiency improvements; the difficulty of implementing new technologies in societies built around old ones; that physical impacts of delivering goods and services are not localized, but are distributed throughout the economies; and that the scale of resource use is growing and not stabilizing.

QUANTITATIVE ASSESSMENT OF SUSTAINABILITY KPIS IN PROCESS DESIGN ACTIVITIES

Sustainable Process Design (SPD)

Conventional process design is based on technical and economic criteria, with the system boundary drawn around the process itself. This often leads to designs which are optimised economically around the plant but under-optimised upstream and downstream of the process. Recognising this deficiency in process design, Azapagic *et al.* have proposed a general methodology for Sustainable

Process Design (SPD) to enable consideration on a life cycle basis of economic, environmental and social criteria, alongside the usual technical criteria.

The methodology consists of the three following steps:

Identification of Sustainability Design Criteria and Stakeholders

The starting point in SPD is identification of sustainability criteria on which the alternatives, and later the whole design, will be assessed. At the initial stage of design when little information on the whole system may be available, it is proposed to use generic sustainability criteria that apply to any process. Further examples are provided by IChemE (2003), who have developed a set of sustainability indicators specifically for the chemical and process industry. Identification of specific sustainability criteria and their prioritisation for a particular process is then carried out in consultation with stakeholders, who may include employees, investors, neighbouring communities and citizens, non-governmental organizations (NGOs) and government.

Table: Examples of sustainability design criteria and indicators.

Economic	Environmental	Social
Micro-economic, *e.g.*, - Capital costs - Operating costs - Profitability - Investments (*e.g.* decommissioning, health and safety and so on). Macro-economic, *e.g.*, - Value added - Taxes paid - Other investments (*e.g.* ethical) - Environmental liabilities	Raw materials Energy Emissions to: - Air, water, and land. Environmental impacts, *e.g.* - Global warming - Ozone depletion - Acidification - Human toxicity - Eco-toxicity - Summer smog - Eutrophication	Provision of employment Health and safety of: - Employees/contractors - Customers - Citizens Nuisance - Odour - Noise - Visual Impact Public acceptability - Process - Product

Identification and Evaluation of Alternatives

Evaluation of process alternatives is first carried out on a qualitative basis by identifying advantages and disadvantages of different alternatives with respect to the sustainability criteria identified in the previous step. Ideally, the outcome of this stage should be identification of the most promising processing route, raw materials, energy sources and so on; however, in practice, it is more likely that there will be several feasible and potentially sustainable alternatives. In that case, further, quantitative, assessment of the alternatives is necessary; this is carried out as part of step 3.

Sustainability Assessment and Design Optimisation

The qualitative sustainability criteria identified are translated into the appropriate economic, environmental and social indicators or metrics. The indicators are largely quantitative, although some are expressed as qualitative statements. They are then used to assess economic, environmental and social sustainability of the design on a life cycle basis. In addition to the traditional microeconomic indicators, such as net present value, discounted cash flow analysis, returns on capital invested and so on, it is important to consider the additional economic indicators, including value added and investments in, for example, pollution prevention and decommissioning. LCA is used as a tool for assessing environmental sustainability. Both the life cycles of the plant and the product are considered.

Social sustainability criteria can be translated into both quantitative and qualitative indicators. For example, provision of employment can be expressed in quantitative terms as 'number of employees'. The design is then optimised to address the sustainability 'hot spots' identified during the assessment.

Strengths

Sustainable Process Design (SPD) methodology shows how sustainability considerations could be integrated into process design on a life cycle basis.

It enables identification of relevant sustainability criteria and indicators, comparison of alternatives, sustainability assessment of the overall design and identification of 'hot spots' in the life cycle of the system. In this way, it is possible to arrive at a design configuration that would ensure the most sustainable performance of the plant and product over their whole life cycles especially by identifying and choosing the alternatives that would not normally be considered in conventional design.

Application of SPD principles at a process design stage would help to improve the level of sustainability of future processes and products.

SPD helps in identifying and addressing the important environmental or social issues at an early stage that otherwise would have remained hidden until the plant is in operation.

Weaknesses/Limitations

Currently, there is little experience in applying SPD in practice. For instance, commercially available flowsheeting packages have yet to incorporate sustainability considerations.

The methodology is quite complex.

Interpretation of results can be difficult, due to numerous criteria that are considered simultaneously.

Opportunities for Broadening and Deepening LCA

The tool proposes quantitative indicators for process-relevant social and economic sustainability criteria. Social indicators include number of employees, occupation exposure limits, potential safety risks from fire and explosion. Economic criteria include the usual micro-economic aspects (capital and operating costs, NPV etc.) and macroeconomic issues (*e.g.* contribution to GDP, value added, etc.). These indicators could also be used to broaden LCA for sustainability analysis.

The drive for broader corporate social responsibility also demands consideration of integrating sustainability criteria at the early stage of product formulation *i.e.* into process design.

Threats for Broadening and Deepening LCA

Some of SPD's weakness, such as complexity could act as barriers for wider use of the tool.

Key Performance Indicators Defined

The first step to making data-driven decisions in the call center is to develop a comprehensive understanding of what a call center KPI is, as well as how they are used in the call center.

Call center KPIs are quantitative metrics that are used to evaluate constructs that are crucial to the success of the call center. They typically assess the performance of the agent, team, department and/or the call center as a whole. They can be used to determine trends and make data-driven decisions that will increase efficiency, optimize customer satisfaction, increase revenue and reduce costs. They can also be used in benchmarking efforts to measure progress towards goals and to compare the call center's performance with an industry standard. Taken together, call center KPIs are metrics that provide information about how a call center is performing and can be used as a foundation to make strategic decisions that will help to drive results.

Now that you have a good understanding of call center KPIs, lets walk through an example of how you can measure certain KPIs to access your call center's performance in one key domain - customer satisfaction.

Top 10 KPIs that Impact Customer Satisfaction

Knowing which call center KPIs to measure consistently can be challenging, especially for more novice call center managers. This chapter of this blog post is a great place to start. Below are the top 10 KPIs that impact customer satisfaction within the call center compiled from a study conducted with 512 call centers:

1. *First call resolution* - The percentage of calls that the agent resolves the caller's issue without having to escalate, transfer or return the call.

2. *Percentage of calls blocked* - The percentage of callers that received a busy tone when they call.
3. *Average time in queue* - The average amount of time callers wait in call queues before an agent responds.
4. *Average after call work time* - The average amount of time an agent spends completing work related to the call after they finish the call.
5. *Service level* - The percentage of calls answered within a specified number of seconds.
6. *Average abandonment rate* - The percentage of callers who hang up before reaching an agent.
7. *Agent turnover rate* - The percentage of agents who leave the call center.
8. *Average speed of answer* - The average amount of time it takes for the call to be answered by an agent or the Automatic Call Distributor (ACD).
9. *Average handle time* - The average amount of time an agent spends speaking with the caller, including hold time.
10. *Schedule adherence* - A measure of an agent's degree of compliance with their assigned schedule.

If you are looking for a starting point for evaluating your call center's impact on customer satisfaction, the 10 aforementioned KPIs are a great option. Consistently measuring them over time will allow you to make data-driven decisions which will help to increase customer satisfaction.

Steps to Selecting Optimal KPIs

Now that you have a comprehensive understanding of call center KPIs as well as commonly measured KPIs in the call center, it is time to select KPIs that will allow for an accurate assessment of your call center's performance. Below is a step-by-step guide to help you do just that.

1. Identify KPIs that are in line with your business objectives and corporate strategy.
2. Identify and define the KPIs that will be measured.
3. Decide which data points will contribute to the KPI calculation and how the data will be acquired.
4. Define each KPI's purpose and make sure that the purpose is clear to your entire team.
5. Ensure that the KPIs are measuring different domains to gain a more comprehensive view of the performance of your call center.
6. Set a specific, quantitative target, a range or both for each KPI.
7. Define concrete steps that are required to meet each target.
8. Develop a specific action plan if the KPI falls outside of the target or range.

9. Continually review KPIs, interpret the data in a meaningful way and consider causes for any trends or outliers.
10. Optimize the definition of the KPI, refine the data points included in the KPI calculation and select new KPIs to measure in an effort to constantly improve your approach to measuring your call center's performance.

In order to effectively engage in call center benchmarking, evaluation and decision-making, managers must follow the steps outlined above to select the most relevant call center KPIs and make more informed decisions based on the acquired data.

Tools to Consistently Measure KPIs

The final step to optimizing your approach to measuring KPIs and acting on the data is to leverage the right tools. Below is information about tools to help you consistently measure call center KPIs so that you and your team can make strategic decisions that will have a significant impact on your call center's performance.

1. Call center software with integrated reporting - The first, and arguably the most important, tool in your team's toolbelt should be call center software with integrated real-time and historical reporting. This will provide your team with the data they need exactly when and how they need it.
2. Real-time metrics dashboard for agents - Agents should be provided with a real-time metrics dashboard so they are able to see KPIs like how many callers are in the queue, their colleagues' agent status, longest wait time and average abandonment time.
3. Real-time metrics dashboard for managers - Managers should have access to KPIs that are most helpful to them in making data-driven decisions. For example, their metrics dashboard should display service level, number of calls in queue, average abandonment time, longest wait time, etc., in real-time.
4. Call monitoring dashboard - Call center managers and supervisors should have access to call monitoring capabilities to analyze interaction quality and obtain data for measuring first call resolution.
5. Historical reporting - The entire team can engage in data-driven decision making when they have access to dashboards that display call center metrics such as call volume, service level, handle time, abandonment time, wait time, etc. from any time point.
6. Workforce management software - Managers should leverage workforce management software to help them measure agent attrition and agent absenteeism.
7. Backoffice tools - Managers should have access to backoffice tools to help them measure KPIs such as first call resolution, cost per call, customer satisfaction and interaction quality.

Consistently measuring call center KPIs is essential to gaining an at-a-glance overview of the call center's performance and making informed decisions based on the results. As such, arm your team with the aforementioned tools so they can access the data they need to succeed.

CRITERIA FOR NORMALIZATION AND AGGREGATION OF THE KPIS

Key performance indicator (KPI) is a set of quantifiable measures that a company or industry uses to gauge or compare performance in terms of meeting their strategic and operational goals. KPIs vary between companies and industries, depending on their priorities or performance criteria. Also referred to as "key success indicators (KSI)".

A performance indicator or KPI is an essential and critical element for organizational performance measurement. KPIs are commonly used by an organization to evaluate its success or the success of a particular activity in which it is engaged. Sometimes success is defined in terms of making progress toward strategic goals, but often, success is simply the repeated achievement of some level of operational goal. Accordingly, choosing the right KPIs is reliant upon having a good understanding of what is important to the organization. 'What is important' often depends on the department measuring the performance - the KPIs useful to finance will be quite different than the KPIs assigned, for example.

Key performance indicators define a set of values used to measure against. These raw sets of values, which are fed to systems in charge of summarizing the information, are called indicators. Indicators identifiable as possible candidates for KPIs can be summarized into the following sub-categories:

Quantitative indicators which can be presented as a number.

Practical indicators that interface with existing company processes.

Directional indicators specifying whether an organization is getting better or not.

Actionable indicators are sufficiently in an organization's control to effect change.

Financial indicators used in performance measurement and when looking at an operating index.

Key performance indicators, in practical terms and for strategic development, are objectives to be targeted that will add the most value to the business. These are also referred to as key success indicators.

Because of the need to develop a good understanding of what is important, performance indicator selection is often closely associated with the use of various techniques to assess the present state of the business, and its key activities. These assessments often lead to the identification of potential improvements; and as a consequence, performance indicators are routinely associated with 'performance improvement' initiatives. A very common method for choosing KPIs is to apply a management framework such as the balanced scorecard.

The performance indicators are organized under two categories - core and additional - to differentiate between indicators that are commonly adopted across companies in the oil and gas industry versus indicators that may not have general applicability to all oil and gas industry companies and/or may not be sufficiently well-defined for common adoption.

Core indicators are typically:

- Considered relevant to almost all oil and gas industry companies;
- Inherent to activities in the oil and gas industry (*e.g.*, upstream and downstream);
- Of common interest to a wide range of local and global stakeholders;
- Generally related to aspects or issues of national or global significance;
- Sufficiently mature in terms of consistent usage and reproducibility by those in the oil and gas industry.

On this basis, the core indicators have been defined to enable generally consistent reporting or aggregation on a global basis. There can be value and benefit in using core indicators to promote consistent performance reporting among companies, encourage best practice sharing and enable industry associations and organizations to generate reasonable overviews of sector performance.

Additional indicators can be of equal or even greater importance to individual companies than core indicators in specific contexts of location, activity or stakeholder group.

Additional indicators are typically:

- Assessed as relevant by the reporting company and its stakeholders;
- Associated with only a subset of the industry;
- Reflective of local regulations or legislation;
- Generally related to issues of local or regional significance;
- Evolving and under development.

Additional indicators may often represent a leading practice in sustainability or non-financial reporting.

Furthermore, some qualitative additional indicators may pertain to issues for which there are currently no generally accepted definitions or performance measurement practices. Additional indicators are typically locally defined and/or relate to local or regional issues. Since indicator definitions may not provide comparable or meaningful descriptors of overall company performance, reporters should exercise caution when interpreting additional indicators on a global basis by either consolidating information or aggregating data. Therefore, reporting at the local (operating unit or country) level is becoming more prevalent for oil and gas industry companies, especially to describe performance in locations where a particular issue has high significance or sensitivity.

Presentation of only consolidated qualitative information or aggregated quantitative data with these indicators may not be as meaningful or useful unless local context, disaggregated data or other explanations are provided by the reporting company. A company may not report all indicators addressed here if it has assessed that the indicator or issue is not relevant across all of its activities or because of insufficient information systems, quality, availability or resources.

Core and additional indicators can be defined either quantitatively or qualitatively. Quantitative indicators are reported as a number with a dimensional unit or some form of a numerical index. Certain indicators, however, do not readily lend themselves to quantification. Many social issues, in particular, are primarily reported in qualitative terms as there is not yet common understanding of appropriate quantitative measures.

When qualitative indicators are appropriate, reporting companies are encouraged to consolidate information and report their performance in terms of underlying policies, commitments, programme initiatives, stakeholder partnerships, industry alliances and case study examples that describe results, benefits and lessons learned from various initiatives.

Over time, qualitative indicators may evolve into more quantitative measures. Companies may start out by describing performance related to operational practices and by using anecdotal examples and local case studies. In time, these anecdotal descriptions may converge into a more objective approach for reporting performance within an organizational segment or operating region of a company. Finally, these practices may emerge as a quantitative index for measuring performance or assessing impacts.

Companies report performance data at varying levels of aggregation ranging from individual facilities to national/regional locations and to global coverage for the entire corporation. Aggregate reporting at the corporate level is most commonly observed for reporting occupational injuries, environmental emissions and incident data as part of both regulated and voluntary public reporting. Reporting companies are encouraged to determine the level of aggregation that is appropriate and provides a meaningful representation of the data being presented.

There are two principal aspects of performance indicators that are of interest to internal and external users of sustainability or non-financial indicator performance data: the absolute quantity of the indicator and the normalized quantity relative to some other measured input or output.

Reporting companies often present raw performance data in terms of absolute quantities that can be expressed in a physical unit of measurement related to weight, volume, energy or financial value. In general, absolute data can be expressed in units of measurement that are readily convertible. Absolute quantities may provide information about the magnitude or size of an output, input, value, or result.

Normalized quantities are relative figures representing ratios between two absolute quantities of the same or different kind. Ratios allow comparisons among operations of different size and facilitate comparisons of similar products or processes. They also help relate the performance and achievements of one company, business unit, or organization to those of another. Ratio indicators can provide information on the efficiency of an activity, on the relative intensity of an output (*e.g.*, energy intensity) or on the relative quality of a value or achievement.

Often, companies measure and report performance based on both absolute and normalized quantities to provide a more complete and balanced representation of sustainability or non financial performance.

For many oil and gas companies, describing the boundaries of reported information or indicators is an important consideration because they often cut across an array of complex operational and organizational relationships, as well as direct and indirect impacts. In the oil and gas industry, two or more parties are often involved in an asset, such as in a joint venture, and work together under a variety of legal forms. In some but not all situations, performance indicators can be aggregated and normalized across a range of dimensions that consider ownership, management accountability, geographic and national locations, industrial sectors, company divisions, business units, facilities and source types

MITIGATING METRICS MADNESS: HOW TO TELL KPIS FROM MERE METRICS

With more ways than ever to capture information, organizations today have a wealth of metrics at their disposal. The rise of dashboards has provided a neat package in which to display metrics; with the ability to create a limitless number of dashboards — from enterprise level, to business unit level, department level, all the way down to the individual — the number of metrics being tracked has skyrocketed. Nowhere has this proliferation become more evident than within IT organizations. The owners of vast amounts of data apparently feel obligated to show others that they are doing something with it.

Yet for many IT organizations, this wealth of metrics yields a decided dearth of useful insight. Metrics have multiplied to the point that they've become not only overwhelming, but also meaningless to those who monitor them. In fact, studies have shown that organizations are tracking as many as *nine times* the effectual number of measures.

While many IT organizations have fallen into this trap — with rationales such as "We've *always* collected this metric" or "We collect it because we *can*" — this abundance of metrics can actually be a detriment to organizational success for a number of reasons. Building on existing transactional data often produces a plethora of metrics that lack strategic context, reflect only past behavior, or don't measure an organization's intended outcomes. When metrics are created simply because the data is available, they often bear little connection to organizational objectives. For example, it can be tempting to create a metric for the *number of network outages per day*. However, that metric alone may not convey enough information if the service-level objective is based on the users' average response time or encourages users to migrate from the call center to self-service.

Historically, organizations have generated reports by running query tools against relational databases and other data stores. Each individual user would generate his or her own report independently, without regard to the common elements. For example, two different reports may each contain revenue, but they might have defined it in different ways (operating vs. cash flow, all products vs. nondiscontinued products, etc.). In recent years, organizations have tried to standardize these elements into metrics by agreeing on definitions and use cases. Unfortunately, the number of metrics in use has ballooned, especially with the proliferation of dashboards. Thus the organization goes from being overwhelmed by reports to being overwhelmed by metrics. At this point many organizations have so many metrics that they no longer know which ones are important.

The Challenge of Metrics

The lack of context surrounding data-driven metrics is compounded by the fact that most metrics being used are activity metrics and represent information about the past. For example, even though *number of trouble tickets closed* and *cost per change request* are two of the most common IT metrics, *% R&D investment leading to service improvement* and *retention of high-potential staff* may be more appropriate metrics for an IT department that is interested in its strategic value to the organization. Relying solely on activity and lagging metrics is akin to driving while looking in the rearview mirror; it limits an organization's ability to drive in its chosen direction.

Moreover, many organizations naturally draw from transactional (*e.g.*, financial, ERP) systems for their data. Unfortunately, this means that the resulting metrics typically reflect an organization's activity or *outputs*, not the *outcomes* it is trying to achieve. Returning to the example above, it might be tempting to measure user satisfaction by creating an activity metric such as *percentage of trouble tickets resolved in the same day*, because this data is already contained in most service management solutions. However, the number of trouble tickets resolved doesn't directly measure user satisfaction. Instead, it may be more appropriate to regularly survey a portion of the users in order to gauge reported satisfaction. This survey metric, while subjective, more directly reflects the intended outcome.

Even when organizations focus on the right metrics, they often struggle to ensure that the metrics are understood – how they are calculated, where they came from, and what they mean – correctly and consistently throughout the organization. Take *profitability* as an example. Just on the face of it, how do you know whether it is gross or net? Fully burdened or raw cost? Operating or cash flow? Once you establish metrics, be sure that everyone in the organization understands what they are, how they are to be used, how they will be calculated, and from what source the data will come. Documentation is essential here, with formal, standard definitions. Be sure you have the capability to include such documentation with your metrics.

Kpis Vs. Metrics: What's the Difference?

With so many metrics, how do you decide which ones are critical? The answer lies in distinguishing which metrics are actually key performance indicators (KPIs). KPIs are performance metrics explicitly linked to a strategic objective that help an organization translate strategy execution into quantifiable terms. Well-designed KPIs provide quick insight into trends and summary information, while supporting drill-down into more detailed metrics. This allows an organization to see where it's doing well and where it requires improvements and/or course adjustments. Think of KPIs as the yardstick by which success and progress are measured; they are the measures most tightly linked to the organization's success or failure in executing strategy.

All KPIs are metrics, but not all metrics are KPIs. An organization will have many metrics, but few KPIs. While metrics can be a measure of just about anything, KPIs are the measures that matter most. KPIs should also be actionable; if they are not actionable, how can they help you achieve your goals? In other words, don't measure something as a KPI that you can't change with specific actions.

So when is a metric a KPI? KPIs are metrics that are:

1. Outcome-oriented – tied to an objective
2. Target-based – have at least one defined time-sensitive target value
3. Rated or graded – have explicit thresholds that grade the difference (or gap) between the actual value and the target

The above three criteria, used in evaluating whether a metric meets KPI status, serve as a "litmus test" to help ensure focus on the measures that truly matter to the success of your organization.

THE KEYS TO KPIs: A THREE-PART LITMUS TEST

Outcome-Oriented

The first – and perhaps most critical – criterion a metric must meet to be deemed a KPI is that it must be *outcome-oriented.* Metrics that track *inputs* (the amount of financial and nonfinancial resources applied to providing a service or producing a product) or *outputs* (the quantity of service or products produced) are just metrics. A KPI tracks *outcomes* that measure progress toward a defined goal so that you can understand impacts.

Another way of looking at this is that a KPI is explicitly tied to an objective. If you can't describe the business goal it's monitoring, it's not a KPI – it's a metric. While it might seem natural to simply start by looking at your current pool of metrics and asking which of them meet this criterion, such an approach focuses only on those measures already being tracked, which may exclude other measures critical to your objectives. For example, if you don't currently measure *employee satisfaction*, then it likely will be excluded from consideration, even though it may be a critical factor in the *retention of high-potential staff* strategic objective.

A more effective approach is to start with your goals. Take existing metrics (and, as much as possible, take organizational politics off the table for a moment) and ask yourselves, "What measures will tell us if we are on track with the objectives in our strategy plan?" In doing so, be sure to consider a balance of measures – operational as well as financial, leading as well as lagging, and subjective (qualitative) as well as objective (quantitative).

While financial and other backward-looking lagging indicators provide an important view of how the organization has performed up until now, they offer little visibility into how the organization will perform moving forward. Leading indicators, on the other hand, help forecast future performance, providing critical insight into how today's decisions will impact tomorrow's performance and giving you an opportunity to address issues and/or shift course if necessary. If your goal is to *maximize customer lifetime value*, a common lagging indicator is *total $ purchased per customer*, while a leading one could be *% customers that buy again within six months of initial purchase*. If a high percentage of customers buy again after six months, then it's likely they will continue to buy in the future, which indicates a high lifetime value. Alternatively, a call center that wants to reach the goal*be a one-stop shop for all customer interactions* might normally use the lagging indicator *% cases closed in first contact*. However, the leading indicator *% of self-help transactions on Web site that are abandoned* often provides clearer insight into whether customers are getting their questions answered.

While people naturally tend to think of metrics as quantitative in nature – pulled from transactional systems, for example – it is important not to overlook qualitative metrics. Qualitative metrics leverage subjective information that is often critical to performance, such as feedback from important constituents (*e.g.*, employees, customers).

In starting with your goal of identifying the right KPIs, you may establish KPIs that aren't currently being tracked. What if you can't get access to the underlying data or if it isn't stored in any database? Rather than relying on the time-consuming and error-prone process of manually collecting data from disparate people, a good performance management system will provide you with a flexible workflow and role-based alerts that manage the process of entering, approving, and publishing KPIs. Be sure not only that you are focusing on the right KPIs, but also that you are tracking all of the KPIs critical to your objectives.

Target-Based

The second point to remember when evaluating whether a metric is truly a KPI is that KPIs are more than just numbers. In order to provide a meaningful gauge of progress toward organizational objectives, a metric must have context. Milestone-based targets provide this context.

As an example, let's say that a metric intended to measure user satisfaction currently has a value of 61. Is 61 cause for celebration or concern? If the target were 100 – as it might be for a test in school (where 90-100 is an A, 80-89 is a B, etc.)

– then 61 would be cause for concern. However, if 61 were one of the best scores in the class, then the teacher might resort to grading on a curve, in which case 61 would be a good value. The only way you can effectively evaluate whether the value is good or bad – and consequently, contributing to or detracting from the achievement of your objectives – is by having targets associated with your key performance indicators. Targets represent the value you would like a KPI to be at a specific moment in time.

Rather than relying on a single value for a target, you'll also want to develop incremental milestone targets along the way. Effective KPIs have targets associated with a specific time frame. For example, for the KPI *reduce the cost of service*, the target for one year might be 12%, with incremental milestones of 0% for the first six months and 2% per month for the remaining six months. The milestone targets can be used during the year to determine whether your organization is on track to reach its ultimate goal target.

Determining appropriate targets for your organization can be a bit of a scavenger hunt. Check business plans, budgets, and product plans for ideas. A manufacturing organization might develop a business plan that includes how many units it plans to produce each month or each production shift. A organization typically gives each rep a quota of how much that person should sell each quarter. Budgets specify how much you are allowed to spend. All of these are targets. In some cases, you may need to start out with best guesses. Using these as a starting point, you can refine the targets over time once you have additional experience.

Whatever their source, targets should motivate. When establishing target values, keep in mind that values that are either too easy or too hard to achieve are not motivating. Use "stretch" targets, where near attainment (85%) is deemed success.

Rated or Graded

In order to actually gauge performance, you'll need to calculate the gap between the actual and the target values of a KPI. In addition to being outcome-oriented and target-based, a true and effective KPI should also rate or grade the size of this gap. Associating a rating system with KPIs provides a quick and easy-to-understand reading of whether a particular KPI status is good or bad, how on or off target it is.

For example, you can use the traditional "letter grade" system, in which A, B, C, D, or F grades are assigned based on the calculated gap. KPIs achieving 90% or more of their target receive an A grade; 80%-89% of target is a B; 70%-79% of target is a C, and so on. The letter grade system implies that an A is a "stretch target" – a B grade would be deemed success.

For some IT organizations, the idea of assigning letter grades may clash with their culture. In that case, the above system can be adapted to use less rigid score names such as "Exceptional," "Very Good," "Good," "Needs Improvement," and "Unacceptable." Another alternative is the popular traffic-light metaphor, which uses "dark green," "light green," "yellow," "light red," and "dark red."

Whatever system you choose, consider the fact that a range of five thresholds is more likely to be accepted than three, due to people's natural tendency to avoid extremes. With five thresholds, even after discarding the extremes, you still have three useful values with which to gauge progress.

The Value of KPIs

While migrating from metrics to KPIs may seem like a time-consuming exercise, the benefits far outweigh the costs. The simple exercise of focusing attention on a smaller number of *key* performance indicators and tying them to the organizational outcomes will help better align IT with its users and customers. In addition, there can be substantial cost savings once the organization is no longer spending resources collecting, transforming, approving, and publishing metrics that aren't as critical as others. Finally, by taking this path, IT organizations can better demonstrate their value and more effectively vocalize their true benefit to the overall organization.

Establishing the right KPIs is just as essential a component of successful strategic execution as defining the right strategy for your organization. Trying to execute without KPIs – or perhaps worse, the wrong ones – is akin to taking a road trip to a destination that you've never been to without using a map. If you're lucky enough to ultimately make it to your intended destination, the odds are it will have cost you excessive cycles, wrong turns, and detours – not to mention frustration – to get there. Key performance indicators are an important part of the roadmap on your journey to better performance.

Chapter 8

Direct-Dispense Polymeric Waveguides Platform for Optical Chemical Sensors

Mohamad Hajj-Hassan[1], Timothy Gonzalez[2], Ebrahim Ghafar-Zadeh[1], Hagop Djeghelian[2], Vamsy Chodavarapu[1,*], Mark Andrews[2] and Daniel Therriault[3]

1 Department of Electrical and Computer Engineering, McGill University, 3480 University Street, Montreal, Quebec, Canada H3A 2A7; E-Mails: mohamad.hajjhassan@mail.mcgill.ca (M. H. H); Ebrahim.ghafarzadeh@mail.mcgill.ca (E. G. Z.)

2 Department of Chemistry, McGill University, 801 Sherbrooke Street West, Montreal, Quebec, Canada H3A 2K6; E-Mails: timothy.gonzalez@mail.mcgill.ca (T. G.); hagop.djeghelian@mail.mcgill.ca (H. D.); mark.andrews@mcgill.ca (M. A.);

3 Department of Mechanical Engineering, Ecole Polytechnique de Montreal, Case postale 6079, Succursale Centre-ville, Quebec, Canada H3C 3A7; E-Mail: daniel.therriault@polymtl.ca (D. T.)

* Author to whom correspondence should be addressed; E-Mail: vamsy.chodavarapu@mcgill.ca (V. C.); Tel.: +1-514-398-7111; Fax: +1-514-398-4470

ABSTRACT

We describe an automated robotic technique called direct-dispense to fabricate a polymeric platform that supports optical sensor arrays. Direct-dispense, which is a type of the emerging direct-write microfabrication techniques, uses fugitive organic inks in combination with cross-linkable polymers to create microfluidic channels and other microstructures. Specifically, we describe an application of direct-dispensing to develop optical biochemical sensors by fabricating planar ridge waveguides that support sol-gel-derived xerogel-based thin films. The xerogel-based sensor materials act as host media to house luminophore biochemical recognition elements. As a prototype implementation, we demonstrate gaseous oxygen (O_2) responsive optical sensors that operate on the basis of monitoring luminescence intensity signals. The optical sensor employs a Light Emitting Diode (LED) excitation source and a standard silicon photodiode as the detector. The sensor operates over the full scale (0%-100%) of O_2 concentrations with a response time of less than 1 second. This work has implications for the development of miniaturized multi-sensor platforms that can be cost-effectively and reliably mass-produced.

Keywords

Direct-Dispense, Direct-Write, Xerogels, Oxygen Sensors, Waveguides, Optical Sensors, Fluorescence, Chemical Sensors, Polymer Waveguides.

1. INTRODUCTION

Multi-sensor systems and sensor microarrays are undergoing rapid development because they can multiplex and de-multiplex biological or chemical information from multi-component samples [1, 2]. Optical sensing based on fluorescence spectroscopy is a well established technique for highly selective and sensitive biochemical sensing [3, 4]. In comparison to competing approaches, optical sensors may be less susceptible to contamination while providing fast multiplexed response and high efficiency. Optical sensor arrays have proven attractive for DNA-chip technology and protein microarrays widely being used for drug discovery, genomics and proteomics [5, 6]. Biochemical sensor arrays are also seeing development in applications like point-of-care diagnostics, medical/ clinical biosensing, personal/food safety, and environmental monitoring [3, 7]. To date, a number of platforms have been demonstrated for optical biochemical sensor arrays. Some platforms use optical fibers [8], planar waveguides [9], some of which have been coupled with micro-LED arrays [10], and arrays on a LED [11].

We work on waveguide based sensors which offer a degree of robustness and that can be integrated relatively easily into microfluidic structures for lab-on-chip applications [12]. The waveguide structures act as a support platform for immobilizing biochemical recognition elements. We describe a direct-dispense fabrication process to implement the waveguide based sensor arrays. Direct-dispense and direct-write techniques are commonly used for advanced micro-electronic packaging, microfabrication, and microelectrodes development [13, 14]. With direct-dispense, select materials can be precisely deposited with features on the order of several micrometers through a microscale nozzle. A longer term objective is to evaluate the potential of direct-dispense techniques for low-cost printable optical sensor arrays. Recently, a number of research groups including ours have employed the direct-dispense technique to create microfluidic channels and other microstructures [15, 16]. In literature, we find several fabrication techniques that have been grouped as direct-write techniques which include laser (optical) lithography, ink-jet printing and screen printing [17]. The fabrication technique described here is one example of the several direct-write techniques and involves mechanical robotic deposition to pattern fugitive organic inks on a substrate. Optically transparent epoxy or polymeric materials are then deposited with direct-dispense system using the previously deposited fugitive ink structures as guide or sacrificial layers in a way that is reminiscent of the use of positive photoresists. Subsequently, the fugitive ink is thermally removed leaving behind micropatterned epoxy or polymeric structures. This process can be repeated to create complex polymeric microstructures that can be used as high-performance, multi-functional sensor platforms which can support microfluidics and optical signal collection and transmission.

Sol-gel derived materials have a good potential for creating sensor arrays in combination with direct-dispense techniques. Several reports describe the use of sol-gel-processed xerogels as encapsulation media for various biochemical recognition elements [7, 18, 19]. Xerogels offer a number of advantages, including low temperature processing, optical transparency, relative long time stability, and straightforward doping procedures to encapsulate guest sensing elements. Xerogel sensor materials have been deposited using pin-printing [7], dip-coating [8], spin-coating [20], and photo-(ultraviolet) patterning [21]. This flexibility makes xerogels attractive to combine with the direct-dispense method. In this paper, we show how modified xerogels can be deposited selectively and integrated with a patterned polymeric support platform using direct-dispensing. Ru(II) diimine complexes have relatively long lived triplet metal-to-ligand charge transfer states that make them useful for quenchometric sensing. We use *tris*(2,2′-bipyridyl) dichlororuthenium(II) hexahydrate ($[Ru(bpy)_3]^{2+} \cdot 6H_2O$) as the guest luminophore in the xerogel, because it is one of the more widely studied complexes for oxygen sensing by luminescence quenching. For the prototype discussed in this paper, we use an optically transparent epoxy Epo-Tek 301-1 (Epoxy Technologies Inc.) to fabricate polymer ridge waveguides. Figure 1 shows a simplified block diagram of the prototype optical sensor platform implementing the polymeric waveguide support platform and xerogel sensor materials. Other configurations common to existing optical chemical sensors are also possible where the excitation source and the photodetector are on the either side of the xerogel sensor array or both the excitation source and the photodetector are on the same side (top or bottom) of the sensor array [12, 22-24].

Following Figure 1, the sensor partly uses direct LED excitation and partly guided evanescent-wave excitation due to the exponentially decaying evanescent field from the multimode waveguides [8, 25]. Subsequently, fluorescence is generated from Ru(II) complexes trapped in the xerogel claddings with an intensity proportional to the target analyte concentration. A part of this generated fluorescence is captured by the waveguides and channeled to a photodetector. In the following sections, we describe the direct-dispense technique to create the waveguide structures, preparation of the modified xerogel sensor materials that are suitable for direct-dispense process, and finally, the integration of a xerogel sensor film with the waveguide structures.

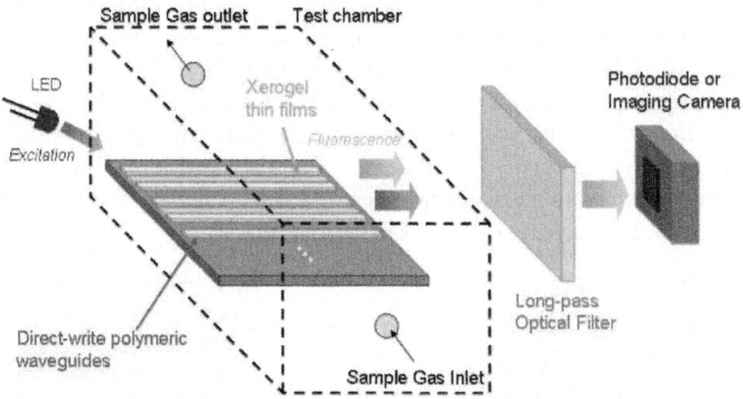

Figure 1. Simplified block diagram of the optical sensor system implementing the polymeric waveguide support platform and xerogel recognition materials.

2. DIRECT-DISPENSE FABRICATION OF WAVEGUIDE STRUCTURES

A robotic deposition tool (Model I&J 2200, FISNAR Inc.) was used for the direct-dispense process, in which a fugitive ink, typically consisting of a composite of 75% of petroleum jelly and 25% of microcrystalline wax, was extruded through a micro-nozzle (Model 2400, EFD Inc.) and deposited onto a substrate. Typical substrates may be selected from plastic, generic microscope borosilicate glass slides, or fragments of silicon wafers [15, 16]. The automated direct-dispense tool is as shown in Figure 2. The pattern to be written is first mapped out as a trajectory by programming it into the robot, which in turn commands and controls the dispensing system to emit the fugitive ink from the nozzle. In the simplest case (Figure 3), straight lines of the fugitive ink are deposited. Next, uncured liquid polymer resin is deposited at room temperature with the direct-dispense apparatus.

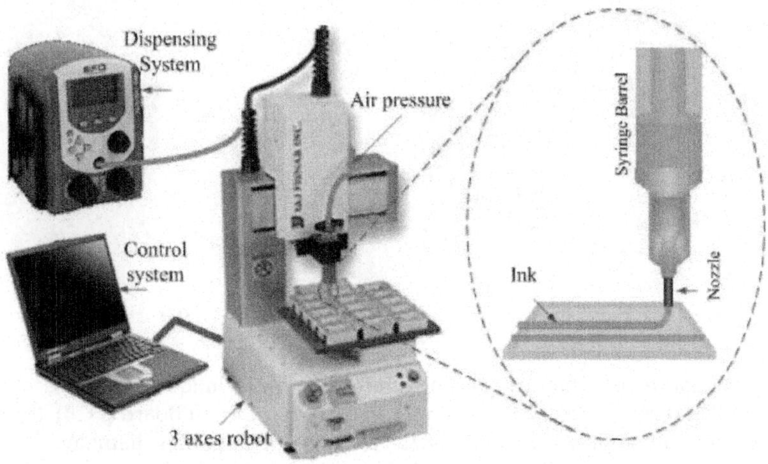

Figure 2. Direct-dispense microfabrication system.

The fugitive ink pattern acts as a periodic reservoir to hold the fluid resin, which is subsequently cured. The ink is then melted at 65°C and extracted, leaving behind patterned polymeric microstructures. Figures 3 (a), (b), and (c) depict the fabrication steps for the direct-dispense process. The fugitive ink layer must have appropriate flow viscosity, solidify and adhere well to the substrate and melt at a sufficiently low temperature to yield uniform quality and reproducible structures from polymer resins. As an example, Figure 4 shows functionality of the direct-dispense system to fabricate different microchannels.

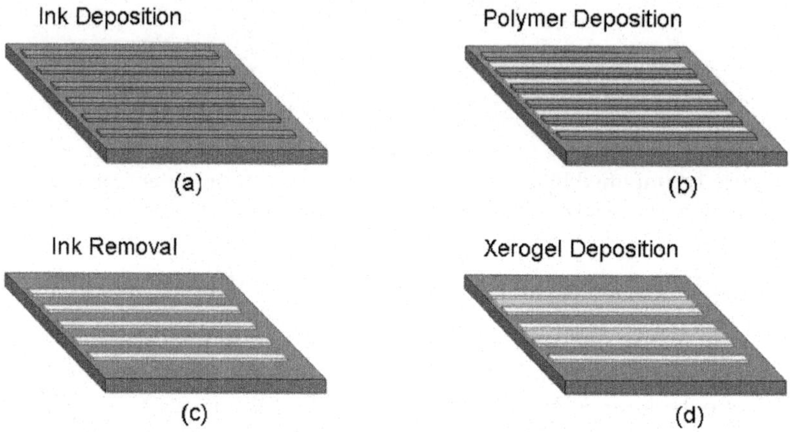

Figure 3. Fabrication process flow for the development of polymeric waveguide support platform using direct-dispense process and the immobilization of xerogel mateirals.

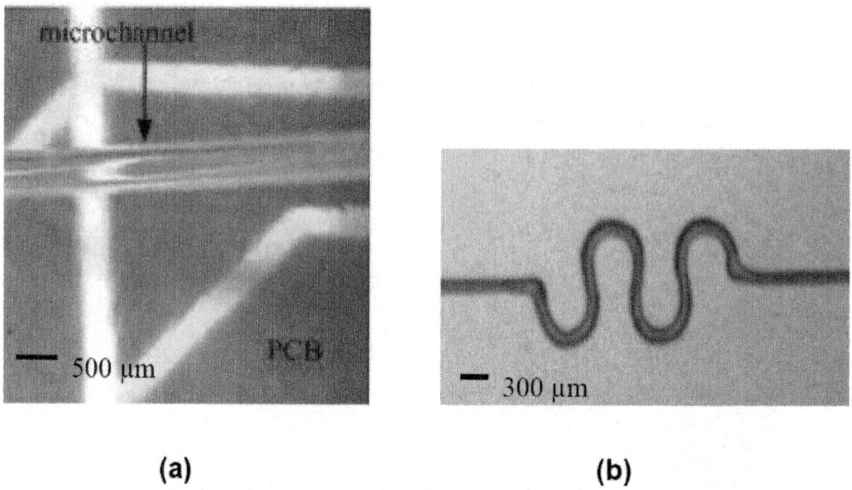

Figure 4. Microstructures fabricated using direct-dispense technique. (a) Microscopic image of a fabricated 500 μm diameter microchannel on a Printed Circuit Board (PCB). (b) Serpentine microchannel with an inner diameter of 150 μm and 200 μm outer diameter.

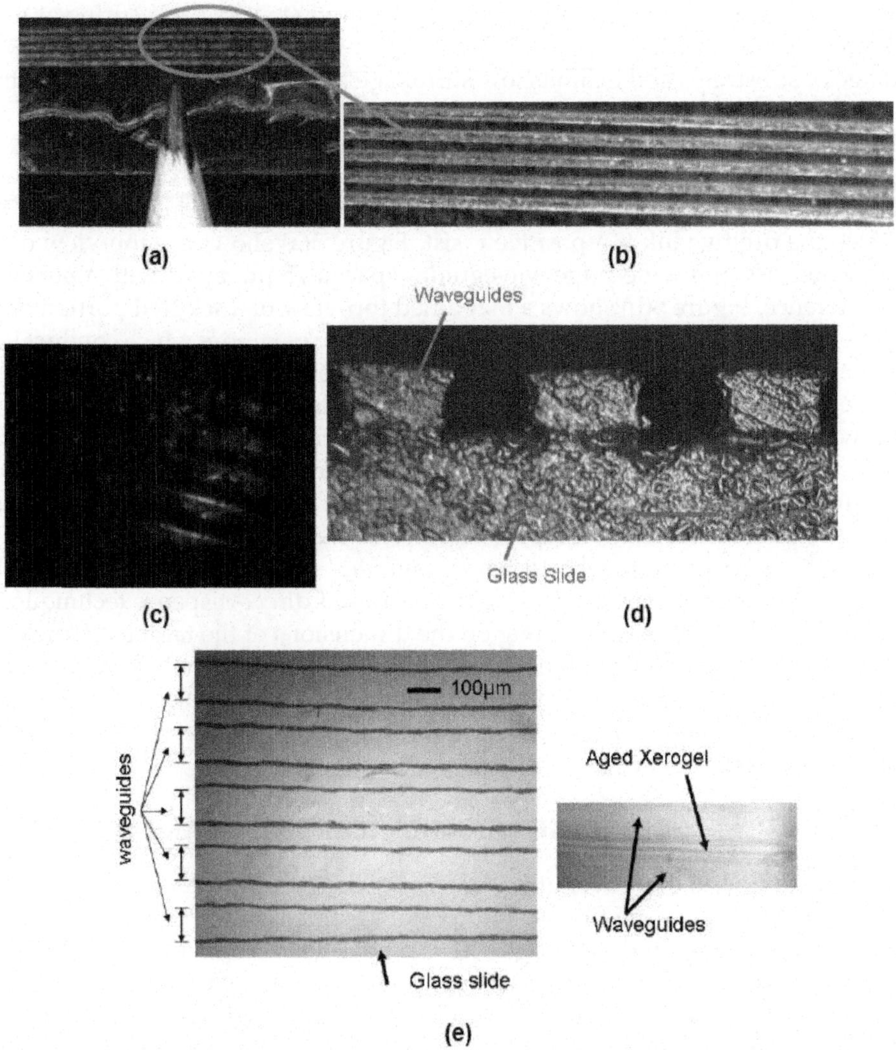

Figure 5. Fabricated epoxy waveguides using the direct-dispense deposition technique. (a) Shows the top-view of the fabricated 100 µm wide ridge waveguides spaced 75 µm apart with a pencil tip as reference; (b) Magnified top-view of a selected portion of the waveguides; (c) Multimode epoxy waveguides which conduct light from a blue (λ_{peak} = 470 nm) LED; (d) Cross-sectional view of the waveguides; (e) Magnified top-view of the transparent waveguides under intense optical illumination which shows the straightness and reproducibility of the structures. Figure (e) also shows the deposited xerogel after aging.

Accordingly, the direct-dispense system is programmed to make multimode ridge waveguides 100 µm in width, 100 µm thick and several millimeters in length. First, 100 µm lines of ink spaced 75 µm apart were deposited on a borosilicate glass slide. A layer of low viscosity, optically transparent epoxy Epo-Tek 301-1 was then deposited in the periodic reservoirs between the lines to form the incipient waveguides. The Epo-Tek 301-1 polymerized after standing and curing at room temperature for 24 hours. Cured Epo-Tek 301-1 adheres tenaciously to a wide range of substrates and exhibits low shrinkage, so that the ink effectively acts as a mold. Epo-Tek 301-1 has an optical transparency of >96% between 400 nm and 1,600 nm wavelengths. The last step in the fabrication of the epoxy waveguides is the removal of ink lines located between the epoxy guides. The stripping step was performed by placing samples in an oven or on a hot plate maintained at 65°C. In effect, the fugitive ink is a positive resist. Figure 5(a) shows the top-view of the fabricated 100 μm wide ridge waveguides spaced 75 μm apart with a pencil tip as reference. Figure 5(b) shows a magnified top-view of a selected portion of the waveguides. Figure 5(c) shows multimode epoxy waveguides which conduct light from a blue (λ_{peak} = 470 nm) LED. Figure 5(d) shows cross-sectional view of the waveguides. In Figure 5(d) we notice a semi-circular structure to the side-walls for the waveguides. This structure is due to the cylindrical structure of the fugitive-ink lines. Figure 5(e) shows a magnified top-view of the waveguides that are transparent under intense optical illumination. This image will help in understanding the straightness and fabrication reproducibility of the waveguides. Figure 5(e) also shows the deposited xerogel between waveguides after aging for 5 days. Typically, the microstructures produced using the proposed direct-dispense technique are about 5-10% inaccurate depending on the dimensions of the targeted structures. This inaccuracy increases as the dimensions of the structures reduce to the system limit of 10 μm. The structural variations are mainly due to ambient vibrations that affect the nano-manipulation system controlling the robotic deposition arm of the direct-dispense system.

3. XEROGELS BASED BIOCHEMICAL SENSOR ARRAYS

In addition to the waveguide element for delivery and receiving optical signals, the xerogel sensing film is an important element of the array. Xerogels are micro-/nano-structured porous glasses that have been extensively researched for the development of sensors for various biochemical analytes including oxygen, glucose, pH, and lactate [7, 10, 19, 26]. The appeal of xerogels for biochemical recognition elements derives from their thermal stability, tunable pore dimensions and distributions, adjustable pH and dipolarity, a broad optical transparency window, and evidence that they can be used to sequester a wide variety of active agents like luminophores in their porous networks. In the area of xerogel based biochemical sensor arrays, Cho *et al.* [7] have explored pin-printing to create xerogel sensor arrays. Aubonnet *et al.* [21] introduced photopatterned waveguides for sensing, but the performance of the sensors was limited to the detection of gaseous samples because waveguide transmission losses were too large in sol-gel samples that readily hydrated. More recently, Burke *et al.* [22, 27] developed embossing techniques to create sol-gel waveguides. They used pin-printing to integrate xerogel materials with the waveguides. The direct-dispense microfabrication technique described here is comparable to sol-gel processing techniques for the waveguides development in terms of ease of fabrication and potential low cost. The use of direct-dispense technique for the waveguide microstructures can further improve the integration and functionality of sensor microsystems by combining the properties of gaseous/fluidic sample manipulation and sensing.

Real-time monitoring of O_2 is important for blood gas analysis, water treatment, and food quality analysis. More than often, optical oxygen sensor systems are based on the measurement of luminescence intensity. Here, we prototype an O_2 sensor developed from the fugitive ink direct-dispense process. The detection scheme invokes fluorescence quenching of physically doped triplet $[Ru(bpy)_3]^{2+}$ [3, 28], whose quenching by oxygen has been quite well-studied [7, 10, 26]. Luminophore quenching in xerogel media is governed by many factors, and in the simplest scenario of a homogeneous distribution of luminophore molecules with a single exponential decay, luminescence quenching is described by the standard Stern-Volmer expression of Equation (1). In this expression, I_0 and τ_0 are the luminescence intensity and lifetime in the absence of analyte respectively, I is the luminescence intensity in the presence of analyte, K_{SV} is the Stern-Volmer constant, k_q is the bimolecular quenching constant and A is the fractional analyte concentration [28].

$$Io/_1 = 1 + K_{SV}[A] = 1 + K_q\tau_0[A] \tag{1}$$

It is recognized that physical doping and even covalent binding of the luminescent probe can give rise to heterogeneities in the local environment of the guest to cause nonlinear SV plots [29]. Two-site models, systematic changes in the host xerogel composition and more complex covalent attachment schemes have all been used to linearize and improve the SV response [30]. These strategies, desirable as they are, lie outside the scope of our current work which is not that of sensor optimization; instead our focus is to demonstrate functionality of a printable sensor array created by the fugitive ink direct-dispense technique.

3.1. Preparation of the Xerogel Oxygen (O_2) Sensors

We prepared what is commonly referred to as a Class II hybrid xerogel. This material abandons the classic combinations of tetraethoxyorthosilicates (TEOS) and tetramethoxyorthosilicates (TMOS) in favor of a silicon alkoxide component that is organically modified with a nonhydrolyzable covalent C-Si bond. Moreover, and anticipating future studies, the organic substituent can be a latently reactive moiety like a polymerizable monomer that can be further used to alter the properties of the host matrix (through crosslink density). Thus, the *sol* was prepared by acid catalysed hydrolysis and polycondensation of 3-methacryloxypropyltrimethoxysilane (MAPTMS) in combination with methacrylic acid and ZrO_2. Accordingly, MAPTMS (10.14 g, 0.040 mol) was combined with 0.05M hydrochloric acid (HCl, 0.54 g). The solution was stirred and aged for 1 hour at room temperature. During this time, a separate solution was prepared from 70% zirconium(IV)propoxide (6 g) in 1-propanol (0.013 mol) and methacrylic acid (1.1 g, 0.013 mol). These were combined and aged for 40 minutes. The methacrylic acid is incorporated to chelate Zr^{4+} in order to suppress the otherwise rapid hydrolysis that would deposit large particles of light scattering zirconia when the zirconium alkoxide is combined with the aqueous MAPTMS. After 1 hour, the MAPTMS solution is added drop-wise to the zirconium solution. The resulting *sol* is further hydrolyzed by the addition of distilled water (0.98 g) and stirred while being aged for at least 16 hours. The final ratio of water to silicon (r-value) is 1.5. The fluorescence sensor was prepared from a 1:4 mixture of 10 mmol/L *tris*(2,2′-bipyridyl) dichlororuthenium(II) hexahydrate in water and MAPTMS *sol*. 400 μL aliquots of $[Ru(bpy)_3]^{2+} \cdot 6H_2O$ were added to MAPTMS *sol* (1,400 μL). Rapid *gel* formation was avoided by adding methanol (1 mL) containing water (200 μL) to the final *sol*.

Prior to xerogel deposition, the waveguide samples were diced using a semi-conductor-dicer to a length of 1 centimeter. The exposed glass substrate and epoxy waveguides were cleaned with isopropanol. The epoxy surface was first coated with the adhesion promoter, 3-aminopropyltriethoxy-silane (APTS). The APTS was dried at 50°C for 2 hours. Once dry, the epoxy surface was again cleaned with isopropanol. Sol-gel sensing layer was subsequently deposited by direct-dispense in the spaces left by the stripped ink between two epoxy waveguide structures. The resulting structure of the sensor is shown in Figure 3(d). The hybrid glass was then dried for 1 hour at 50°C before a second layer was applied using the same procedure. The sensor waveguide structures were dried at 50°C for five days.

4. EXPERIMENTAL RESULTS AND DISCUSSION

The experimental setup for testing the proposed planar ridge waveguides is similar to the system shown in Figure 1. The setup consists of a blue LED (Digikey, λ_{peak}: 470 nm) for exciting the xerogel coated waveguide structures on the glass slide. The LED is sinusoidally modulated at 400 Hz with peak-peak sinusoidal amplitude of 1 V and DC offset of 2.7 V. The resulting optical signal collected by the epoxy waveguides is passed through a long pass (Thorlabs: FGL590, $\lambda_{cut\text{-}off}$: 590 nm) optical filter to remove the excitation light and allow only the fluorescence emission to pass through. The fluorescence signal is detected by a standard silicon photodiode (Thorlabs: PDA100A) and the output of the photodiode is recorded by a lock-in amplifier (Stanford Research Systems, SRS830). Both the excitation LED and photodetector are placed as close as possible (less than 5 mm) to the waveguide structures to maximize the efficiency in the sensing process. The sensor response is measured as a function of gaseous O_2 concentration at room temperature (25°C). The O_2 concentration is varied by changing its relative percentage with respect to gaseous N_2 concentration. The gases are mixed within a custom built gas mixing manifold, consisting of a matched pair of flow controllers (Cole-Parmer: EW-03229) connected to O_2 and N_2 gas cylinders.

As a prototype using the direct-dispense process, we fabricated a linear array of five polymer waveguides which housed two xerogel waveguides. As explained earlier, this configuration was used mainly for two reasons (i) xerogels are in aqueous form and hence two waveguide structures are used to contain the solution between them during deposition and (ii) the waveguide spacing between consecutive sensors is maintained in order to clearly distinguish the response from each sensor in the array. The sensor platform is scalable and can be used to implement large scale arrays for multi-sensor systems. As introduced earlier, the sensors operate on the basis of evanescent wave sensing [8, 25]. In the prototype system we used a photodiode as the detector which can only measure the cumulative fluorescence response from the sensor array. Alternatively, using a Charge Coupled Device (CCD) camera as the detector would enable us to distinguish the response from each sensor in the array.

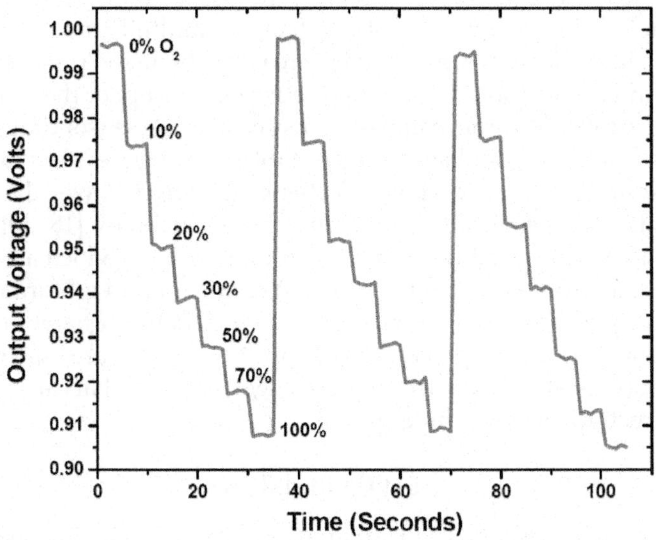

Figure 6. Response of the O_2 responsive xerogel sensor materials.

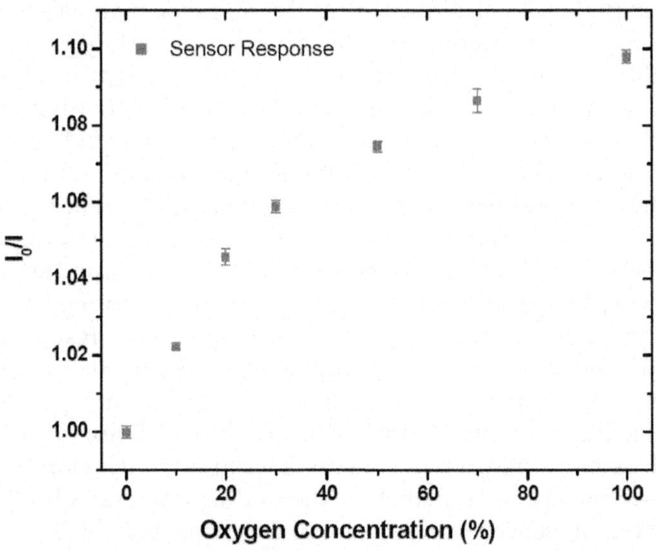

Figure 7. Stern-Volmer response of the xerogel O_2 sensors.

Figure 6 shows the response of the sensor as continually recorded by the lock-in-amplifier from the photodiode output by varying the oxygen concentration with respect to nitrogen gas. From the graph, we notice that the sensor system has repeatable performance with a response time on the order of 1 second. Measuring the exact response of the xerogel sensor material is difficult and the measured time is the experimental system response time which includes the time taken by the gas concentration to flow from the tanks through the controllers and tubing and fill the test chamber. Figure 7 shows the Stern-Volmer plot for the xerogel sensors. Each datum is the mean of results from multiple measurements. While the Stern-Volmer plot is nonlinear [28] for reasons discussed previously in Section 3, the data in Figures 6 and 7 show that the basic concept of the sensor array has been demonstrated. We notice that the sensitivity of the sensor film is comparable to optical sensors that are based on printable technologies developed by other research groups [21, 22, 27]. However, the sensitivity is lower than sol-gel sensor materials that specifically aim for increased sensitivity [18, 30]. The xerogel formulation has been modified from traditional recipes (which are typically immobilized with dip coating or spin coating) to avoid rapid *gel* formation from *sol* state. This modification is a novel step that renders the xerogel immobilization compatible with direct-dispense fabrication process. We are currently working towards improving the sensitivity of the xerogel materials that can be immobilized with the direct- dispense technique.

5. CONCLUSIONS

We have described the implementation of direct-dispense patterning to develop polymeric waveguide support platform for optical multi-sensor systems. Specifically, we demonstrated a waveguide array structure made with direct-dispense to support sol-gel derived xerogel recognition materials which are made responsive to oxygen concentration. The direct-dispense fabrication process suggests a scalable, versatile, cost-effective technology towards printable optical sensor arrays. The direct-dispense process may in the future lead to a high-performance, multi-functional sensor platform that can support gaseous/fluidic sample handling and optical signal collection and transmission. Advantages of this technique as compared to other soft-microfabrication techniques, are that one needs only a single direct-dispense setup (no requirement for combining multiple fabrication strategies or the creation of micro-contact printing masters), and ability to develop structures on–the-fly. The developed xerogel sensors performed for the full-scale range (0%-100%) of gaseous oxygen concentration with a response time of less than 1 second. The sensors show comparable sensitivity to other competent optical chemical sensors based on printable and patternable technologies. However, the sensitivity needs to be improved for practical field-usable applications. This is not unexpected, as more research in general is required to develop high sensitivity xerogel sensor materials. Nonetheless, the findings suggest a bright outlook for miniaturized multi-sensor platforms by direct-dispense.

Acknowledgements

We would like to thank the financial support given by Natural Sciences and Engineering Research Council of Canada (NSERC) and Le Fonds Québécois de la Recherche sur la Nature et les Technologies (FQRNT).

REFERENCES

1. Bally, M.; Halter, M.; Voros, J.; Grandin, H.M. Optical microarray biosensing techniques. *Surf. Interface Anal.* **2006**, *38*, 1442-1458.
2. Ligler, F.S.; Sapsford, K.E.; Golden, J.P.; Shriver-Lake, L.C.; Taitt, C.R.; Dyer, M.A.; Barone, S.; Myatt, C.J. The array biosensor: Portable, automated systems. *Anal. Sci.* **2007**, *23*, 5-10.
3. McDonagh, C.; Burke, C.S.; MacCraith, B.D. Optical chemical sensors. *Chem. Rev.* **2008**, *108*, 400-422.
4. Narayanaswamy, R.; Wolfbeis, O.S. *Optical Sensors: Industrial, Enviro nmental and Diagnostic Applications*, 1st, Ed.; Springer: New York, USA, 2004.
5. Heller, M.J. DNA microarray technology: Devices, systems, and applications. *Annu. Rev. Biomed. Eng.* **2002**, *4*, 129-153.
6. Kuswandi, B.; Tombelli, S.; Marazza, G.; Mascini, M. Recent advances in optical DNA biosensors technology. *Chimia* **2005**, *59*, 236-242.
7. Cho, E.J.; Tao, Z.Y.; Tehan, E.C.; Bright, F.V. Multianalyte pin-printed biosensor arrays based on protein-doped xerogels. *Anal. Chem.* **2002**, *74*, 6177-6184.
8. Wolfbeis, O.S. Fiber-optic chemical sensors and biosensors. *Anal. Chem.* **2006**, *78*, 3859-3873.
9. Sapsford, K.E.; Shubin, Y.S.; Delehanty, J.B.; Golden, J.P.; Taitt, C.R.; Shriver-Lake, L.C.; Ligler, F.S. Fluorescence-based array biosensors for detection of biohazards. *J. Appl. Microbiol.* **2004**, *96*, 47-58.
10. Chodavarapu, V.P.; Bukowski, R.M.; Kim, S.J.; Titus, A.H.; Cartwright, A.N.; Bright, F.V. Multi-sensor system based on phase detection, an LED array, and luminophore-doped xerogels. *Electron. Lett.* **2005**, *41*, 1031-1033.
11. Cho, E.J.; Bright, F.V. Optical sensor array and integrated light source. *Anal. Chem.* **2001**, *73*, 3289-3293.
12. Lee, K.S.; Lee, H.L.T.; Ram, R.J. Polymer waveguide backplanes for optical sensor interfaces in microfluidics. *Lab Chip* **2007**, *7*, 1539-1545.
13. Kuhn, M.; Napporn, T.; Meunier, M.; Vengallatore, S.; Therriault, D. Direct-write microfabrication of single-chamber micro solid oxide fuel cells. *J. Micromech. Microeng.* **2008**, *18*, 015005-015012.
14. Lewis, J.A. Direct-write assembly of ceramics from colloidal inks. *Curr. Opin. Solid State Mat. Sci.* **2002**, *6*, 245-250.
15. Therriault, D.; Shepherd, R.F.; White, S.R.; Lewis, J.A. Fugitive inks for direct-write assembly of three-dimensional microvascular networks. *Advan. Mater.* **2005**, *17*, 395-399.
16. Ghafar-Zadeh, E.; Sawan, M.; Therriault, D. Novel direct-write CMOS-based laboratory-on-chip: Design, assembly and experimental results. *Sensor. Actuator. A-Phys.* **2007**, *134*, 27-36.
17. Pique, A.; Chrisey, D.B. *Direct Write Technologies for Rapid Prototyping Applications: In Sensors, Electronics and Integrated Power Sources*; Academic Press: San Diego, USA, 2002.
18. MacCraith, B.D.; McDonagh, C.; Mcevoy, A.K.; Butler, T.; OKeeffe, G.; Murphy, V. Optical chemical sensors based on sol-gel materials: Recent advances and critical issues. *J. Sol-Gel Sci. Technol.* **1997**, *8*, 1053-1061.

19. Jeronimo, P.C.A.; Araujo, A.N.; Montenegro, M.C.B.S.M. Optical sensors and biosensors based on sol-gel films. *Talanta* **2007**, *72*, 13-27.

20. Chodavarapu, V.P.; Shubin, D.O.; Bukowski, R.M.; Titus, A.H.; Cartwright, A.N.; Bright, F.V. CMOS-based phase fluorometric oxygen sensor system. *IEEE Trans. Circuit. Syst.-I*, **2007**, *54*, 111-118.

21. Aubonnet, S.; Barry, H.F.; von Bultzingslowen, C.; Sabattie, J.M.; MacCraith, B.D. Photo-patternable optical chemical sensors based on hybrid sol-gel materials. *Electron. Lett.* **2003**, *39*, 913-914.

22. Burke, C.S.; Polerecky, L.; MacCraith, B.D. Design and fabrication of enhanced polymer waveguide platforms for absorption-based optical chemical sensors. *Meas. Sci. Technol.* **2004**, *15*, 1140-1145.

23. Dumais, P.; Callender, C.L.; Ledderhof, C.J.; Noad, J.P. Monolithic integration of microfluidic channels, liquid-core waveguides, and silica waveguides on silicon. *Appl. Opt.* **2006**, *45*, 9182-9190.

24. Lavers, C.R.; Itoh, K.; Wu, S.C.; Murabayashi, M.; Mauchline, I.; Stewart, G.; Stout, T.; Planar optical waveguides for sensing applications. *Sensor. Actuator. B-Chem.* **2000**, *69*, 85-95.

25. Boisde, G.; Harmer, A. *Chemical and Biochemical Sensing With Optical Fibers and Waveguides*; Artech House Publishers: Boston, USA, 1996.

26. Maccraith, B.D.; Mcdonagh, C.M.; Okeeffe, G.; Mcevoy, A.K.; Butler, T.; Sheridan, F.R. Sol-Gel Coatings for Optical Chemical Sensors and Biosensors. *Sensor. Actuator. B-Chem.* **1995**, *29*, 51-57.

27. Burke, C.S.; McGaughey, O.; Sabattie, J.M.; Barry, H.; McEvoy, A.K.; McDonagh, C.; MacCraith, B.D. Development of an integrated optic oxygen sensor using a novel, generic platform. *Analyst* **2005**, *130*, 41-45.

28. Lakowicz, J.R. *Principles of Fluorescence Spectroscopy*. 2[ed] Ed.; Kluwer Academic/Plenum Publishers: New York, USA, 1999.

29. Lei, B.F.; Li, B.; Zhang, H.R.; Lu, S.Z.; Zheng, Z.H.; Li, W.L.; Wang, Y. Mesostructured silica chemically doped with Ru-II as a superior optical oxygen sensor. *Adv. Funct. Mater.* **2006**, *16*, 1883-1891.

30. Tang, Y.; Tehan, E.C.; Tao, Z.Y.; Bright, F.V. Sol-gel-derived sensor materials that yield linear calibration plots, high sensitivity, and long-term stability. *Anal. Chem.* **2003**, *75*, 2407-2413.

INDEX

D

E

F

G

This page left intentionally blank.